BETZ HANDBOOK OF

INDUSTRIAL WATER CONDITIONING

NINTH EDITION 1991

Library of Congress Catalog Card Number: 91-61770
ISBN 0-913641-00-6

Copyright 1991
Betz Laboratories, Inc.
Trevose, PA 19053

Previous Editions

First Edition, 1942
Second Edition, 1945
Third Edition, 1950
Fourth Edition, 1953
Fifth Edition, 1957
Sixth Edition, 1962
Seventh Edition, 1976
Eighth Edition, 1980

Printed in U.S.A.

The 9th Edition of the Betz Handbook of
Industrial Water Conditioning is printed on
recycled paper that meets the minimum 50%
waste paper requirements established by the
Federal EPA in its guidelines for recycled
paper products.

ACKNOWLEDGMENT

This handbook was authored by the engineering and technical staff of Betz Laboratories, Inc. We thank the following water treatment specialists for their editorial contributions to this Ninth Edition.

A.M. Agree	C.R. Istre
C.R. Ascolese	N.N. Johnston
W.E. Bornak	J.M. Knapp
C.E. Callow	M.C. Kogut
J.S. Clavin	D.U. Laub
I.J. Cotton	D.M. Lusardi
D.P. Davis	D. Marturana
P.J. DiFranco	J.P. McAlary
E.R. Domino	M.S. Morgan
D.B. Dyer	O.J. Nussbaum
K.S. Eble	B. Nyzio
G.E. Geiger	B.J. Peters
C.J. Gerth	G.M. Reggiani
M. Goldblatt	J.O. Robinson
J.W. Griffin	J.T. Rose
R.E. Hargrave	E.L. Schorpp
R.W. Hartung	E.A. Thungstrom
L.A. Huchler	D.B. Way

We would also like to recognize the following members of the Editorial Board

B.C. Moore	J.O. Robinson
R.J. Palangio	

and acknowledge the following for reviewing editiorial contributions.

P.R. Burgmayer	R.C. May
A.B. Carlisle	A.D. Ricci
G.W. Delaney	J.C. Rodgers
M.J. Esmacher	W.R. Snyder
E.W. James	F. Termine
F.C. Klaessig	S. Vasconcellos
J.B. Lord	J.M. Young
D.A. Lusardi	A.D. Zsolnay
L.A. Lyons	

Many Betz employees contributed to this edition and previous editions. We wish to particularly acknowledge the following for their efforts:

H. Bartle	M. Fralin
R. Bauers	G. E. Pearson
L. Borek	F.S. Spoerle
A.F. Camerota	J.M. Straub
M.M. Donadieu	

TABLE OF CONTENTS

CONTROL WATER ANALYSES
AND THEIR INTERPRETATION (CONT'D)

DEDICATED TO WORLDWIDE PRESERVATION AND ENHANCEMENT OF OUR WATER RESOURCES

Our natural environment is delicately balanced. It is a system of complex global cycles working in harmony to reuse limited resources. And every effort must be made to preserve this elemental balance for generations to come.

Betz recognizes the need to preserve the environment worldwide, and dedicates this Ninth Edition of the Betz Handbook to that endeavor.

It is with this dedication to the environment that we donate all profits from this new Betz Handbook to the global preservation and enhancement of our natural water resources.

Directors, Officers, and Employees of Betz Laboratories, Inc.

Introduction

Water discharged to oceans, lakes, and rivers must meet increasingly stringent environmental standards.

WATER SOURCES, IMPURITIES, AND CHEMISTRY

Abundant supplies of fresh water are essential to the development of industry. Enormous quantities are required for the cooling of products and equipment, for process needs, for boiler feed, and for sanitary and potable water supply.

THE PLANETARY WATER CYCLE

Industry is a small participant in the global water cycle. The finite amount of water on the planet participates in a very complicated recycling scheme that provides for its reuse. This recycling of water is termed the "Hydrologic Cycle" (see Figure 1-1).

Evaporation under the influence of sunlight takes water from a liquid to a gaseous phase. The water may condense in clouds as the temperature drops in the upper atmosphere. Wind transports the water over great distances before releasing it in some form of precipitation. As the water condenses and falls to the ground, it absorbs gases from the environment. This is the principal cause of acid rain and acid snow.

WATER AS A SOLVENT

Pure water (H_2O) is colorless, tasteless, and odorless. It is composed of hydrogen and oxygen. Because water becomes contaminated by the substances with which it comes into contact, it is not available for use in its pure state. To some degree, water can dissolve every naturally occurring substance on the earth. Because of this property, water has been termed a "universal solvent."

Although beneficial to mankind, the solvency power of water can pose a major threat to industrial equipment. Corrosion reactions cause the slow dissolution of metals by water. Deposition reactions, which produce scale on heat transfer surfaces, represent a change in the solvency power of water as its temperature is varied. The control of corrosion and scale is a major focus of water treatment technology.

WATER IMPURITIES

Water impurities include dissolved and suspended solids. Calcium bicarbonate is a soluble salt. A solution of calcium bicarbonate is clear, because the calcium and bicarbonate are present as atomic-sized ions which are not large enough to reflect light. Some soluble minerals impart a color to the solution. Soluble iron salts produce pale yellow or green solutions; some copper salts form intensely blue solutions. Although colored, these solutions are clear.

Suspended solids are substances that are not completely soluble in water and are present as particles. These particles usually impart a visible turbidity to the water.

Dissolved and suspended solids are present in most surface waters. Seawater is very high in soluble sodium chloride; suspended sand and silt make it slightly cloudy. An extensive list of soluble and suspended impurities found in water is given in Table 1-1.

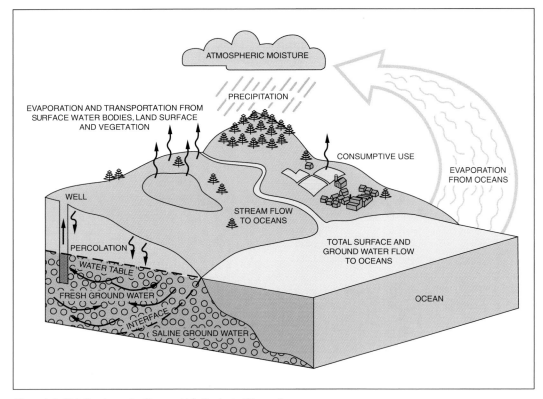

Figure 1-1. Global water cycle. (Source: U.S. Geological Survey.)

Table 1-1. Common impurities found in fresh water.

Constituent	Chemical Formula	Difficulties Caused	Means of Treatment
Turbidity	none—expressed in analysis as units	imparts unsightly appearance to water; deposits in water lines, process equipment, etc.; interferes with most process uses	coagulation, settling, and filtration
Hardness	calcium and magnesium salts, expressed as $CaCO_3$	chief source of scale in heat exchange equipment, boilers, pipe lines, etc.; forms curds with soap, interferes with dyeing, etc.	softening; demineralization; internal boiler water treatment; surface-active agents
Alkalinity	bicarbonate (HCO_3^-), carbonate (CO_3^{2-}), and hydroxide (OH^-), expressed as $CaCO_3$	foam and carryover of solids with steam; embrittlement of boiler steel; bicarbonate and carbonate produce CO_2 in steam, a source of corrosion in condensate lines	lime and lime–soda softening; acid treatment; hydrogen zeolite softening; demineralization dealkalization by anion exchange
Free Mineral Acid	H_2SO_4, HCl, etc., expressed as $CaCO_3$	corrosion	neutralization with alkalies
Carbon Dioxide	CO_2	corrosion in water lines, particularly steam and condensate lines	aeration, deaeration, neutralization with alkalies
pH	hydrogen ion concentration defined as: $$pH = \log \frac{1}{[H^+]}$$	pH varies according to acidic or alkaline solids in water; most natural waters have a pH of 6.0–8.0	pH can be increased by alkalies and decreased by acids
Sulfate	SO_4^{2-}	adds to solids content of water, but in itself is not usually significant; combines with calcium to form calcium sulfate scale	demineralization, reverse osmosis, electrodialysis, evaporation

Table 1-1, continued

Constituent	Chemical Formula	Difficulties Caused	Means of Treatment
Chloride	Cl^-	adds to solids content and increases corrosive character of water	demineralization, reverse osmosis, electrodialysis, evaporation
Nitrate	NO_3^-	adds to solids content, but is not usually significant industrially; high concentrations cause methemoglobinemia in infants; useful for control of boiler metal embrittlement	demineralization, reverse osmosis, electrodialysis, evaporation
Fluoride	F^-	cause of mottled enamel in teeth; also used for control of dental decay; not usually significant industrially	adsorption with magnesium hydroxide, calcium phosphate, or bone black; alum coagulation
Sodium	Na^+	adds to solids content of water; when combined with OH^-, causes corrosion in boilers under certain conditions	demineralization, reverse osmosis, electrodialysis, evaporation
Silica	SiO_2	scale in boilers and cooling water systems; insoluble turbine blade deposits due to silica vaporization	hot and warm process removal by magnesium salts; adsorption by highly basic anion exchange resins, in conjunction with demineralization, reverse osmosis, evaporation
Iron	Fe^{2+}(ferrous) Fe^{3+} (ferric)	discolors water on precipitation; source of deposits in water lines, boilers, etc.; interferes with dyeing, tanning, papermaking, etc.	aeration; coagulation and filtration; lime softening; cation exchange; contact filtration; surface-active agents for iron retention
Manganese	Mn^{2+}	same as iron	same as iron
Aluminum	Al^{3+}	usually present as a result of floc carryover from clarifier; can cause deposits in cooling systems and contribute to complex boiler scales	improved clarifier and filter operation
Oxygen	O_2	corrosion of water lines, heat exchange equipment, boilers, return lines, etc.	deaeration; sodium sulfite; corrosion inhibitors
Hydrogen Sulfide	H_2S	cause of "rotten egg" odor; corrosion	aeration; chlorination; highly basic anion exchange
Ammonia	NH_3	corrosion of copper and zinc alloys by formation of complex soluble ion	cation exchange with hydrogen zeolite; chlorination; deaeration
Dissolved Solids	none	refers to total amount of dissolved matter, determined by evaporation; high concentrations are objectionable because of process interference and as a cause of foaming in boilers	lime softening and cation exchange by hydrogen zeolite; demineralization, reverse osmosis, electrodialysis, evaporation
Suspended Solids	none	refers to the measure of undissolved matter, determined gravimetrically; deposits in heat exchange equipment, boilers, water lines, etc.	subsidence; filtration, usually preceded by coagulation and settling
Total Solids	none	refers to the sum of dissolved and suspended solids, determined gravimetrically	see "Dissolved Solids" and "Suspended Solids"

Surface Water

The ultimate course of rain or melting snow depends on the nature of the terrain over which it flows. In areas consisting of hard-packed clay, very little water penetrates the ground. In these cases, the water generates "runoff." The runoff collects in streams and rivers. The rivers empty into bays and estuaries, and the water ultimately returns to the sea, completing one major phase of the hydrologic cycle shown in Figure 1-1.

As water runs off along the surface, it stirs up and suspends particles of sand and soil, creating silt in the surface water. In addition, the streaming action erodes rocky surfaces, producing more sand. As the surface water cascades over rocks, it is aerated. The combination of oxygen, inorganic nutrients leached from the terrain, and sunlight supports a wide variety of life forms in the water, including algae, fungi, bacteria, small crustaceans, and fish.

Often, river beds are lined with trees, and drainage areas feeding the rivers are forested. Leaves and pine needles constitute a large percentage of the biological content of the water. After it dissolves in the water, this material becomes a major cause of fouling of ion exchange resin used in water treatment.

The physical and chemical characteristics of surface water contamination vary considerably over time. A sudden storm can cause a dramatic short-term change in the composition of a water supply. Over a longer time period, surface water chemistry varies with the seasons. During periods of high rainfall, high runoff occurs. This can have a favorable or unfavorable impact on the characteristics of the water, depending on the geochemistry and biology of the terrain.

Surface water chemistry also varies over multi-year or multidecade cycles of drought and rainfall. Extended periods of drought severely affect the availability of water for industrial use. Where rivers discharge into the ocean, the incursion of salt water up the river during periods of drought presents additional problems. Industrial users must take surface water variability into account when designing water treatment plants and programs.

Groundwater

Water that falls on porous terrains, such as sand or sandy loam, drains or percolates into the ground. In these cases, the water encounters a wide variety of mineral species arranged in

Table 1-2. A comparison of surface water and groundwater characteristics.

Characteristic	Surface Water	Groundwater
Turbidity	high	low
Dissolved minerals	low–moderate	high
Biological content	high	low
Temporal variability	very high	low

complex layers, or strata. The minerals may include granite, gneiss, basalt, and shale. In some cases, there may be a layer of very permeable sand beneath impermeable clay. Water often follows a complex three-dimensional path in the ground. The science of groundwater hydrology involves the tracking of these water movements.

In contrast to surface supplies, groundwaters are relatively free from suspended contaminants, because they are filtered as they move through the strata. The filtration also removes most of the biological contamination. Some groundwaters with a high iron content contain sulfate-reducing bacteria. These are a source of fouling and corrosion in industrial water systems.

Groundwater chemistry tends to be very stable over time. A groundwater may contain an undesirable level of scale-forming solids, but due to its fairly consistent chemistry it may be treated effectively.

Mineral Reactions. As groundwater encounters different minerals, it dissolves them according to their solubility characteristics. In some cases chemical reactions occur, enhancing mineral solubility.

A good example is the reaction of groundwater with limestone. Water percolating from the surface contains atmospheric gases. One of these gases is carbon dioxide, which forms carbonic acid when dissolved in water. The decomposition of organic matter beneath the surface is another source of carbon dioxide. Limestone is a mixture of calcium and magnesium carbonate. The mineral, which is basic, is only slightly soluble in neutral water.

The slightly acidic groundwater reacts with basic limestone in a neutralization reaction that forms a salt and a water of neutralization. The salt formed by the reaction is a mixture of calcium and magnesium bicarbonate. Both bicarbonates are quite soluble. This reaction is the source of the most common deposition and corrosion problems faced by industrial users. The calcium and magnesium (hardness) form scale on heat transfer surfaces if the groundwater is not treated before use in industrial cooling and boiler systems. In boiler feedwater applications, the thermal breakdown of the bicarbonate in the boiler leads to high levels of carbon dioxide in condensate return systems. This can cause severe system corrosion.

Structurally, limestone is porous. That is, it contains small holes and channels called "interstices." A large formation of limestone can hold vast quantities of groundwater in its structure. Limestone formations that contain these large quantities of water are called aquifers, a term derived from Latin roots meaning water-bearing.

If a well is drilled into a limestone aquifer, the water can be withdrawn continuously for decades and used for domestic and industrial applications. Unfortunately, the water is very hard, due to the neutralization/dissolution reactions described above. This necessitates extensive water treatment for most uses.

CHEMICAL REACTIONS

Numerous chemical tests must be conducted to ensure effective control of a water treatment program. Most of these tests are addressed in detail in Chapters 39–71. Because of their significance in many systems, three tests, pH, alkalinity, and silica, are discussed here as well.

pH Control

Good pH control is essential for effective control of deposition and corrosion in many water systems. Therefore, it is important to have a good understanding of the meaning of pH and the factors that affect it.

Pure H_2O exists as an equilibrium between the acid species, H^+ (more correctly expressed as a protonated water molecule, the hydronium ion, H_3O^+) and the hydroxyl radical, OH^-. In neutral water the acid concentration equals the hydroxyl concentration and at room temperature they both are present at 10^{-7} gram equivalents (or moles) per liter.

The "p" function is used in chemistry to handle very small numbers. It is the negative logarithm of the number being expressed. Water that has 10^{-7} gram equivalents per liter of hydrogen ions is said to have a pH of 7. Thus, a neutral solution exhibits a pH of 7. Table 1-3 lists the concentration of H^+ over 14 orders of magnitude. As it varies, the concentration of OH^- must also vary, but in the opposite direction, such that the product of the two remains constant.

Confusion regarding pH arises from two sources:

■ the inverse nature of the function

■ the pH meter scale

It is important to remember that as the acid

Table 1-3. pH relationships.

pH[a]	H⁺ Concentration Exponential Notation, gram moles/L	H⁺ Concentration, Normality	OH⁻ Concentration, Normality	OH⁻ Concentration, Exponential Notation, gram moles/L	pOH[a]
0	10^0	1	0.00000000000001	10^{-14}	14
1	10^{-1}	0.1	0.0000000000001	10^{-13}	13
2	10^{-2}	0.01	0.000000000001	10^{-12}	12
3	10^{-3}	0.001	0.00000000001	10^{-11}	11
4	10^{-4}	0.0001	0.0000000001	10^{-10}	10
5	10^{-5}	0.00001	0.000000001	10^{-9}	9
6	10^{-6}	0.000001	0.00000001	10^{-8}	8
7	10^{-7}	0.0000001	0.0000001	10^{-7}	7
8	10^{-8}	0.00000001	0.000001	10^{-6}	6
9	10^{-9}	0.000000001	0.00001	10^{-5}	5
10	10^{-10}	0.0000000001	0.0001	10^{-4}	4
11	10^{-11}	0.00000000001	0.001	10^{-3}	3
12	10^{-12}	0.000000000001	0.01	10^{-2}	2
13	10^{-13}	0.0000000000001	0.1	10^{-1}	1
14	10^{-14}	0.00000000000001	1	10^0	0

[a] pH + pOH = 14.

Table 1-4. Comparative pH levels of common solutions.

12	OH⁻ alkalinity 500 ppm as $CaCO_3$
11	OH⁻ alkalinity 50 ppm as $CaCO_3$; Columbus, OH, drinking water[a]
10	OH⁻ alkalinity 5 ppm as $CaCO_3$
9	strong base anion exchanger effluents
8	phenolphthalein end point
7	neutral point at 25 °C
6	Weymouth, MA, drinking water[a]
5	methyl orange end point
4	FMA 4 ppm as $CaCO_3$
3	FMA 40 ppm as $CaCO_3$; strong acid cation exchanger effluents
2	FMA 400 ppm as $CaCO_3$

[a] Extremes of drinking water pH.

concentration increases, the pH value decreases (see Table 1-4).

The pH meter can be a source of confusion, because the pH scale on the meter is linear, extending from 0 to 14 in even increments. Because pH is a logarithmic function, a change of 1 pH unit corresponds to a 10-fold change in acid concentration. A decrease of 2 pH units represents a 100-fold change in acid concentration.

Alkalinity

Alkalinity tests are used to control lime–soda softening processes and boiler blowdown and to predict the potential for calcium scaling in cooling water systems. For most water systems, it is important to recognize the sources of alkalinity and maintain proper alkalinity control.

Carbon dioxide dissolves in water as a gas. The dissolved carbon dioxide reacts with solvent water molecules and forms carbonic acid according to the following reaction:

$$CO_2 \;+\; H_2O \;\rightarrow\; H_2CO_3$$

Only a trace amount of carbonic acid is formed, but it is acidic enough to lower pH from the neutral point of 7. Carbonic acid is a weak acid, so it does not lower pH below 4.3. However, this level is low enough to cause significant corrosion of system metals.

If the initial loading of CO_2 is held constant and the pH is raised, a gradual transformation into the bicarbonate ion HCO_3^- occurs. This is shown in Figure 1-2. The transformation is complete at pH 8.3. Further elevation of the pH forces a second transformation—into carbonate, CO_3^{2-}. The three species—carbonic acid, bicarbonate, and carbonate—can be converted from one to another by means of changing the pH of the water.

Variations in pH can be reduced through "buffering"—the addition of acid (or caustic). When acid (or caustic) is added to a water containing carbonate/bicarbonate species, the pH of the system does not change as quickly as it does in pure water. Much of the added acid (or caustic) is consumed as the carbonate/bicarbonate (or bicarbonate/carbonic acid) ratio is shifted.

Alkalinity is the ability of a natural water to neutralize acid (i.e., to reduce the pH depression expected from a strong acid by the buffering mechanism mentioned above). Confusion arises in that alkaline pH conditions exist at a pH above

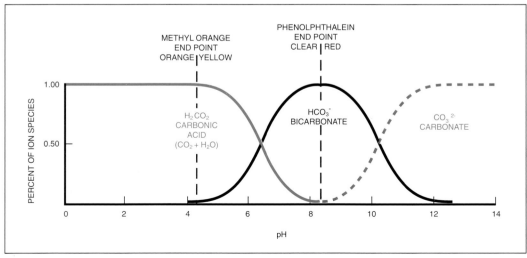

Figure 1-2. Carbonic acid, bicarbonate, and carbonate distribution as a function of pH.

7, whereas alkalinity in a natural water exists at a pH above 4.4.

Alkalinity is measured by a double titration; acid is added to a sample to the Phenolphthalein end point (pH 8.3) and the Methyl Orange end point (pH 4.4). Titration to the Phenolphthalein end point (the P-alkalinity) measures OH^- and $\frac{1}{2}CO_3^{2-}$; titration to the Methyl Orange end point (the M-alkalinity) measures OH^-, CO_3^{2-} and HCO_3^-.

Silica

When not properly controlled, silica forms highly insulating, difficult-to-remove deposits in cooling systems, boilers, and turbines. An understanding of some of the possible variations in silica testing is valuable.

Most salts, although present as complicated crystalline structures in the solid phase, assume fairly simple ionic forms in solution. Silica exhibits complicated structures even in solution.

Silica exists in a wide range of structures, from a simple silicate to a complicated polymeric material. The polymeric structure can persist when the material is dissolved in surface waters.

The size of the silica polymer can be substantial, ranging up to the colloidal state. Colloidal silica is rarely present in groundwaters. It is most commonly present in surface waters during periods of high runoff.

The polymeric form of silica does not produce color in the standard molybdate-based colorimetric test for silica. This form of silica is termed "nonreactive." The polymeric form of silica is not thermally stable and when heated in a boiler reverts to the basic silicate monomer, which is reactive with molybdate.

As a result, molybdate testing of a boiler feedwater may reveal little or no silica, while boiler blowdown measurements show a level of silica that is above control limits. High boiler water silica and low feedwater values are often a first sign that colloidal silica is present in the makeup.

One method of identifying colloidal silica problems is the use of atomic emission or absorption to measure feedwater silica. This method, unlike the molybdate chemistry, measures total silica irrespective of the degree of polymerization.

ENVIRONMENTAL CONSIDERATIONS

Concern for the environment is not a new issue, as evidenced in Figure 2-1 by the notice printed in the January, 1944 issue of *The Betz Indicator.*

In the 1960's it became evident that there could be a dark side to the economic development that resulted from the decades of rapid industrial growth following World War II. During this period the general public became aware of the consequences of improper waste material handling and industrial accidents. Frightening incidents at Love Canal, Seveso, and Bhopal in the 1970's and 1980's had tragic effects on members of the general public beyond the fence line of the facilities. In the past few decades, public awareness has grown concerning many other important environmental issues:

■ acid rain

■ global warming ("greenhouse effect")

■ stratospheric ozone depletion

■ tropical deforestation

■ the urban trash crisis

■ pesticides in groundwater

■ hazardous waste disposal

■ natural and synthetic carcinogens

Figure 2-1. Concern for the environment and the potential for recovery of valuable materials has been a concern for decades. The above notice appeared in the January, 1944 issue of "The Betz Indicator."

Focus on environmental considerations has shifted from a single-medium approach (air, water, land) to a holistic approach. Early regulations permitted the removal of a solvent, such as trichloroethane (methyl chloroform), from contaminated groundwater by counter-current air stripping. It was soon realized that while the water was no longer contaminated, an air pollutant had been created in the process. Today's regulations address the fact that moving a pollutant from one medium to another does not eliminate the problem. In the example given above, the solvent removed from the water must be condensed or adsorbed by activated carbon and recovered or incinerated.

Another change is a recognition that city sewers are an appropriate means of disposal only for those industrial wastes that are removed or degraded to environmentally compatible products in the municipal treatment plant. Industrial wastes that cause a degradation of effluent water quality or render the sewage sludge hazardous must be managed in ways that are environmentally acceptable. The accomplishment of this goal will require the continuing, long-term efforts of all concerned.

The cost of manufacturing a product now includes factors for waste disposal and pollution prevention. Often, it is more economical to alter processes to produce less waste or more benign wastes, and to recover usable materials from waste streams, than to make a contaminated waste stream suitable for disposal.

THE INDUSTRIAL USE OF WATER

It is becoming increasingly apparent that fresh water is a valuable resource that must be protected through proper management, conservation, and use.

Although two-thirds of the Earth's surface is covered by water, most of it is seawater, which is not readily usable for most needs. All fresh water comes from rainfall, which percolates into the soil or runs off into rivers and streams. The hydrologic cycle is dynamic, as shown in Chapter 1.

In order to ensure an adequate supply of high-quality water for industrial use, the following practices must be implemented (see Figure 2-2):

- purification and conditioning prior to consumer (potable) or industrial use

- conservation (and reuse where possible)

- wastewater treatment

Cooling systems are being modified in industrial applications to reduce the use of fresh water makeup. The operation of cooling towers at high cycles of concentration and the reuse of waste streams (including municipal plant effluent for cooling tower makeup) can contribute significantly to reduced water consumption.

Both groundwater and surface waters can become polluted as a result of the improper management of wastes (Figure 2-3). Because of the increasing demands for fresh water, there is a continuing need to share resources. Regulations will require the increasing treatment of all domestic and industrial wastewaters in order to remove industrial and priority pollutants and restore the

Figure 2-2. Effective water purification is essential to ensure that all users have an adequate supply of high-quality water.

effluent water to the quality required by the next user. Facilities that treat domestic waste must also control the more conventional pollutants, such as BOD (biological oxygen demand), ammonia, and nitrates, and restore the pH if it is out of the neutral zone.

Concerns about the safety of drinking water supplies are widespread. Although there are many pollutants that degrade water quality (including natural pollutants), those that attract the greatest public attention result from industrial activity and the use of agricultural pesticides and fertilizers.

Environmental regulations establish quality criteria for both industrial and domestic waste treatment discharges. Although some countries have more comprehensive laws and permit regulations than others, stringent pollution control standards will probably be adopted globally in the coming years.

AIR QUALITY

Geographic boundaries are not recognized by the winds. Air quality issues are complicated by the fact that they are usually of multinational concern. Significant issues such as acid rain, stratospheric ozone depletion, and the greenhouse effect require a degree of international cooperation that is difficult to achieve (see Figure 2-4). Technologies available today can have a positive and measurable impact on these issues. Several chapters in this handbook describe technologies that increase boiler and industrial cooling efficiency. In paper mills, generating plants, steel mills, refineries, and other major energy consumers, each incremental increase in energy efficiency represents a reduction in required fuel. As a result of reduced fuel consumption, less carbon dioxide is produced, and where coal- or other sulfur-containing fuels are used there is also a decrease in sulfur oxide emissions. Fluidized bed boilers are being used increasingly to reduce the presence of acidic gasses (SO_x and NO_x) in the boiler flue gas.

One of the problems faced by governments is the amount of energy required to accomplish wet scrubbing (to remove acid gases) and electrostatic precipitation of particulates. These processes, combined, consume up to 30% of the energy released by the burning of coal (see Figure 2-5). While these processes reduce the contaminants thought to cause acid rain, they increase the amount of coal burned and thereby increase the production of carbon dioxide, one of the gases

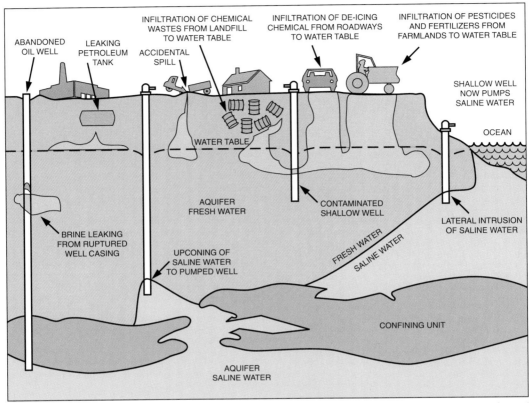

Figure 2-3. Groundwater (and surface water) can become contaminated in a variety of ways. (Source: U.S. Geological Survey.)

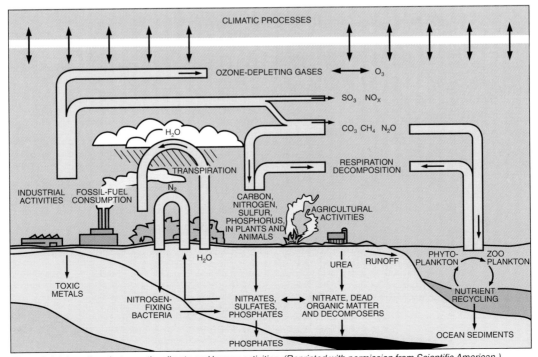

Figure 2-4. Interactions between the climate and human activities. (Reprinted with permission from Scientific American.)

Figure 2-5. Scrubbers reduce atmospheric emissions.

thought to cause the "greenhouse effect."

Many of the air pollutants of concern could be greatly reduced through the use of alternative energy sources, such as nuclear fission (and at some point, probably nuclear fusion), geothermal, wind, hydroelectric, photovoltaic, biomass, and solar. At this time, many of the alternatives are significantly more expensive than the use of fossil fuels, and each has its own problems. There are no clear and simple solutions; no source of energy has been developed that is both economically attractive and without environmental drawbacks.

Over the past several years, most industrialized countries have passed laws addressing air pollution concerns and industrial and power plant emissions. Nations have begun to come together in a cooperative fashion to formulate agreements and protocols to deal with global atmospheric concerns. There has been a multinational agreement to phase out the use of certain chlorofluorocarbon compounds (used as refrigerant gases and for other purposes) because they have been linked with a reduction of ozone in the stratosphere. There is reason to believe that a reduction in stratospheric ozone will allow a higher level of UV radiation to reach the earth's surface, and this is expected to cause an increase in the incidence of skin cancer along with other undesirable effects.

There are movements to establish multinational agreements that provide incentives to allow economic progress to occur in developing countries without the destruction of their rain forests. The rain forests should be preserved not only for the sake of conservation but also because they remove vast quantities of atmospheric carbon dioxide through photosynthesis and thus have a favorable effect on global warming and the greenhouse effect.

Human understanding of atmospheric chemistry is far from complete. As our understanding grows there will undoubtedly be many changes in direction and emphasis regarding atmospheric pollutants. Because a sizeable amount of atmospheric pollution results from industrial activity and power generation, the scope and stringency of industrial air pollution regulations will continue to increase.

INDUSTRIAL WASTE REDUCTION AND ENERGY CONSERVATION

In the 20th century, industrialized nations evolved from exploiters of bountiful natural resources to conservators of scarce resources. In the early 1900's, the consumption of industrial products was modest and natural resources appeared to be limitless. As the demand for electric power and industrial products grew, the limitations of the Earth's natural resources became an increasing concern. Today, even developing countries are very interested in the controlled development and utilization of their resources.

In addition to producing a desired output at a certain cost, industrial producers must now consider the following objectives:

■ to consume a minimum of raw materials and energy

■ to minimize waste through efficient use of resources

■ to recover useful materials from production waste

■ to treat any residual waste so that it can be converted to an environmentally acceptable form before disposal

In addition to concerns about the depletion of natural resources, there are widespread concerns about waste disposal practices. The burying of untreated industrial wastes, whether classified as hazardous or nonhazardous, is no longer an acceptable practice. Landfill of stabilized residues from the incineration, thermal treatment, or biological oxidation/degradation of industrial wastes is the approach accepted by most countries today.

Certain materials that are the waste products of one process can be recovered for reuse in another application. For example, boiler blowdown may be used as cooling tower makeup in certain instances. Other waste products may contain valuable components that can be extracted. As the cost of waste disposal has escalated, it has become economically feasible to use alternative raw materials and to alter processes so that less waste or less hazardous waste is produced. The treatment of waste and wastewater so that it can be successfully reused is an increasing need (Figure 2-6).

The most efficient driving force for the selection of alternative, waste-reducing raw materials and processes is the marketplace. Because of the high cost of waste treatment and disposal, certain processes can offset higher initial costs with reduced operating expenses. For example, membrane systems (reverse osmosis, electro-dialysis reversal, etc.) have been used successfully to treat boiler makeup water and reduce the total level of contamination in the waste discharge in comparison with ion exchange systems. Membrane treatment of cooling tower blowdown has also been used to reduce the total quantity of wastewater. The stripping of carbon dioxide and ammonia from process condensate streams has made it feasible to reuse them as boiler feedwater. The reduction of cooling tower blowdown by the use of side stream softeners and/or filters, along with effective deposit control and corrosion inhibition programs, is also increasing.

Although global efforts are being made to ensure that the wastes from industrial processes are properly managed, the cost of remedying the damage from past practices must also be addressed. Injudicious burial of industrial wastes in the past has resulted in significant groundwater contamination (leaching) problems. Because the underground movement of chemicals leaching from dumping areas is extremely difficult to monitor and track, this form of pollution is of major concern to the general public. A large percentage of the world's population relies on groundwater from wells or springs for its potable water supply.

Because the turnover of an aquifer can take years, or even decades, any contamination can be serious. Fortunately, certain natural processes, including microbiological digestion, may break down leaching pollutants to nonharmful materials. One remedy that is gaining acceptance is the addition of certain nutrients and inoculum cultures to contaminated soils to accelerate the biological degradation of pollutants. This process is referred to as bioremediation and has many useful variants.

Industrial and commercial producers have an obligation to minimize consumption of the Earth's natural resources and to generate a minimum of pollutants and waste.

The term "zero risk" is often used to represent the ultimate goal of generating products without any possibility of producing environmental effects. As zero risk is approached (although in most cases it can never be fully attained), the cost to the producer and to society in general becomes increasingly larger for each increment of risk avoided (see Figure 2-7).

It has become clear to all nations ·that the protection of the environment is an immediate and ongoing concern. It will take a great deal of time and effort to redesign industrial processes to minimize wastes produced. Deposit and corrosion control treatments that are effective under demanding conditions and also environmentally acceptable are necessary. Efficient treatment, handling, feeding, and control systems are essential to ensure optimum system performance with minimum impact on the environment.

Figure 2-6. Effective treatment can make some wastewaters suitable for reuse.

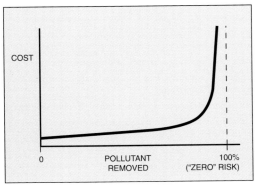

Figure 2-7. The cost of each increment of risk abatement becomes increasingly greater as "zero risk" is approached.

APPLYING QUALITY METHODS TO WATER TREATMENT[1]

Typical utility water systems are subject to considerable variation. Makeup water characteristics can change over time. The abruptness and degree of change depend on the source of the water. Water losses from a recirculating system, changes in production rates, and chemical feed rates all introduce variation into the system and thereby influence the ability to maintain proper control of the system. Other variables inherent in utility water systems include:

- water flow/velocity
- water temperature
- process/skin temperature
- process demands
- evaporation rates
- operator skill/training
- water characteristics (suspended solids, hardness, pH swings)
- treatment product quality

These variables are considered and introduced during the applications and pilot plant testing of new products for the treatment of various water systems. Pilot plant simulation of actual operational variation is a challenging task. Every industrial water system is unique, not only in the production operations it supports and the sources of water it receives, but also in the degree of inherent variation encountered due to the factors listed above. While a very sensitive treatment program that must operate within a narrow control range may be suitable for one system, another system requiring the same degree of protection may be incapable of maintaining the required control. Consequently, inferior results must be accepted unless the system is improved to support the sensitive program.

In operating systems, proper treatment of influent, boiler, cooling, and effluent waters often requires constant adjustment of the chemistry to meet the requirements of rapidly changing system conditions. A well designed program is essential to maintaining proper control. The program should include proper control limits and the ability to troubleshoot problems that interfere with control of water chemistry. Success in troubleshooting depends on the knowledge, logic, and skills of the troubleshooter. In order to improve operations it is necessary to recognize the importance of continuous improvement and to be familiar with some tools and procedures necessary to support this effort.

Adequate and reliable data are essential if variation in a system is to be measured and reduced. Specialized computer software can assist efforts to manage, summarize, and use data effectively. Process data can be stored in a database and retrieved and analyzed as needed in a variety of formats. Computers provide nearly instantaneous access to many months or years of process data that would require several filing cabinets if stored on paper log sheets. The computer can be used to graph and analyze the data in a variety of formats, such as statistical process control (SPC), trend analysis, and histograms (see Figure 3-1). The operator is able to troubleshoot the system based on these analyses without spending large amounts of time manually researching and analyzing the data.

In his classic book *Managerial Breakthrough* (McGraw Hill: New York, 1964, pp 1–14), Dr. J. M. Juran develops the important distinction between quality control and quality improvement, and describes the elements of effective problem-solving in each case. These distinctions and relationships are summarized in Figure 3-2.

[1] Derived from Wilton, C.T. "Quality Improvement Tools," Juran Institute, Inc., 1989.
Reproduced here with permission of the copyright holder.

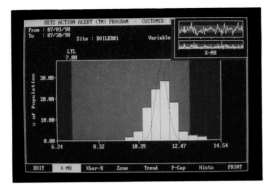

Figure 3-1. The ability to access data and review it in various formats contributes to faster and more effective problem-solving.

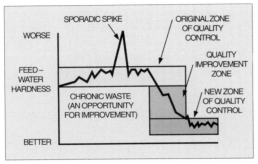

Figure 3-2. Quality improvement program sets new standard for feedwater hardness. (Derived from "Quality Improvement Tools," Wilton, C.T., Juran Institute, Inc., 1989. Reproduced here with permission of the copyright holder.)

QUALITY CONTROL ZONE

Although the performance of a process varies from day to day, the average performance and the range of variation are fairly constant over time. This level of performance is inherent in the process and is provided for in the system design. The Quality Control Zone in Figure 3-2 depicts the accepted average and accepted range of variation in feedwater hardness. This zone is often adopted as the standard of performance.

Sometimes, performance falls outside the accepted, or standard, range of variation in the Quality Control Zone. This is depicted in Figure 3-2 by the sporadic spike. The goal of problem-solving in the Quality Control Zone is to reestablish performance within the standard. This involves the following steps:

■ detecting the change (sporadic spike)

■ identifying the cause of the change

■ taking corrective action to restore the status quo

QUALITY IMPROVEMENT ZONE

Problem solving in the Quality Improvement Zone (also depicted in Figure 3-2) can have an even greater impact. The goal of quality improvement is to reject the status quo as the standard and reach a level of performance never before achieved. This level, the "New Zone of Quality Control," represents the achievement of lower costs and/or better performance. In this case, significantly lower feedwater hardness decreases scaling potential and improves boiler reliability.

This step extends the scope of problem solving beyond the correction of obvious problems. While it is important to "make the system work," it is often more important to view the entire system to identify areas of potential improvement. Some systems are poorly planned; others have not been updated to keep pace with changing requirements and progressing technology. In either case, it is often the system that causes control and operational problems—not the people working within the system.

Quality Improvement Tools

While a proper mindset must exist for continuous improvement, certain problem-solving procedures and tools can add structure and consistency to the effort. The following quality improvement tools provide the means to summarize and present meaningful data in a way that adds significantly to the successful resolution of chronic problems.

Flow Diagrams. A flow diagram provides a graphic presentation of the steps required to produce a desired result. For example, this tool may be used to clarify the procedures used to regenerate a softener or the steps to be taken in the event of an upset in a cooling tower. Flow diagrams are used in problem-solving to give all parties a common understanding of the overall process.

Brainstorming. In diagnosing a problem, new and useful ideas can result when all of the people familiar with the process meet to share their experiences and ideas. Possible causes are discussed and possible solutions are presented and evaluated.

Cause–Effect Diagrams. An important first step in quality improvement is the identification of the root causes of a problem. A cause–effect diagram provides an effective way to organize and display the various ideas of what those root causes might be. Figure 3-3 graphically presents possible causes for reduced demineralizer throughput.

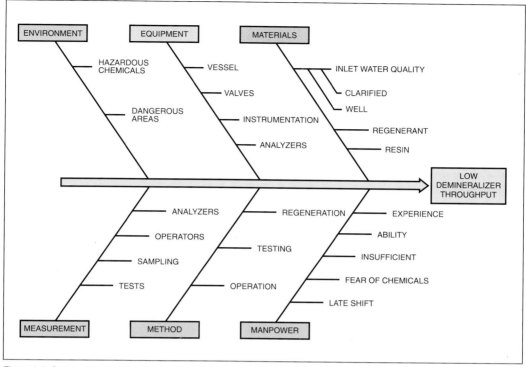

Figure 3-3. Cause–effect diagram presents possible causes for reduced demineralizer throughput.

Scatter Diagrams. A scatter diagram is useful in providing a clear, graphic representation of the relationship between two variables. For example, boiler feedwater iron levels might be plotted as a function of feedwater pH to confirm or rule out a cause–effect relationship.

Pareto Analysis. Pareto analysis is a ranked comparison of factors related to a quality problem, or a ranking of the cost of various problems. It is an excellent graphic means of identifying and focusing on the vital few factors or problems. Figure 3-4 represents an analysis of the calculated cost of various problems interfering with the successful management of a utility water system.

Meaningful Data Collection. Meaningful collection of data and facts is fundamental to every quality improvement effort. Quality improvement is an information-intensive activity. In many cases, problems remain unsolved for long periods of time due to a lack of relevant information. A good data collection system must be carefully planned in order to provide the right information with a minimum of effort and with minimal chance of error.

In order to plan for data collection, it is necessary to identify potential sources of bias and

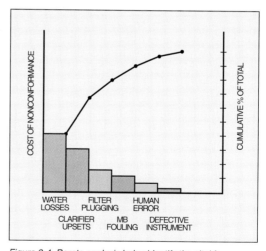

Figure 3-4. Pareto analysis helps identify the vital few factors that have the greatest impact on water system performance.

develop procedures to address them:

■ *Exclusion bias.* If a part of the process being investigated has been left out, the result will be biased if the data is intended to represent the entire process. For example, if data on attemperating water purity is not included in an evaluation of a steam turbine fouling problem, the cause could be missed.

■ *Interaction bias.* The process of collecting the data itself can affect the process being studied. For example, if an operator knows that cooling tower treatment levels are being monitored by the central laboratory, he may be more careful conducting his own tests.

■ *Perception bias.* The attitudes and beliefs of the data collectors can influence what they perceive and how they record it. If an operator believes that swings in steam header pressure are his responsibility, he may record that operation was normal at the

time of boiler water carryover.

■ *Operational bias.* Failure to follow the established procedures is a common operational bias. For example, failure to cool a boiler water sample to 25 °C (77 °F) often leads to an erroneous pH measurement.

Graphs and Charts. Pictorial representations of quantitative data, such as line charts, pie charts, and bar graphs, can summarize large amounts of data in a small area and communicate complex situations concisely and clearly.

Histograms. The pictorial nature of a histogram

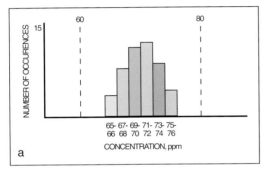

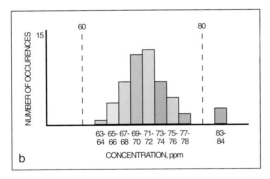

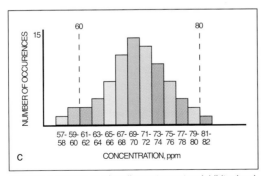

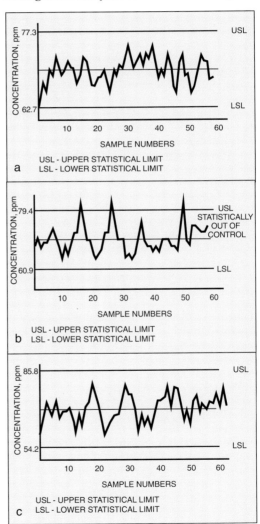

Figure 3-5. Histogram of cooling water system inhibitor level. (a) normal distribution within specified control limits; (b) generally normal distribution with a few isolated data points outside of engineering control limits; (c) normal distribution but too much variation in the system to stay within engineering control limits.

Figure 3-6. Statistical Process Control (SPC) charts present the cooling system inhibitor level data points used to construct the histograms in Figure 3-5.

(a graphic summation of variation in a set of data) reveals patterns that are difficult to see in a simple table of numbers. Figure 3-5(a) is a histogram that shows the variation of inhibitor level in a cooling water system. Each bar along the horizontal axis represents a specific range of inhibitor concentration, in parts per million. The scale on the vertical axis represents the number of occurrences within each range of concentration. The shape of this particular histogram indicates a normal and predictable pattern of distribution. There are no incidents of nonconformance outside of the specified tolerance limits of 60–80 ppm, represented by the dotted lines.

In contrast, the patterns of variation depicted in Figure 3-5(b) and (c) represent problems which must be corrected. The pattern of distribution in Figure 3-5(b) is relatively normal, but a few incidents of nonconformance occur outside of the engineering limits, departing significantly from the otherwise normal distribution. The cause of these occurrences must be investigated, and the process corrected to a more predictable pattern. Figure 3-5(c) represents a normal and predictable pattern, but reveals several occurrences that fall outside of the specified 60–80 ppm limits, indicating that there is too much natural variation in the process.

Statistical Process Control. Statistical process control (SPC) is the use of statistical methods to study, analyze, and control the variation in any process. It is a vehicle through which one can extract meaningful information about a process so that corrective action, where necessary, can be implemented. While a histogram is a pictorial representation of patterns of variation, SPC is used to quantify this variation and determine mathematically whether the process is stable or unstable, predictable or erratic. Figure 3-6 shows three SPC charts of the individual values of measurement used to construct the histograms in Figure 3-5. In these cases, the data is plotted chronologically and used interactively to determine whether a value falls outside of the statistical (predictability) limits.

With statistical process control, the actual historical data is used to calculate the upper and lower statistical limits as a guideline for future operation. Anything falling outside of the statistical limits is considered to be a special cause of variation requiring immediate attention. Of course, if the common causes of variation are excessive for either engineering or economic reasons, as is the case in Figures 3-5(c) and 3-6(c), improvement to the process is necessary until the statistical limits are narrowed to the point of acceptability.

External Treatment

All fresh water comes from rainfall, which percolates into the soil or runs off into rivers and streams.

AERATION

Aeration is a unit process in which air and water are brought into intimate contact. Turbulence increases the aeration of flowing streams (Figure 4-1). In industrial processes, water flow is usually directed countercurrent to atmospheric or forced-draft air flow. The contact time and the ratio of air to water must be sufficient for effective removal of the unwanted gas.

Aeration as a water treatment practice is used for the following operations:

- carbon dioxide reduction (decarbonation)

- oxidation of iron and manganese found in many well waters (oxidation tower)

- ammonia and hydrogen sulfide reduction (stripping)

Aeration is also an effective method of bacteria control.

METHODS OF AERATION

Two general methods may be used for the aeration of water. The most common in industrial use is the water-fall aerator. Through the use of spray nozzles, the water is broken up into small droplets or a thin film to enhance countercurrent air contact.

In the air diffusion method of aeration, air is diffused into a receiving vessel containing counter-current flowing water, creating very small air bubbles. This ensures good air–water contact for "scrubbing" of undesirable gases from the water.

Water-Fall Aerators

Many variations of the water-fall principle are used for this type of aeration. The simplest configuration employs a vertical riser that discharges water by free fall into a basin (Figure 4-2). The riser usually operates on the available head of water. The efficiency of aeration is improved as the fall distance is increased. Also, steps or

shelves may be added to break up the fall and spread the water into thin sheets or films, which increases contact time and aeration efficiency.

Coke tray and wood or plastic slat water-fall aerators are relatively similar in design and have the advantage of small space requirements.

Coke tray aerators are widely used in iron and manganese oxidation because a catalytic effect is secured by contact of the iron/manganese-bearing water with fresh precipitates. These units consist of a series of coke-filled trays through which the water percolates, with additional aeration obtained during the free fall from one tray to the next.

Wood or plastic slat tray aerators are similar to small atmospheric cooling towers. The tray slats are staggered to break up the free fall of the water and create thin films before the water finally drops into the basin.

Forced draft water-fall aerators (see Figure 4-3) are used for many industrial water conditioning purposes. Horizontal wood or plastic slat trays, or towers filled with packing of various shapes and materials, are designed to maximize disruption of

Figure 4-1. Natural aeration in streams.

Figure 4-2. Multicone aerator. (Courtesy of Infilco Degremont, Inc.)

the falling water into small streams for greater air–water contact. Air is forced through the unit by a blower which produces uniform air distribution across the entire cross section, cross current or countercurrent to the fall of the water. Because of these features, forced draft aerators are more efficient for gas removal and require less space for a given capacity.

Air Diffusion Aerators

Air diffusion systems aerate by pumping air into water through perforated pipes, strainers, porous plates, or tubes. Aeration by diffusion is theoretically superior to water-fall aeration because a fine bubble of air rising through water is continually exposed to fresh liquid surfaces, providing maximum water surface per unit volume of air. Also, the velocity of bubbles ascending through the water is much lower than the velocity of free-falling drops of water, providing a longer contact time. Greatest efficiency is achieved when water flow is countercurrent to the rising air bubbles.

APPLICATIONS

In industrial water conditioning, one of the major objectives of aeration is to remove carbon dioxide. Aeration is also used to oxidize soluble iron and manganese (found in many well waters) to insoluble precipitates. Aeration is often used to reduce the carbon dioxide liberated by a treatment process. For example, acid may be fed to the effluent of sodium zeolite softeners for boiler alkalinity control. Carbon dioxide is produced as

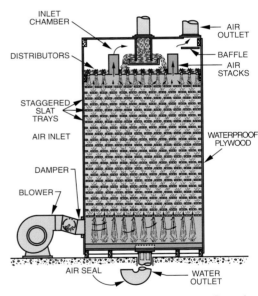

Figure 4-3. Forced draft aerator. (Courtesy of the Permutit Company, Inc.)

a result of the acid treatment, and aeration is employed to rid the water of this corrosive gas. Similarly, when the effluents of hydrogen and sodium zeolite units are blended, the carbon dioxide formed is removed by aeration.

In the case of cold lime softening, carbon dioxide may be removed from the water before the water enters the equipment. When carbon dioxide removal is the only objective, economics usually favor removal of high concentrations of carbon dioxide by aeration rather than by chemical precipitation with lime.

Air stripping may be used to reduce concentrations of volatile organics, such as chloroform, as well as dissolved gases, such as hydrogen sulfide and ammonia. Air pollution standards must be considered when air stripping is used to reduce volatile organic compounds.

Iron and Manganese Removal

Iron and manganese in well waters occur as soluble ferrous and manganous bicarbonates. In the aeration process, the water is saturated with oxygen to promote the following reactions:

$$4Fe(HCO_3)_2 + O_2 + 2H_2O = 4Fe(OH)_3\downarrow + 8CO_2$$

ferrous   oxygen water   ferric   carbon
bicarbonate   hydroxide dioxide

$$2Mn(HCO_3)_2 + O_2 = 2MnO_2 + 4CO_2\uparrow + 2H_2O$$

manganese oxygen manganese carbon   water
bicarbonate   dioxide   dioxide

The oxidation products, ferric hydroxide and manganese dioxide, are insoluble. After aeration, they are removed by clarification or filtration.

Occasionally, strong chemical oxidants such as chlorine (Cl_2) or potassium permanganate ($KMnO_4$) may be used following aeration to ensure complete oxidation.

Dissolved Gas Reduction

Gases dissolved in water follow the principle that the solubility of a gas in a liquid (water) is directly proportional to the partial pressure of the gas above the liquid at equilibrium. This is known as Henry's Law and may be expressed as follows:

$$C_{total} = kP$$

where
C_{total} = total concentration of the gas in solution
P = partial pressure of the gas above the solution
k = a proportionality constant known as Henry's Law Constant

However, the gases frequently encountered in water treatment (with the exception of oxygen) do not behave in accordance with Henry's Law because they ionize when dissolved in water. For example:

$$H_2O + CO_2 \rightleftharpoons H^+ + HCO_3^-$$

water   carbon   hydrogen   bicarbonate
dioxide   ion   ion

$$H_2S \rightleftharpoons H^+ + HS^-$$

hydrogen   hydrogen hydrosulfide
sulfide   ion   ion

$$H_2O + NH_3 \rightleftharpoons NH_4^+ + OH^-$$

water ammonia ammonium hydroxide
ion   ion

Carbon dioxide, hydrogen sulfide, and ammonia are soluble in water under certain conditions to the extent of 1,700, 3,900, and 531,000 ppm, respectively. Rarely are these concentrations encountered except in certain process condensates. In a normal atmosphere, the partial pressure of each of these gases is practically zero. Consequently, the establishment of a state of equilibrium between water and air by means of aeration results in saturation of the water with nitrogen and oxygen and nearly complete removal of other gases.

As the equations above show, ionization of the gases in water is a reversible reaction. The common ion effect may be used to obtain almost complete removal of these gases by aeration. If the concentration of one of the ions on the right side of the equation is increased, the reaction is driven to the left, forming the gas. In the case of carbon dioxide and hydrogen sulfide, hydrogen ion concentration may be increased by the addition of an acid. Bicarbonate and carbonate ions in the water will form carbon dioxide, which can be removed by aeration.

In a similar manner, an increase in hydroxyl ion concentration through the addition of caustic soda aids in the removal of ammonia.

Figures 4-4, 4-5, and 4-6 show the percentage of gas removal that may be obtained at various pH levels.

Gas removal by aeration is achieved as the level of gas in the water approaches equilibrium

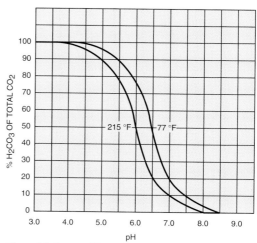

Figure 4-4. Percent CO_2 available for removal.

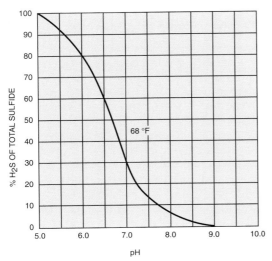

Figure 4-5. Percent H₂S available for removal.

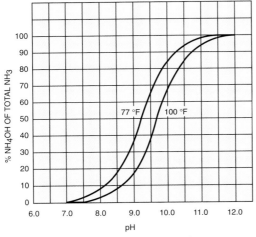

Figure 4-6. Percent NH₃ available for removal.

with the level of the gas in the surrounding atmosphere. The process is improved by an increase in temperature, aeration time, the volume of air in contact with the water, and the surface area of water exposed to the air. As previously indicated, pH is an important consideration. The efficiency of aeration is greater where the concentration of the gas to be removed is high in the water and low in the atmosphere.

LIMITATIONS

Temperature significantly affects the efficiency of air stripping processes. Therefore, these processes may not be suitable for use in colder climates. Theoretically, at 68 °F the carbon dioxide content of the water can be reduced to 0.5 ppm by aeration to equilibrium conditions. This is not always practical from an economic standpoint, and reduction of carbon dioxide to 10 ppm is normally considered satisfactory.

Although removal of free carbon dioxide increases the pH of the water and renders it less corrosive from this standpoint, aeration also results in the saturation of water with dissolved oxygen. This does not generally present a problem when original oxygen content is already high. However, in the case of a well water supply that is high in carbon dioxide but devoid of oxygen, aeration simply exchanges one corrosive gas for another.

The efficiency of aeration increases as the initial concentration of the gas to be removed increases above its equilibrium value. Therefore, with waters containing only a small amount of carbon dioxide, neutralization by alkali addition is usually more cost-effective.

The complete removal of hydrogen sulfide must be combined with pH reduction or chemical oxidation.

Nonvolatile organic compounds cannot be removed by air stripping. For example, phenols and creosols are unaffected by the aeration process alone.

CLARIFICATION

Suspended matter in raw water supplies is removed by various methods to provide a water suitable for domestic purposes and most industrial requirements. The suspended matter can consist of large solids, settleable by gravity alone without any external aids, and nonsettleable material, often colloidal in nature. Removal is generally accomplished by coagulation, flocculation, and sedimentation. The combination of these three processes is referred to as conventional clarification.

Coagulation is the process of destabilization by charge neutralization. Once neutralized, particles no longer repel each other and can be brought together. Coagulation is necessary for the removal of the colloidal-sized suspended matter.

Flocculation is the process of bringing together the destabilized, or "coagulated," particles to form a larger agglomeration, or "floc."

Sedimentation refers to the physical removal from suspension, or settling, that occurs once the particles have been coagulated and flocculated. Sedimentation or subsidence alone, without prior coagulation, results in the removal of only relatively coarse suspended solids.

STEPS OF CLARIFICATION

Finely divided particles suspended in surface water repel each other because most of the surfaces are negatively charged. The following steps in clarification are necessary for particle agglomeration:

■ *Coagulation.* Coagulation can be accomplished through the addition of inorganic salts of aluminum or iron. These inorganic salts neutralize the charge on the particles causing raw water turbidity, and also hydrolyze to form insoluble precipitates, which entrap particles. Coagulation can also be effected by the addition of water-soluble organic polymers with numerous ionized sites for particle charge neutralization.

■ *Flocculation.* Flocculation, the agglomeration of destabilized particles into large particles, can be enhanced by the addition of high-molecular-weight, water-soluble organic polymers. These polymers increase floc size by charged site binding and by molecular bridging.

Therefore, coagulation involves neutralizing charged particles to destabilize suspended solids. In most clarification processes, a flocculation step then follows. Flocculation starts when neutralized or entrapped particles begin to collide and fuse to form larger particles. This process can occur naturally or can be enhanced by the addition of polymeric flocculant aids.

INORGANIC COAGULANTS

Table 5-1 lists a number of common inorganic coagulants. Typical iron and aluminum coagulants are acid salts that lower the pH of the treated water by hydrolysis. Depending on initial raw water alkalinity and pH, an alkali such as lime or caustic must be added to counteract the pH depression of the primary coagulant. Iron and aluminum hydrolysis products play a significant role in the coagulation process, especially in cases where low-turbidity influent waters benefit from the presence of additional collision surface areas.

Variation in pH affects particle surface charge and floc precipitation during coagulation. Iron and aluminum hydroxide flocs are best precipitated at pH levels that minimize the coagulant solubility. However, the best clarification performance may not always coincide with the optimum pH for hydroxide floc formation. Also, the iron and aluminum hydroxide flocs increase volume

Table 5-1. Common inorganic coagulants.

Name	Typical Formula	Typical Strength	Typical Forms Used in Water Treatment	Density	Typical Uses
Aluminum sulfate	$Al_2(SO_4)_3 \cdot$ 14 to 18H_2O	17% Al_2O_3	lump, granular, or powder	60–70 lb/ft³	primary coagulant
Alum		8.25% Al_2O_3	liquid	11.1 lb/gal	
Aluminum chloride	$AlCl_3 \cdot 6H_2O$	35% $AlCl_3$	liquid	12.5 lb/gal	primary coagulant
Ferric sulfate	$Fe_2(SO_4)_3 \cdot 9H_2O$	68% $Fe_2(SO_4)_3$	granular	70–72 lb/ft³	primary coagulant
Ferric-floc	$Fe_2(SO_4)_3 \cdot 5H_2O$	41% $Fe_2(SO_4)_3$	solution	12.3 lb/gal	primary coagulant
Ferric chloride	$FeCl_3$	60% $FeCl_3$, 35–45% $FeCl_3$	crystal, solution	60–64 lb/ft³ 11.2–12.4 lb/gal	primary coagulant
Sodium aluminate	$Na_2Al_2O_4$	38–46% $Na_2Al_2O_4$	liquid	12.3–12.9 lb/gal	primary coagulant; cold/hot precipitation softening

requirements for the disposal of settled sludge.

With aluminum sulfate, optimum coagulation efficiency and minimum floc solubility normally occur at pH 6.0 to 7.0. Iron coagulants can be used successfully over the much broader pH range of 5.0 to 11.0. If ferrous compounds are used, oxidation to ferric iron is needed for complete precipitation. This may require either chlorine addition or pH adjustment. The chemical reactions between the water's alkalinity (natural or supplemented) and aluminum or iron result in the formation of the hydroxide coagulant as in the following:

$$Al_2(SO_4)_3 + 6NaHCO_3 = 2Al(OH)_3\downarrow + 3Na_2SO_4 + 6CO_2$$

aluminum sulfate / sodium bicarbonate / aluminum hydroxide / sodium sulfate / carbon dioxide

$$Fe_2(SO_4)_3 + 6NaHCO_3 = 2Fe(OH)_3\downarrow + 3Na_2SO_4 + 6CO_2$$

ferric sulfate / sodium bicarbonate / ferric hydroxide / sodium sulfate / carbon dioxide

$$Na_2Al_2O_4 + 4H_2O = 2Al(OH)_3\downarrow + 2NaOH$$

sodium aluminate / water / aluminum hydroxide / sodium hydroxide

POLYELECTROLYTES

The term polyelectrolytes refers to all water-soluble organic polymers used for clarification, whether they function as coagulants or flocculants.

Water-soluble polymers may be classified as follows:

■ *anionic*—ionize in water solution to form negatively charged sites along the polymer chain

■ *cationic*—ionize in water solution to form positively charged sites along the polymer chain

■ *nonionic*—ionize in water solution to form very slight negatively charged sites along the polymer chain

Polymeric primary coagulants are cationic materials with relatively low molecular weights (under 500,000). The cationic charge density (available positively charged sites) is very high. Polymeric flocculants or coagulant aids may be anionic, cationic, or nonionic. Their molecular weights may be as high as 50,000,000. Table 5-2 describes some typical organic polyelectrolytes.

For any given particle there is an ideal molecular weight and an ideal charge density for optimum coagulation. There is also an optimum charge density and molecular weight for the most efficient flocculant.

Because suspensions are normally nonuniform, specific testing is necessary to find the coagulants and flocculants with the broadest range of performance.

Primary Coagulant Polyelectrolytes

The cationic polyelectrolytes commonly used as primary coagulants are polyamines and poly-(DADMACS). They exhibit strong cationic ionization and typically have molecular weights of less than 500,000. When used as primary coagulants, they adsorb on particle surfaces, reducing the repelling negative charges. These polymers may also bridge, to some extent, from one particle to another but are not particularly effective flocculants. The use of polyelectrolytes permits water clarification without the precipitation of additional hydroxide solids formed by inorganic coagulants. The pH of

Table 5-2. Common organic polyelectrolytes.

Polymer Type	Typical Formula	Typical Molecular Weight	Available Forms	Typical Uses
Nonionic	Polyacrylamide $$\left[\begin{array}{c} -CH_2-CH- \\ \mid \\ C=O \\ \mid \\ NH_2 \end{array} \right]_n$$	1×10^6 to 2×10^7	powder, emulsion, solution	flocculant in clarification with inorganic or organic coagulants
Anionic	Hydrolyzed polyacrylamide $$\left[\begin{array}{c} -CH_2-CH- \\ \mid \\ C=O \\ \mid \\ NH_2 \end{array} \right]_n \left[\begin{array}{c} CH_2-CH- \\ \mid \\ C=O \\ \mid \\ ONa \end{array} \right]_y$$	1×10^6 to 2×10^7	powder, emulsion, solution	flocculant in clarification with inorganic or organic coagulants
Cationic	Poly(DADMAC) or poly(DMDAAC) polymers $$\left[\begin{array}{c} -CH_2-CH-CH-CH_2- \\ \mid \qquad \mid \\ CH_2 \quad CH_2 \\ \diagdown \diagup \\ N \;\;+\;\; Cl^- \\ \diagup \diagdown \\ CH_3 \quad CH_3 \end{array} \right]_n$$	250 to 500×10^3	solution	primary coagulant alone or in combination with inorganics in clarification
Cationic	Quaternized polyamines $$\left[\begin{array}{c} \qquad\qquad CH_3 \\ \qquad\qquad \mid \\ -CH_2-CH-CH_2-N- \\ \mid \qquad\qquad \mid \\ OH \qquad\quad CH_3 \end{array} \right]_n$$	10 to 500×10^3	solution	primary coagulant alone or in combination with inorganics in clarification
Cationic	Polyamines $$\left[-CH_2-CH_2-NH_2- \right]_n$$	10^3 to 10^6	solution	primary coagulant alone or in combination with inorganics in clarification

the treated water is unaffected.

The efficiency of primary coagulant polyelectrolytes depends greatly on the nature of the turbidity particles to be coagulated, the amount of turbidity present, and the mixing or reaction energies available during coagulation. With lower influent turbidities, more turbulence or mixing is required to achieve maximum charge neutralization.

Raw waters of less than 10 NTU (Nephelometric Turbidity Units) usually cannot be clarified with a cationic polymer alone. Best results are obtained by a combination of an inorganic salt and cationic polymer. In-line clarification should be considered for raw waters with low turbidities.

Generally, waters containing 10 to 60 NTU are most effectively treated with an inorganic coagulant and cationic polymer. In most cases, a significant portion of the inorganic coagulant demand can be met with the cationic polyelectrolyte. With turbidity greater than 60 NTU, a polymeric primary coagulant alone is normally sufficient.

In low-turbidity waters where it is desirable to avoid using an inorganic coagulant, artificial turbidity can be added to build floc. Bentonite clay is used to increase surface area for adsorption and entrapment of finely divided turbidity. A polymeric coagulant is then added to complete the coagulation process.

The use of organic polymers offers several advantages over the use of inorganic coagulants:

■ *The amount of sludge produced during clarification can be reduced by 50–90%.* The approximate dry weight of solids removed per pound of dry alum and ferric sulfate are approximately 0.25 and 0.5 lb, respectively.

■ *The resulting sludge contains less chemically bound water and can be more easily dewatered.*

■ *Polymeric coagulants do not affect pH.* Therefore, the need for supplemental alkalinity, such as lime, caustic, or soda ash, is reduced or eliminated.

■ *Polymeric coagulants do not add to the total dissolved solids concentration.* For example, 1 ppm of alum adds 0.45 ppm of sulfate ion (expressed as $CaCO_3$). The reduction in sulfate can significantly extend the capacity of anion exchange systems.

■ *Soluble iron or aluminum carryover in the clarifier effluent may result from inorganic coagulant use.* Therefore, elimination of the inorganic coagulant can minimize the deposition of these metals in filters, ion exchange units, and cooling systems.

Coagulant Aids (Flocculants)

In certain instances, an excess of primary coagulant (whether inorganic, polymeric, or a combination of both) may be fed to promote large floc size and to increase settling rate. However, in some waters, even high doses of primary coagulant will not produce the desired effluent clarity. A polymeric coagulant aid added after the primary coagulant may, by developing a larger floc at low treatment levels, reduce the amount of primary coagulant required.

Generally, very high-molecular-weight, anionic polyacrylamides are the most effective coagulant aids. Nonionic or cationic types have proven successful in some clarifier systems. Essentially, the polymer bridges the small floc particles and causes them to agglomerate rapidly into larger, more cohesive flocs that settle quickly. The higher-molecular-weight polymers bridge suspended solids most effectively.

Coagulant aids have proven quite successful in precipitation softening and clarification to achieve improved settling rates of precipitates and finished water clarity.

COLOR REDUCTION

Frequently, the objective of clarification is the reduction of color. Swamps and wetlands introduce color into surface waters, particularly after heavy rainfalls. Color-causing materials can cause various problems, such as objectionable taste, increased microbiological content, fouling of anion exchange resins, and interference with coagulation and stabilization of silt, soluble iron, and manganese.

Most organic color in surface waters is colloidal and negatively charged. Chemically, color-producing compounds are classified as humic and fulvic acids. Color can be removed by chlorination and coagulation with aluminum or iron salts or organic polyelectrolytes. Chlorine oxidizes color compounds, while the inorganic coagulants can physically remove many types of organic color by neutralization of surface charges. The use of chlorine to oxidize organic color bodies may be limited due to the production of chlorinated organic by-products, such as trihalomethanes. Additional color removal is acheived by chemical interaction with aluminum or iron hydrolysis products. Highly charged cationic organic polyelectrolytes can also be used to coagulate some types of color particles.

Coagulation for color reduction is normally carried out at pH 4.5 to 5.5. Optimum pH for turbidity removal is usually much higher than that for color reduction. The presence of sulfate ions can interfere with coagulation for color reduction, whereas calcium and magnesium ions can improve the process and broaden the pH range in which color may be reduced effectively.

CONVENTIONAL CLARIFICATION EQUIPMENT

The coagulation/flocculation and sedimentation process requires three distinct unit processes:

■ high shear, rapid mix for coagulation

■ low shear, high retention time, moderate mixing for flocculation

■ liquid and solids separation

Horizontal Flow Clarifiers

Originally, conventional clarification units consisted of large, rectangular, concrete basins divided into two or three sections. Each stage of the clarification process occurred in a single section of the basin. Water movement was horizontal with plug flow through these systems.

Because the design is suited to large-capacity basins, horizontal flow units are still used in some large industrial plants and for clarifying municipal water. The retention time is normally long (up to 4–6 hr), and is chiefly devoted to settling. Rapid mix is typically designed for 3–5 min and slow mix for 15–30 min. This design affords great flexibility in establishing proper chemical addition points. Also, such units are relatively insensitive to sudden changes in water throughput.

The long retention also allows sufficient reaction time to make necessary adjustments in chemical and polymer feed if raw water conditions suddenly change. However, for all but very large treated water demands, horizontal units require high construction costs and more land space per unit of water capacity.

Upflow Clarifiers

Compact and relatively economical, upflow clarifiers provide coagulation, flocculation, and sedimentation in a single (usually circular) steel or concrete tank. These clarifiers are termed "upflow" because the water flows up toward the effluent launders as the suspended solids settle. They are characterized by increased solids contact through internal sludge recirculation. This is a key feature in maintaining a high-clarity effluent and a major difference from horizontal clarifiers.

Because retention time in an upflow unit is approximately 1–2 hr, upflow basins can be much smaller in size than horizontal basins of equal throughput capacity. A rise rate of 0.70–1.25 gpm/ft² of surface area is normal for clarification. Combination softening–clarification units may operate at up to 1.5 gpm/ft² of surface area due to particle size and densities of precipitated hardness.

In order to achieve high throughput efficiency, upflow units are designed to maximize the linear overflow weir length while minimizing the opportunity for short-circuiting through the settling zone. In addition, the two mixing stages for coagulation and flocculation take place within the same clarification tank.

Although upflow units may provide more efficient sedimentation than horizontal designs, many upflow clarifiers compromise on the rapid and slow mix sequences. Some types provide rapid, mechanical mixing and rely on flow turbulence for flocculation; others eliminate the rapid mix stage and provide only moderate turbulence for

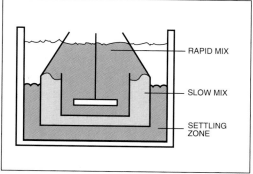

Figure 5-1. Clarifier and zones.

flocculation. However, in most cases, users can overcome rapid mix deficiencies by adding the primary coagulant further upstream of the clarifier. Figure 5-1 shows the rapid mix, slow mix, and settling zones of a typical upflow, solids-contact clarifier.

Sludge Blanket and Solids-Contact Clarification. Most upflow designs are called either "sludge blanket" or "solids-contact" clarifiers. After coagulation and/or flocculation in the sludge blanket units, the incoming water passes through the suspended layer of previously formed floc. Figure 5-2 shows an upflow sludge blanket clarifier.

Because the centerwell in these units is often shaped like an inverted cone, the rise rate of the water decreases as it rises through the steadily enlarging cross section. When the rise rate decreases enough to equal the settling rate of the suspended floc exactly, a distinct sludge/liquid interface forms.

Sludge blanket efficiency depends on the filtering action as the freshly coagulated or flocculated water passes through the suspended floc. Higher sludge levels increase the filtration efficiency. In practice, the top sludge interface is carried at the highest safe level to prevent upsets that might result in large amounts of floc carryover into the overflow. Excessive sludge withdrawal or blowdown should also be avoided. The sludge blanket level is often highly sensitive to changes in throughput, coagulant addition, and changes in raw water chemistry and temperature.

"Solids-contact" refers to units in which large volumes of sludge are circulated internally. The term also describes the sludge blanket unit and simply means that prior to and during sedimentation the chemically treated water contacts previously

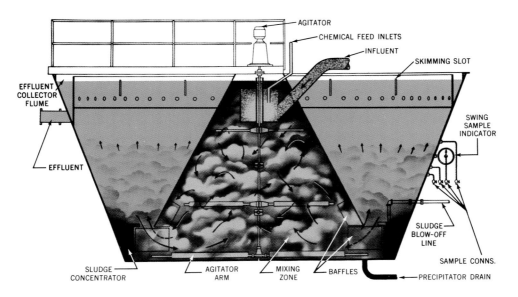

Figure 5-2. Upflow sludge blanket clarifier. (Courtesy of the Permutit Company, Inc.)

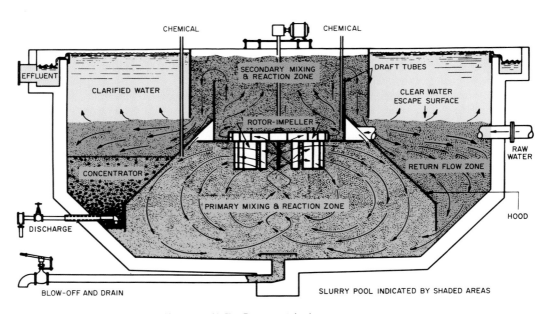

Figure 5-3. Solids-contact clarifier. (Courtesy of Infilco Degremont, Inc.)

coagulated solids. Solids-contact, slurry pool units do not rely on filtration as in sludge blanket designs.

Solids-contact units often combine clarification and precipitation softening. Bringing the incoming raw water into contact with recirculated sludge improves the efficiency of the softening reactions and increases the size and density of the floc particles. Figure 5-3 illustrates a typical solids-contact unit.

IN-LINE CLARIFICATION

In-line clarification is the process of removing raw water turbidity through the addition of coagulant just prior to filtration. In-line clarification is generally limited to raw waters with typical turbidities of less than 20 NTU, although upflow filters may tolerate higher loading. Polyelectrolytes and/or inorganic coagulants are used to improve filtration efficiency and run length. Polymers are favored because they do not create additional suspended solids loading, which can shorten filter run length.

Filter design may be downflow or upflow, depending on raw water turbidity and particle size. The downflow dual-media unit generally consists of layers of various grades of anthracite and sand supported on a gravel bed. After backwashing, the larger anthracite particles separate to the top of the bed, while the more dense, smaller sand particles are at the bottom. The purpose is to allow bed penetration of the floc, which reduces the potential for excessive pressure drops due to blinding off the top portion of filter media. Thus, higher filtration rates are realized without a significant loss in effluent quality. Normal filtration rates are 5–6 gpm/ft².

Coagulant Selection and Feeding for In-Line Clarification

The choice of a polymer coagulant and feed rate depends on equipment design and influent water turbidity. Initially, in-line clarification was used in the treatment of low-turbidity waters, but it is now being used on many types of surface waters. For most waters, the use of a polymeric cationic coagulant alone is satisfactory. However, the addition of a high-molecular-weight, anionic polymer may improve filtration efficiency.

Polymer feed rates are usually lower than those used in conventional clarification, given the same raw water characteristics. Complete charge neutralization and bridging are not necessary and should be avoided, because total coagulation or flocculation may promote excessive entrapment of suspended solids in the first portion of the filter media. This can cause blinding of the media, high pressure drops, and short operating runs.

Sufficient polymer is applied only to initiate neutralization, which allows attraction and adsorption of particles through the entire bed. Often, polymer feed rates are regulated by trial and error on the actual units to minimize effluent turbidity and maximize service run length.

Because optimum flocculation is undesirable, polymers are injected just upstream of the units. Normally, a short mixing period is required to achieve the degree of reaction most suitable for unit operation. Dilution water may be recommended to disperse the polymer properly throughout the incoming water. However, it may be necessary to move the polymer injection point several times to improve turbidity removal. Due to the nature of operation, a change of polymer feed rate will typically show a change in effluent turbidity in a relatively short period of time.

COAGULATION TESTING

Raw water analyses alone are not very useful in predicting coagulation conditions. Coagulation chemicals and appropriate feed rates must be selected according to operating experience with a given raw water or by simulation of the clarification step on a laboratory scale.

Jar testing is the most effective way to simulate clarification chemistry and operation. A multiple-paddle, beaker arrangement (Figure 5-4) permits the comparison of various chemical combinations, all of which are subjected to identical hydraulic conditions. The effects of rapid and slow mix intensity and duration may also be observed.

In addition to determining the optimum chemical program, it is possible to establish the correct order of addition. The most critical measurements in the jar test are coagulant and/or flocculant dosages, pH, floc size and settling characteristics, floc-forming time, and finished water clarity. To simulate sludge circulation, sludge formed in one series of jar tests (or a sludge sample from an operating clarifier) may be added to the next jar test. Results of jar tests are only relative, and frequent adjustments are necessary in full-scale plant operation. Monitoring and control units, such as a streaming current detector, can be used

Figure 5-4. Jar test coagulation study.

for on-line feedback control.

Zeta potential measurements have been used experimentally to predict coagulant requirements and optimum pH levels. Because the measurement technique requires special apparatus and a skilled technician, zeta potential has never become practical for controlling industrial water clarification plants. Also, because zeta potential measures only one aspect of the entire process, it may not reflect all conditions leading to coagulation efficiency.

Chemical Additions

The most efficient method for adding coagulation chemicals varies according to the type of water and system used, and must be checked by means of jar testing. However, there is a usual sequence:

1. chlorine

2. bentonite (for low-turbidity waters)

3. primary inorganic and/or polymer coagulant

4. pH-adjusting chemicals

5. coagulant aid

Waters with a high organic content exhibit an increased primary coagulant demand. Chlorine may be used to assist coagulation by oxidizing organic contaminants which have dispersing properties. Chlorination prior to primary coagulant feed also reduces the coagulant dosage. When an inorganic coagulant is used, the addition of pH-adjusting chemicals prior to the coagulant establishes the proper pH environment for the primary coagulant.

All treatment chemicals, with the exception of coagulant aids, should be added during very turbulent mixing of the influent water. Rapid mixing while the aluminum and iron coagulants are added ensures uniform cation adsorption onto the suspended matter.

High shear mixing is especially important when cationic polymers are used as primary coagulants. In general, it is advisable to feed them as far ahead of the clarifier as possible. However, when a coagulant aid is added, high shear mixing must be avoided to prevent interference with the polymer's bridging function. Only moderate turbulence is needed to generate floc growth.

FILTRATION

Filtration is used in addition to regular coagulation and sedimentation for removal of solids from surface water or wastewater. This prepares the water for use as potable, boiler, or cooling make-up. Wastewater filtration helps users meet more stringent effluent discharge permit requirements.

Filtration, usually considered a simple mechanical process, actually involves the mechanisms of adsorption (physical and chemical), straining, sedimentation, interception, diffusion, and inertial compaction.

Filtration does not remove dissolved solids, but may be used together with a softening process, which does reduce the concentration of dissolved solids. For example, anthracite filtration is used to remove residual precipitated hardness salts remaining after clarification in precipitation softening.

In most water clarification or softening processes where coagulation and precipitation occur, at least a portion of the clarified water is filtered. Clarifier effluents of 2–10 NTU may be improved to 0.1–1.0 NTU by conventional sand filtration. Filtration ensures acceptable suspended solids concentrations in the finished water even when upsets occur in the clarification processes.

TYPICAL CONSTRUCTION

Conventional gravity and pressure rapid filters operate downflow. The filter medium is usually a 15–30 in. deep bed of sand or anthracite. Single or multiple grades of sand or anthracite may be used.

A large particle bed supports the filter media to prevent fine sand or anthracite from escaping into the underdrain system. The support bed also serves to distribute backwash water. Typical support beds consist of ⅛–1½ in. gravel or anthracite in graded layers to a depth of 12–16 in.

TYPES OF MEDIA

Quartz sand, silica sand, anthracite coal, garnet, magnetite, and other materials may be used as filtration media. Silica sand and anthracite are the most commonly used types. When silica is not suitable (e.g., in filters following a hot process softener where the treated water is intended for boiler feed), anthracite is usually used.

The size and shape of the filter media affect the efficiency of the solids removal. Sharp, angular media form large voids and remove less fine material than rounded media of equivalent size. The media must be coarse enough to allow solids to penetrate the bed for 2–4 in. Although most suspended solids are trapped at the surface or in the first 1–2 in. of bed depth, some penetration is essential to prevent a rapid increase in pressure drop.

Sand and anthracite for filters are rated by effective particle size and uniformity. The effective size is such that approximately 10% of the total grains by weight are smaller and 90% are larger. Therefore, the effective size is the minimum size of most of the particles. Uniformity is measured by comparison of effective size to the size at which 60% of the grains by weight are smaller and 40% are larger. This latter size, divided by the effective size, is called the uniformity coefficient— the smaller the uniformity coefficient, the more uniform the media particle sizes.

Finer sands result in shallower zones for the retention of suspended matter. The most desirable media size depends on the suspended solids characteristics as well as the effluent quality requirements and the specific filter design. In general, rapid sand filters use sand with an effective size of 0.35–0.60 mm (0.014–0.024 in.) and a maximum uniformity coefficient of 1.7. Coarse media, often 0.6–1.0 mm (0.024–0.04 in.), are used for closely controlled coagulation and sedimentation.

MIXED MEDIA FILTER BEDS

The terms "multilayer," "in-depth," and "mixed media" apply to a type of filter bed which is graded by size and density. Coarse, less dense particles are at the top of the filter bed, and fine, more dense particles are at the bottom. Downflow filtration allows deep, uniform penetration by particulate matter and permits high filtration rates and long service runs. Because small particles at the bottom are also more dense (less space between particles), they remain at the bottom. Even after high-rate backwashing, the layers remain in their proper location in the mixed media filter bed.

Table 6-1 lists four media that are used in multilayer filtration. Several other mixed media combinations have also been tested and used effectively. The use of too many different media layers can cause severe backwashing difficulties. For example, if all four materials listed in Table 6-1 were used in the same filter, a wash rate high enough to expand the magnetite layer might wash the anthracite from the filter. High wash water requirements would also result.

Anthracite/sand filter beds normally provide all of the advantages of single-media filtration but require less backwash water than sand or anthracite alone. Similar claims have been made for anthracite/sand/garnet mixed units. The major advantages of dual-media filtration are higher rates and longer runs. Anthracite/sand/garnet beds have operated at normal rates of approximately 5 gpm/ft^2 and peak rates as high as 8 gpm/ft^2 without loss of effluent quality.

Table 6-1. Media used in multilayer filtration.

Media	Effective Size, mm (in.)	Specific Gravity
Anthracite	0.7–1.7 (0.03–0.07)	1.4
Sand	0.3–0.7 (0.01–0.03)	2.6
Garnet	0.4–0.6 (0.016–0.024)	3.8
Magnetite	0.3–0.5 (0.01–0.02)	4.9

CAPPING OF SAND FILTERS

Rapid sand filters can be converted for mixed media operation to increase capacity by 100%. The cost of this conversion is much lower than that of installing additional rapid sand filters.

Capping involves the replacement of a portion of the sand with anthracite. In this conversion, a 2–6 in. layer of 0.4–0.6 mm (0.016–0.024 in.) sand

is removed from the surface of a bed and replaced with 4–8 in. of 0.9 mm (0.035 in.) anthracite. If an increase in capacity is desired, a larger amount of sand is replaced. Pilot tests should be run to ensure that a reduction in the depth of the finer sand does not reduce the quality of the effluent.

GRAVITY FILTERS

Gravity filters (see Figure 6-1) are open vessels that depend on system gravity head for operation. Apart from the filter media, the essential components of a gravity filter include the following:

■ *The filter shell*, which is either concrete or steel and can be square, rectangular, or circular. Rectangular reinforced concrete units are most widely used.

■ *The support bed*, which prevents loss of fine sand or anthracite through the underdrain system. The support bed, usually 1–2 ft deep, also distributes backwash water.

■ *An underdrain system*, which ensures uniform collection of filtered water and uniform distribution of backwash water. The system may consist of a header and laterals, with perforations or strainers spaced suitably. False tank bottoms with appropriately spaced strainers are also used for underdrain systems.

■ *Wash water troughs*, large enough to collect backwash water without flooding. The troughs are spaced so that the horizontal travel of backwash water does not exceed 3–3½ ft. In conventional sand bed units, wash troughs are placed approximately 2 ft above the filter surface. Sufficient freeboard must be provided to prevent loss of a portion of the filter media during operation at maximum backwash rates.

■ *Control devices* that maximize filter operation efficiency. Flow rate controllers, operated by venturi tubes in the effluent line, automatically maintain uniform delivery of filtered water. Backwash flow rate controllers are also used. Flow rate and head loss gauges are essential for efficient operation.

PRESSURE FILTERS

Pressure filters are typically used with hot process softeners to permit high-temperature operation and to prevent heat loss. The use of pressure filters eliminates the need for repumping of

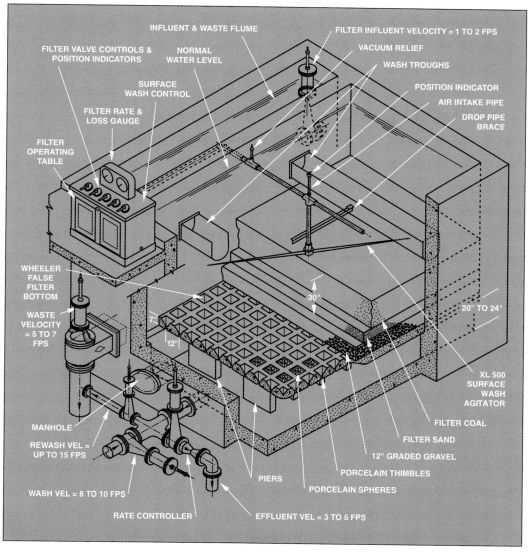

Figure 6-1. Typical gravity filter unit. (Courtesy of Roberts Filter Manufacturing Co.)

filtered water. Pressure filters are similar to gravity filters in that they include filter media, supporting bed, underdrain system, and control device; however, the filter shell has no wash water troughs.

Pressure filters, designed vertically or horizontally, have cylindrical steel shells and dished heads. Vertical pressure filters (see Figure 6-2) range in diameter from 1 to 10 ft with capacities as great as 300 gpm at filtration rates of 3 gpm/ft². Horizontal pressure filters, usually 8 ft in diameter, are 10–25 ft long with capacities from 200 to 600 gpm. These filters are separated into compartments to allow individual backwashing.

Backwash water may be returned to the clarifier or softener for recovery.

Pressure filters are usually operated at a service flow rate of 3 gpm/ft². Dual or multimedia filters are designed for 6–8 gpm/ft². At ambient temperature, the recommended filter backwash rate is 6–8 gpm/ft² for anthracite and 13–15 gpm/ft² for sand. Anthracite filters associated with hot process softeners require a backwash rate of 12–15 gpm/ft² because the water is less dense at elevated operating temperatures. Cold water should not be used to backwash a hot process filter. This would cause expansion and contraction of the system metallurgy, which would lead to

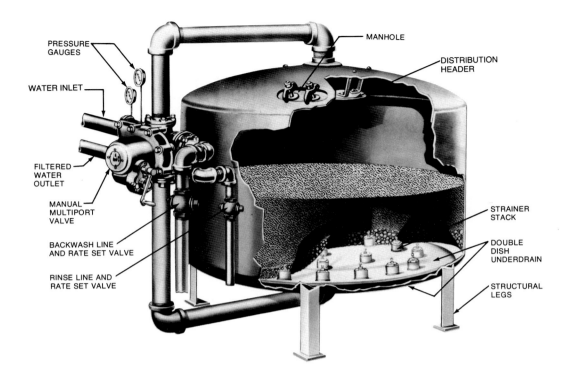

PRESSURE GAUGES

WATER INLET

FILTERED WATER OUTLET

MANUAL MULTIPORT VALVE

BACKWASH LINE AND RATE SET VALVE

RINSE LINE AND RATE SET VALVE

MANHOLE

DISTRIBUTION HEADER

STRAINER STACK

DOUBLE DISH UNDERDRAIN

STRUCTURAL LEGS

Figure 6-2. Vertical-type pressure sand filter. (Courtesy of The Permutit Company, Inc.)

metal fatigue. Also, the oxygen-laden cold water would accelerate corrosion.

UPFLOW FILTERS

Upflow units contain a single filter medium—usually graded sand. The finest sand is at the top of the bed with the coarsest sand below. Gravel is retained by grids in a fixed position at the bottom of the unit. The function of the gravel is to ensure proper water distribution during the service cycle. Another grid above the graded sand prevents fluidization of the media. Air injection during cleaning (not considered backwash because the direction of flow is the same as when in-service) assists in the removal of solids and the reclassification of the filter media. During operation, the larger, coarse solids are removed at the bottom of the bed, while smaller solids particles are allowed to penetrate further into the media. Typical service flow rates are 5–10 gpm/ft². An example of this unit is shown in Figure 6-3.

AUTOMATIC GRAVITY FILTERS

Several manufacturers have developed gravity

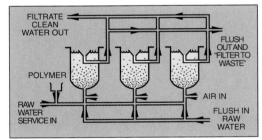

FILTRATE CLEAN WATER OUT

POLYMER

RAW WATER SERVICE IN

FLUSH OUT AND "FILTER TO WASTE"

AIR IN

FLUSH IN RAW WATER

Figure 6-3. Upflow in-line filter. (Courtesy of L'Eau Claire Systems, Inc.)

filters that are backwashed automatically at a preset head loss. Head loss (water level above the media) actuates a backwash siphon and draws wash water from storage up through the bed and out through the siphon pipe to waste. A low level in the backwash storage section breaks the siphon, and the filter returns to service.

Automatic gravity filters are available in diameters of up to 15 ft. When equipped with a high-rate, multilayer media, a single large-diameter unit can filter as much as 1,000 gpm. An example is shown in Figure 6-4.

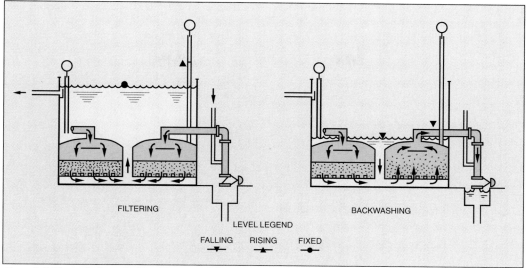

Figure 6-4. Monovalve® gravity filter. (Courtesy of Graver Div., Ecodyne Corporation.)

CONTINUOUS CLEANING FILTERS

Continuous cleaning filter systems eliminate off-line backwash periods by backwashing sections of the filter or portions of the filter media continuously, on-line. Various designs have been introduced. An example is shown in Figure 6-5.

FILTER WASHING—GRAVITY FILTERS

Periodic washing of filters is necessary for the removal of accumulated solids. Inadequate cleaning permits the formation of permanent clumps, gradually decreasing filter capacity. If fouling is severe, the media must be cleaned chemically or replaced.

For cleaning of rapid downflow filters, clean water is forced back up and through the media. In conventional gravity units, the backwash water lifts solids from the bed into wash troughs and carries them to waste. Either of two backwash techniques can be used, depending on the design of the media support structure and the accessory equipment available:

■ *High-rate backwash*, which expands the media by at least 10%. Backwash rates of 12–15 gpm/ft² or higher are common for sand, and rates for anthracite may range from 8 to 12 gpm/ft².

■ *Low-rate backwash*, with no visible bed expansion, combined with air scouring.

Where only water is used for backwash, the backwash may be preceded by surface washing.

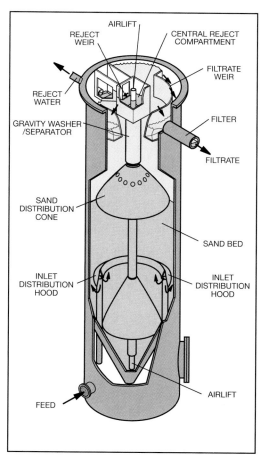

Figure 6-5. DynaSand® continuous cleaning filter. (Courtesy of Parkson Corp.)

In surface washing, strong jets of high-pressure water from fixed or revolving nozzles assist in breaking the filter surface crust. After the surface wash (when there is provision for surface washing), the unit is backwashed for approximately 5–10 min. Following backwash, a small amount of rinse water is filtered to waste, and the filter is returned to service.

High-rate backwash can cause the formation of mud balls inside the filter bed. A high backwash rate and resulting bed expansion can produce random currents in which certain zones of the expanded bed move upward or downward. Encrusted solids from the surface can be carried down to form mud balls. Efficient surface washing helps prevent this condition.

Air scouring with low-rate backwashing can break up the surface crust without producing random currents, if the underdrain system is designed to distribute air uniformly. Solids removed from the media collect in the layer of water between the media surface and wash channels. After the air is stopped, this dirty water is normally flushed out by increased backwash water flow rate or by surface draining. Wash water consumption is approximately the same whether water-only or air/water backwashing is employed.

IN-LINE CLARIFICATION

In-line clarification is the removal of suspended solids through the addition of in-line coagulant followed by rapid filtration. This process is also referred to as in-line filtration, or contact filtration. The process removes suspended solids without the use of sedimentation basins. Coagulation may be achieved in in-line clarification by either of two methods:

- an inorganic aluminum or iron salt used alone or with a high molecular weight polymeric coagulant

- a strongly cationic organic polyelectrolyte

Because metal hydroxides form precipitates, only dual-media filters should be used with inorganic coagulant programs. Floc particles must be handled in filters with coarse-to-fine graded media to prevent rapid blinding of the filter and eliminate backwashing difficulties. Where a high molecular weight polymeric coagulant is used, feed rates of less than 0.1 ppm maximize solids removal by increasing floc size and promoting particle absorption within the filter. This filtration technique readily yields effluent turbidities of less than 0.5 NTU.

The second method of coagulant pretreatment involves the use of a single chemical, a strongly charged cationic polyelectrolyte. This treatment forms no precipitation floc particles, and usually no floc formation is visible in the filter influent.

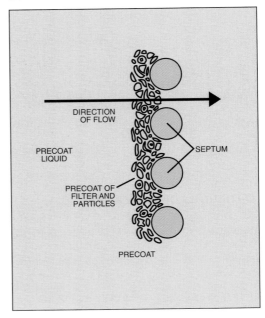

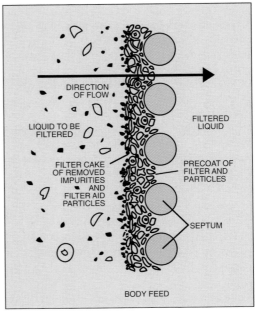

Figure 6-6. Principles of diatomite filtration. (Courtesy of Johns-Manville Corp.)

Solids are removed within the bed by adsorption and by flocculation of colloidal matter directly onto the surface of the sand or anthracite media. The process may be visualized as seeding of the filter bed surfaces with positive cationic charges to produce a strong pull on the negatively charged particles. Because gelatinous hydroxide precipitates are not present in this process, single-media or upflow filters are suitable for poly-electrolyte clarification.

In-line clarification provides an excellent way to improve the efficiency of solids removal from turbid surface waters. Effluent turbidity levels of less than 1 NTU are common with this method.

PRECOAT FILTRATION

Precoat filtration is used to remove very small particulate matter, oil particles, and even bacteria from water. This method is practical only for relatively small quantities of water which contain low concentrations of contaminants.

Precoat filtration may be used following conventional clarification processes to produce water of very low suspended solids content for specific application requirements. For example, precoat filters are often used to remove oil from contaminated condensate.

In precoat filtration, the precoat media, typically diatomaceous earth, acts as the filter media and forms a cake on a permeable base or septum. The base must prevent passage of the precoat media without restricting the flow of filtered water and must be capable of withstanding high pressure differentials. Filter cloths, porous stone tubes, porous paper, wire screens, and wire-wound tubes are used as base materials.

The supporting base material is first precoated with a slurry of precoat media. Additional slurry (body feed) is usually added during the filter run. When the accumulation of matter removed by filtration generates a high pressure drop across the filter, the filter coating is sloughed off by backwashing. The filter bed is then precoated and returned to service. Chemical coagulants are not usually needed but have been used where an ultrapure effluent is required.

PRECIPITATION SOFTENING

Precipitation softening processes are used to reduce raw water hardness, alkalinity, silica, and other constituents. This helps prepare water for direct use as cooling tower makeup or as a first-stage treatment followed by ion exchange for boiler makeup or process use. The water is treated with lime or a combination of lime and soda ash (carbonate ion). These chemicals react with the hardness and natural alkalinity in the water to form insoluble compounds. The compounds precipitate and are removed from the water by sedimentation and, usually, filtration. Waters with moderate to high hardness and alkalinity concentrations (150–500 ppm as $CaCO_3$) are often treated in this fashion.

CHEMISTRY OF PRECIPITATION SOFTENING

In almost every raw water supply, hardness is present as calcium and magnesium bicarbonate, often referred to as carbonate hardness or temporary hardness. These compounds result from the action of acidic, carbon dioxide laden rain water on naturally occurring minerals in the earth, such as limestone. For example:

$$CO_2 + H_2O = H_2CO_3$$
$$\text{carbon dioxide} \quad \text{water} \quad \text{carbonic acid}$$

$$H_2CO_3 + CaCO_3\downarrow = Ca(HCO_3)_2$$
$$\text{carbonic acid} \quad \text{calcium carbonate} \quad \text{calcium bicarbonate}$$

Hardness may also be present as a sulfate or chloride salt, referred to as noncarbonate or permanent hardness. These salts are caused by mineral acids present in rain water or the solution of naturally occurring acidic minerals.

The significance of "carbonate" or "temporary" hardness as contrasted to "noncarbonate" or "permanent" hardness is that the former may be reduced in concentration simply by heating. In effect, heating reverses the solution reaction:

$$Ca(HCO_3)_2 + \text{Heat} = CaCO_3\downarrow + H_2O + CO_2$$
$$\text{calcium bicarbonate} \quad \text{calcium carbonate} \quad \text{water} \quad \text{carbon dioxide}$$

The reduction of noncarbonate hardness, by contrast, requires chemical addition. A combination of lime and soda ash, along with coagulant and flocculant chemicals, is added to raw water to promote a precipitation reaction. This allows softening to take place.

COLD LIME SOFTENING

Precipitation softening accomplished at ambient temperatures is referred to as cold lime softening. When hydrated lime, $Ca(OH)_2$, is added to the water being treated, the following reactions occur:

$$CO_2 + Ca(OH)_2 = CaCO_3\downarrow + H_2O$$
$$\text{carbon dioxide} \quad \text{calcium hydroxide} \quad \text{calcium carbonate} \quad \text{water}$$

$$Ca(HCO_3)_2 + Ca(OH)_2 = 2CaCO_3\downarrow + 2H_2O$$
$$\text{calcium bicarbonate} \quad \text{calcium hydroxide} \quad \text{calcium carbonate} \quad \text{water}$$

$$Mg(HCO_3)_2 + 2Ca(OH)_2 =$$
$$\text{magnesium bicarbonate} \quad \text{calcium hydroxide}$$
$$Mg(OH)_2\downarrow + 2CaCO_3\downarrow + 2H_2O$$
$$\text{magnesium hydroxide} \quad \text{calcium carbonate} \quad \text{water}$$

If the proper chemical control is maintained on lime feed, the calcium hardness may be reduced to 35–50 ppm. Magnesium reduction is a function of the amount of hydroxyl (OH^-) alkalinity excess maintained. Figures 7-1 and 7-2 show these relationships.

Noncarbonate or permanent calcium hardness, if present, is not affected by treatment with lime alone. If noncarbonate magnesium hardness is

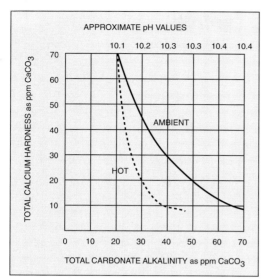

Figure 7-1. Calcium reduction vs. carbonate alkalinity.

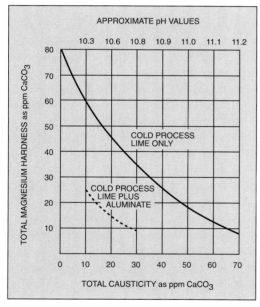

Figure 7-2. Magnesium reduction vs. causticity.

present in an amount greater than 70 ppm and an excess hydroxyl alkalinity of about 5 ppm is maintained, the magnesium will be reduced to about 70 ppm, but the calcium will increase in proportion to the magnesium reduction.

For example, in cold lime treatment of a water containing 110 ppm of calcium, 95 ppm of magnesium, and at least 110 ppm of alkalinity (all expressed as calcium carbonate), calcium could theoretically be reduced to 35 ppm and the

magnesium to about 70 ppm. However, an additional 25 ppm of calcium would be expected in the treated water due to the following reactions:

$$\underset{\substack{\text{magnesium}\\\text{sulfate}}}{MgSO_4} + \underset{\substack{\text{calcium}\\\text{hydroxide}}}{Ca(OH)_2} = \underset{\substack{\text{magnesium}\\\text{hydroxide}}}{Mg(OH)_2\downarrow} + \underset{\substack{\text{calcium}\\\text{sulfate}}}{CaSO_4}$$

$$\underset{\substack{\text{magnesium}\\\text{chloride}}}{MgCl_2} + \underset{\substack{\text{calcium}\\\text{hydroxide}}}{Ca(OH)_2} = \underset{\substack{\text{magnesium}\\\text{hydroxide}}}{Mg(OH)_2\downarrow} + \underset{\substack{\text{calcium}\\\text{chloride}}}{CaCl_2}$$

To improve magnesium reduction, which also improves silica reduction in cold process softening, sodium aluminate may be used. The sodium aluminate provides hydroxyl ion (OH^-) needed for improved magnesium reduction, without increasing calcium hardness in the treated water. In addition, the hydrolysis of sodium aluminate results in the formation of aluminum hydroxide, which aids in floc formation, sludge blanket conditioning, and silica reduction. The reactions are as follows:

$$\underset{\substack{\text{sodium}\\\text{aluminate}}}{Na_2Al_2O_4} + \underset{\text{water}}{4H_2O} = \underset{\substack{\text{aluminum}\\\text{hydroxide}}}{2Al(OH)_3\downarrow} + \underset{\substack{\text{sodium}\\\text{hydroxide}}}{2NaOH}$$

$$\underset{\substack{\text{magnesium}\\\text{sulfate/}\\\text{chloride}}}{Mg\begin{bmatrix}SO_4\\Cl_2\end{bmatrix}} + \underset{\substack{\text{sodium}\\\text{hydroxide}}}{2NaOH} = \underset{\substack{\text{magnesium}\\\text{hydroxide}}}{Mg(OH)_2\downarrow} + \underset{\substack{\text{sodium}\\\text{sulfate/}\\\text{chloride}}}{\begin{bmatrix}Na_2SO_4\\2NaCl\end{bmatrix}}$$

Soda ash (Na_2CO_3) may be used to improve hardness reduction. It reacts with noncarbonate calcium hardness according to the following:

$$\underset{\substack{\text{calcium}\\\text{sulfate}}}{CaSO_4} + \underset{\substack{\text{sodium}\\\text{carbonate}}}{Na_2CO_3} = \underset{\substack{\text{calcium}\\\text{carbonate}}}{CaCO_3\downarrow} + \underset{\substack{\text{sodium}\\\text{sulfate}}}{Na_2SO_4}$$

$$\underset{\substack{\text{calcium}\\\text{chloride}}}{CaCl_2} + \underset{\substack{\text{sodium}\\\text{carbonate}}}{Na_2CO_3} = \underset{\substack{\text{calcium}\\\text{carbonate}}}{CaCO_3\downarrow} + \underset{\substack{\text{sodium}\\\text{chloride}}}{2NaCl}$$

However, noncarbonate magnesium hardness reduction in cold process softening requires added lime. The reactions are as follows:

$$\underset{\substack{\text{magnesium}\\\text{sulfate}}}{MgSO_4} + \underset{\substack{\text{calcium}\\\text{hydroxide}}}{Ca(OH)_2} + \underset{\substack{\text{sodium}\\\text{carbonate}}}{Na_2CO_3} =$$

$$\underset{\substack{\text{magnesium}\\\text{hydroxide}}}{Mg(OH)_2\downarrow} + \underset{\substack{\text{calcium}\\\text{carbonate}}}{CaCO_3\downarrow} + \underset{\substack{\text{sodium}\\\text{sulfate}}}{Na_2SO_4}$$

$$\underset{\substack{\text{magnesium}\\\text{chloride}}}{MgCl_2} + \underset{\substack{\text{calcium}\\\text{hydroxide}}}{Ca(OH)_2} + \underset{\substack{\text{sodium}\\\text{carbonate}}}{Na_2CO_3} =$$

$$\underset{\substack{\text{magnesium}\\\text{hydroxide}}}{Mg(OH)_2\downarrow} + \underset{\substack{\text{calcium}\\\text{carbonate}}}{CaCO_3\downarrow} + \underset{\substack{\text{sodium}\\\text{chloride}}}{2NaCl}$$

In these reactions, dissolved solids are not reduced because a solution reaction product (sodium sulfate or sodium chloride) is formed.

WARM LIME SOFTENING

The warm lime softening process operates in the temperature range of 120–140 °F (49–60 °C). The solubilities of calcium, magnesium, and silica are reduced by increased temperature. Therefore, they are more effectively removed by warm lime softening than by cold lime softening. This process is used for the following purposes:

- *To recover waste heat as an energy conservation measure.* The water to be treated is heated by a waste stream, such as boiler blowdown or low-pressure exhaust steam, to recover the heat content.

- *To prepare feed to a demineralization system.* The lower levels of calcium, magnesium, and especially silica reduce the ionic loading on the demineralizer when warm lime-softened water is used rather than cold lime-softened water. This may reduce both the capital and operating costs of the demineralizer. However, most strong base anion resins have a temperature limitation of 140 °F (60 °C); therefore, additional increases in temperature are not acceptable for increasing the effectiveness of contaminant reduction.

- *To lower the blowdown discharge from cooling systems.* Cooling tower blowdown may be treated with lime and soda ash or caustic to reduce calcium and magnesium levels so that much of the blowdown may be returned to the cooling system. Silica levels in the recirculating cooling water are also controlled in this manner.

In any warm lime or warm lime–soda ash process, temperature control is critical because temperature variations of as little as 4 °F/hr (2 °C/hr) can cause gross carryover of the softener precipitates.

HOT PROCESS SOFTENING

Hot process softening is usually carried out under pressure at temperatures of 227–240 °F (108–116 °C). At the operating temperature, hot process softening reactions go essentially to completion. This treatment method involves the same reactions described above, except that raw water CO_2 is vented and does not participate in the lime reaction. The use of lime and soda ash permits hardness reduction down to 0.5 gr/gal, or about 8 ppm, as calcium carbonate.

Magnesium is reduced to 2–5 ppm because of the lower solubility of magnesium hydroxide at the elevated temperatures.

SILICA REDUCTION

Hot process softening can also provide very good silica reduction. The silica reduction is accomplished through adsorption of the silica on the magnesium hydroxide precipitate. If there is insufficient magnesium present in the raw water to reduce silica to the desired level, magnesium compounds (such as magnesium oxide, magnesium sulfate, magnesium carbonate, or dolomitic lime) may be used. Figure 7-3 is a plot of magnesium oxide vs. raw water silica (in ppm), which may be used to estimate the quantity of magnesium oxide required to reduce silica to the levels indicated. Magnesium oxide is the preferred chemical because it does not increase the dissolved solids concentration of the water.

Good sludge contact enhances silica reduction. To ensure optimum contact, sludge is frequently recirculated back to the inlet of the unit.

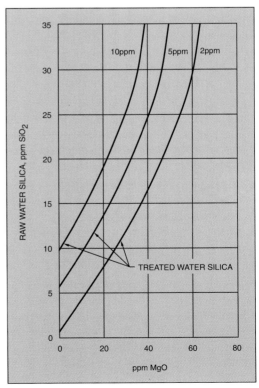

Figure 7-3. Approximate magnesium oxide requirements for silica removal in a hot process softener.

Cold or warm process softening is not as effective as hot process softening for silica reduction. However, added magnesium oxide and good sludge contact will improve results.

Predicted analyses of a typical raw water treated by various lime and lime–soda softening processes are presented in Table 7-1.

ALKALINITY REDUCTION

Treatment by lime precipitation reduces alkalinity. However, if the raw water alkalinity exceeds the total hardness, sodium bicarbonate alkalinity is present. In such cases, it is usually necessary to reduce treated water alkalinity in order to reduce condensate system corrosion or permit increased cycles of concentration.

Treatment by lime converts the sodium bicarbonate in the raw water to sodium carbonate as follows:

$$2NaHCO_3 + Ca(OH)_2 =$$
sodium bicarbonate   calcium hydroxide

$$CaCO_3\downarrow + Na_2CO_3 + 2H_2O$$
calcium carbonate   sodium carbonate   water

Calcium sulfate (gypsum) may be added to reduce the carbonate to required levels. The reaction is as follows:

$$Na_2CO_3 + CaSO_4 = CaCO_3\downarrow + Na_2SO_4$$
sodium carbonate   calcium sulfate   calcium carbonate   sodium sulfate

This is the same reaction involved in the reduction of noncarbonate calcium hardness previously discussed. Table 7-2 shows the treated water

Table 7-2. Alkalinity relationships as determined by titrations.

	Hydroxide	Carbonate	Bicarbonate
P = O	O	O	M
P = M	P	O	O
2P = M	O	2P	O
2P < M	O	2P	M – 2P
2P > M	2P – M	2(M – P)	O

alkalinity relationships to be expected in lime–soda ash softened water.

REDUCTION OF OTHER CONTAMINANTS

Lime softening processes, with the usual filters, will reduce oxidized iron and manganese to about 0.05 and 0.01 ppm, respectively. Raw water organics (color-contributing colloids) are also reduced.

Turbidity, present in most surface supplies, is reduced to about 1.0 NTU with filtration following chemical treatment. Raw water turbidity in excess of 100 NTU may be tolerated in these systems; however, it may be necessary to coagulate raw water solids with a cationic polymer before the water enters the softener vessel to assist liquid-solids separation.

Oil may also be removed by adsorption on the precipitates formed during treatment. However, oil in concentrations above about 30 ppm should be reduced before lime treatment because higher concentrations of oil may exert a dispersing influence and cause floc carryover.

PRECIPITATION PROCESS (CHEMICAL) CONTROL

Lime or lime–soda softener control is usually based on treated water alkalinity and hardness. Samples are tested to determine the alkalinity to

Table 7-1. Typical softener effluent analyses.

	Raw Water	Removal of Calcium Alkalinity Cold-Lime	Lime–soda Softening (Cold)	Lime–soda Softening (Hot)[a]	Lime Softening (Hot)[a]
Total Hardness (as CaCO_3), ppm	250	145	81	20	120
Calcium Hardness (as CaCO_3), ppm	150	85	35	15	115
Magnesium Hardness (as CaCO_3), ppm	100	60	46	5	5
"P" Alkalinity (as CaCO_3), ppm	0	27	37	23	18
"M" Alkalinity (as CaCO_3), ppm	150	44	55	40	28
Silica (as SiO_2), ppm	20	19	18	1–2	1–2
pH	7.5	10.3	10.6	10.5	10.4

[a] Removal of SiO_2 by the hot process, to the levels shown, may require the feed of supplemental magnesium oxide. Sludge recirculation is necessary. All raw water constituents will be diluted by the steam used for heating by approximately 15% if the process is hot.

the P (phenolphthalein, pH 8.3) and M (methyl orange or methyl purple, pH 4.3) end points. The following relationships apply:

$$P \text{ (ppm as CaCO}_3) = \underset{\text{hydroxyl}}{OH^-} + \underset{\text{carbonate}}{^1/_2\, CO_3^{2-}}$$

$$M \text{ (ppm CaCO}_3) = \underset{\text{hydroxyl}}{OH^-} + \underset{\text{carbonate}}{CO_3^{2-}} + \underset{\text{bicarbonate}}{HCO_3^-}$$

In the presence of hydroxyl ion (OH^-), bicarbonate concentration is so low that it may be assumed to be zero.

In the precipitation process, it is advisable to ensure that all of the bicarbonate has been converted to carbonate (the least soluble form of the calcium); therefore, a slight excess of hydroxyl ion should be maintained in the treated water. When the equations above are combined, it can be shown that when 2P – M is positive, hydroxyl ion is present. The usual control range is:

$$2P - M = 5\text{--}15 \text{ ppm}$$

This corresponds to a pH of approximately 10.2.

If soda ash is also used, the control is on the excess carbonate ion. As shown in Figure 7-1 (above), excess carbonate will depress the calcium to the level desired. The usual control range for hot lime–soda units is:

$$M \text{ (alkalinity)} - TH \text{ (total hardness)} = 20\text{--}40 \text{ ppm}$$

For cold lime–soda softening, where effluent magnesium hardness is significantly greater than in hot lime or soda, the control range above may be inappropriate. For cold lime–soda units, soda ash can be controlled such that:

$$2(M - P) - \text{Calcium hardness} = 20\text{--}40 \text{ ppm}$$

Care must be exercised in the specification of soda ash control ranges. If the softened water is to be used as boiler feedwater, hardness removal by the addition of soda ash may not be worth the cost of the resulting increase in steam condensate system corrosion. This corrosion is caused by the higher levels of carbon dioxide in the steam resulting from the higher carbonate alkalinity of the feedwater.

Coagulants/Flocculants/ Sludge Conditioners

Organic polymer flocculants and coagulants are preferred over inorganic salts of aluminum or iron. Polymers add minimal dissolved solids to the water and their use results in reduced sludge quantity compared to the use of inorganic coagulants. Inorganic coagulants must react with raw water alkalinity to form the metallic precipitate that aids in clarification and sludge bed conditioning. For example, alum reacts as follows:

$$\underset{\substack{\text{calcium} \\ \text{bicarbonate}}}{3Ca(HCO_3)_2} + \underset{\substack{\text{aluminum} \\ \text{sulfate}}}{Al_2(SO_4)_3} =$$

$$\underset{\substack{\text{calcium} \\ \text{sulfate}}}{3CaSO_4} + \underset{\substack{\text{aluminum} \\ \text{hydroxide}}}{2Al(OH)_3\downarrow} + \underset{\substack{\text{carbon} \\ \text{dioxide}}}{6CO_2}$$

The precipitated aluminum hydroxide is incorporated within the sludge produced by the softening reactions. This increases the fluidity of the softener sludge, which allows for increased solids contact, improving softening and effluent clarity.

Waters producing high calcium-to-magnesium precipitation ratios usually need sludge bed conditioning chemical feed for proper operation. Specialized organic polymers are available for proper conditioning of the sludge bed without the use of inorganic salts.

Four potentially adverse effects of using inorganic salts may be noted:

- *The inorganic salt reduces the alkalinity.* This converts the hardness to noncarbonate hardness, which is not affected by lime. As a result, inorganic salts increase hardness in water that is naturally deficient in bicarbonate alkalinity.

- *When the water is to be treated further by ion exchange, regenerant consumption is increased.* This is due to the higher hardness and the added soluble sulfate/chloride load.

- *The carbon dioxide generated by the reaction has a lime demand which is twice that of the bicarbonate.* Therefore, increased chemical addition is required.

- *Soluble aluminum in the softener effluent interferes with softened water alkalinity titrations, even when very low levels of soluble aluminum exist.* This interference, which necessitates an increase in lime feed, causes falsely low (2P – M) readings and may be partly responsible for the additional removal of magnesium seen when aluminum salts are used.

EQUIPMENT EMPLOYED

Cold Process

The first cold lime–soda softening was carried out in "batch" fashion. An excess of treating chemicals was mixed with the water in a large basin. After approximately 4 hr, the treated water was decanted from the basin, leaving the settled precipitates in the basin.

Today, continuous sludge-contact softeners (see Figures 7-4 and 7-5) are used to provide a constant flow with effluent quality superior to

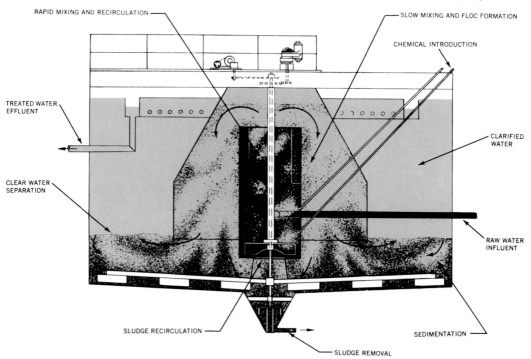

RAPID MIXING AND RECIRCULATION

SLOW MIXING AND FLOC FORMATION

CHEMICAL INTRODUCTION

TREATED WATER EFFLUENT

CLARIFIED WATER

CLEAR WATER SEPARATION

RAW WATER INFLUENT

SLUDGE RECIRCULATION

SEDIMENTATION

SLUDGE REMOVAL

Figure 7-4. Sludge-contact softener. (Courtesy of Graver Water Division, Ecodyne Corporation.)

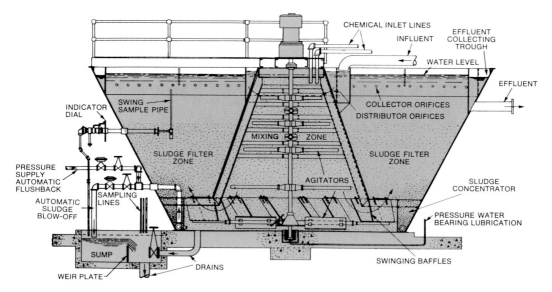

CHEMICAL INLET LINES

INFLUENT

EFFLUENT COLLECTING TROUGH

WATER LEVEL

EFFLUENT

INDICATOR DIAL

SWING SAMPLE PIPE

COLLECTOR ORIFICES

DISTRIBUTOR ORIFICES

MIXING ZONE

SLUDGE FILTER ZONE

SLUDGE FILTER ZONE

PRESSURE SUPPLY AUTOMATIC FLUSHBACK

AGITATORS

SLUDGE CONCENTRATOR

AUTOMATIC SLUDGE BLOW-OFF

SAMPLING LINES

PRESSURE WATER BEARING LUBRICATION

SUMP

SWINGING BAFFLES

WEIR PLATE

DRAINS

Figure 7-5. Sludge-contact softener. (Courtesy of the Permutit Company, Inc.)

that obtained through batch treatment. Treating chemicals are added as a function of flow rate and water quality to the rapid mix zone of the unit. Sludge, recirculated either internally or externally to the unit, may be returned to this rapid mix zone for improved softening, softened water clarity, and silica reduction.

The water then flows to the slow mix zone of the unit. Here, the precipitation reactions continue and the precipitates formed become large enough to begin settling. In the sludge-contact unit, the water flows through a bed of sludge for additional contact. The sludge level is maintained by the proper combination of sludge bed conditioning chemicals, mechanical agitation, hydraulic suspension, and sludge blowdown. A discernible line of separation between clarified water and slurry pool should exist in a properly operated unit. Effluent turbidity is usually less than 10 NTU.

Flow rate is usually limited to less than 1.5 gpm/ft^2 of settling area. A retention time of $1^1/_2$ hr is required to allow the softening reactions to come as close to completion as possible.

Because the reactions in cold process softening are not complete, the water contaminant levels leaving the unit are unstable. With additional time and/or increased temperature, further precipitation will occur downstream of the unit. Frequently, acid or carbon dioxide is added to stabilize the water. The pH is reduced from about 10.2 to between 8.0 and 9.0, which converts the carbonate to the more soluble bicarbonate. Ionically, the reaction is:

$$H^+ \quad + \quad CO_3^{2-} \quad = \quad HCO_3^-$$

$$\text{hydrogen} \qquad \text{carbonate} \qquad \text{bicarbonate}$$
$$\text{ion} \qquad\qquad \text{ion} \qquad\qquad \text{ion}$$

A typical cold lime softener system is shown in Figure 7-6.

Hot Process

Two hot process softener designs are illustrated in Figures 7-7 and 7-8. The former, the simplest in design and fabrication, is referred to as a "downflow" unit. The latter, which incorporates additional features, is referred to as an "upflow" unit. Many variations in design of both units exist, but the principle of operation is quite similar.

In each unit, water is admitted to the top of the vessel designed to operate at 5–15 psig saturated steam pressure (227–240 °F, 108–116 °C). An inlet valve is used to control the inlet water flow as a function of the operating level of the vessel. The water is sprayed into the steam space of the unit and is heated to within 2 or 3 degrees of the

Figure 7-6. Typical precipitator with associated gravity filters.

saturation temperature of the steam. Heating reduces the noncondensible gas content of the water. Oxygen and carbon dioxide are released and vented to the atmosphere with a controlled loss of heating steam. Although they are not deaerators, hot process units reduce oxygen to about 0.3 ppm (0.21 cm^3/L) and carbon dioxide to 0.

This residual oxygen level in the high-temperature water is aggressive and will attack downstream equipment such as filters and zeolites. Therefore, users should consider feeding a chemical oxygen scavenger to the effluent of hot process softeners.

Treatment chemicals are introduced into the top of the vessel as a function of flow and raw water analysis. Although the reactions go essentially to completion quite rapidly, a minimum of 1 hr of retention is designed into the unit. Also, flow rate through the unit is limited to 1.7–2.0 gpm/ft^2. Filter backwash water may be withdrawn from the outlet of the unit, from the filtered water header, or from internal or external storage. Internal storage compartments are illustrated in Figure 7-8. Filter backwash water is usually returned to the unit for recovery.

In the downflow design, the water leaves the vessel after reversing direction and enters the internal hood. Precipitates separate from the water at the hood and continue downward into the cone for removal by blowdown. Sludge blowdown is proportioned to raw water flow.

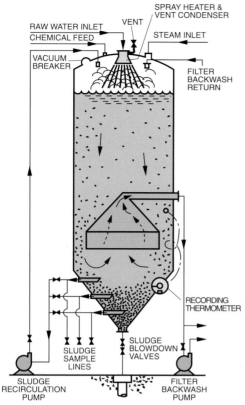

Figure 7-7. Downflow design of hot process softener.

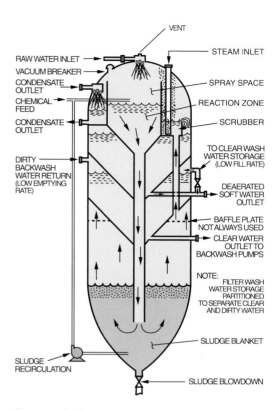

Figure 7-8. Sludge-contact hot process softener.

For improved silica reduction, sludge is recirculated from the cone back to the top of the unit.

For optimum silica reduction, a sludge-contact unit (shown in Figure 7-8) is used. Water and chemicals enter the top of the unit and flow to the bottom of the softener through a downcomer. The sludge level is maintained in such a way that the downcomer always discharges into the sludge bed. This ensures good contact with the sludge, which is rich in magnesium hydroxide. Also, the sludge bed acts as a filter, entrapping finer solids before the water exits near the top of the vessel. Sludge recycle may also be used.

The upflow design also lends itself to easier incorporation of internal compartments for filter backwash storage and return, and condensate or treated water deaeration.

LIMITATIONS

Given proper consideration of raw water quality and ultimate end use of the treated water, the application of precipitation processes has few limitations. However, operational difficulties may be encountered unless the following factors are controlled:

■ *Temperature.* Cold and warm units are subject to carryover if the temperature varies more than 4 °F/hr (2 °C/hr). Hot process units are less sensitive to slight temperature variations. However, a clogged or improper spray pattern can prevent proper heating of the water, and carryover can result.

■ *Hydraulics.* In any system, steady-state operation within design limits optimizes the performance of the equipment. Rapid flow variations can cause severe system upsets. Suitable treated water storage capacity should be incorporated into the total system design to minimize load swings on the softener.

■ *Chemical Control.* This should be as precise as possible to prevent poor water quality. Because of the comparatively constant quality of most well waters, changes in chemical feed rates are largely a function of flow only. However, surface water quality may vary hourly. Therefore, for proper control, it is imperative that users perform frequent testing of the raw water as well as the treated effluent, and adjust chemical feed accordingly.

Table 7-3. Chemicals commonly used in lime or lime–soda softening.

Chemical	Formula of Active Ingredient	Molecular Weight	Equivalent Weight	Approximate Solubility % by Weight, 68 °F (20 °C)	Approximate Specification
Burnt lime, quicklime (calcium oxide)	CaO	56.1	28.0	0.2	for softening; lump, pebble, ground, pulverized; 90% CaO
Hydrated lime, slaked lime (calcium hydroxide)	$Ca(OH)_2$	74.1	37.1	0.2	for softening, 93% $Ca(OH)_2$
Dolomitic lime (calcium oxide and magnesium oxide)	CaO, MgO	variable	variable	variable	for softening and silica reduction; lump, pebble, ground, pulverized; 58% CaO, 40% MgO
Hydrated dolomitic lime (calcium hydroxide and magnesium oxide)	$Ca(OH)_2$, MgO	variable	variable	variable	for softening and silica reduction; 62% $Ca(OH)_2$, 32% MgO
Magnesium oxide	MgO	40.3	20.2	very insoluble (0.002)	for silica reduction; 95% MgO
Soda ash (sodium carbonate)	Na_2CO_3	106	53	21.5	for softening; 58% light, 58% Na_2O, 98-99% Na_2CO_3
Caustic soda (sodium hydroxide)	$NaOH$	40.0	40.0	109.0	for softening; flake or granular; 76% Na_2O, 98% NaOH; liquid form (50%) available
Calcium chloride	$CaCl_2$	111.0	55.5	74.5	for alkalinity reduction; flake–78% $CaCl_2$, granular–75% $CaCl_2$; liquid form available
Gypsum (calcium sulfate)	$CaSO_4 \cdot 2H_2O$	172.0	86.1	0.2	for alkalinity reduction; ground; 100% (approx.) $CaSO_4 \cdot 2H_2O$
Alum (aluminum sulfate)	$Al_2(SO_4)_3 \cdot 18H_2O$	666.0	111.0	71.0	for cold or hot process coagulation/sludge conditioning; 14.5% to 17.5% Al_2O_3; liquid form available
Ferric sulfate	$Fe_2(SO_4)_3$	400.0	66.7	30.0 (approx.)	for cold or hot process coagulation/sludge conditioning; granular; 70% to 90% $Fe_2(SO_4)_3$
Sodium aluminate	$Na_2Al_2O_4$	164	164	40.0	for cold or hot process coagulation/sludge conditioning; granular or powder; 46% Al_2O_3, 31% Na_2O; liquid form available
Polymers (coagulant aids)	variable	variable	variable	variable	for cold or hot process coagulation/sludge conditioning; anionic, cationic, nonionic; powder or liquid
Ferrous sulfate copperas	$FeSO_4 \cdot 7H_2O$	278	139	48.5	for cold process coagulation/sludge conditioning; crystals, granules; 100% (approx.) $FeSO_4 \cdot 7H_2O$

ION EXCHANGE

All natural waters contain, in various concentrations, dissolved salts which dissociate in water to form charged ions. Positively charged ions are called cations; negatively charged ions are called anions. Ionic impurities can seriously affect the reliability and operating efficiency of a boiler or process system. Overheating caused by the buildup of scale or deposits formed by these impurities can lead to catastrophic tube failures, costly production losses, and unscheduled downtime. Hardness ions, such as calcium and magnesium, must be removed from the water supply before it can be used as boiler feedwater. For high-pressure boiler feedwater systems and many process systems, nearly complete removal of all ions, including carbon dioxide and silica, is required. Ion exchange systems are used for efficient removal of dissolved ions from water.

Ion exchangers exchange one ion for another, hold it temporarily, and then release it to a regenerant solution. In an ion exchange system, undesirable ions in the water supply are replaced with more acceptable ions. For example, in a sodium zeolite softener, scale-forming calcium and magnesium ions are replaced with sodium ions.

HISTORY

In 1905, Gans, a German chemist, used synthetic aluminosilicate materials known as zeolites in the first ion exchange water softeners. Although aluminosilicate materials are rarely used today, the term "zeolite softener" is commonly used to describe any cation exchange process.

The synthetic zeolite exchange material was soon replaced by a naturally occurring material called Greensand. Greensand had a lower exchange capacity than the synthetic material, but its greater physical stability made it more suitable for industrial applications. Capacity is defined as the amount of exchangeable ions a unit quantity of resin will remove from a solution. It is usually expressed in kilograins per cubic foot as calcium carbonate.

The development of a sulfonated coal cation exchange medium, referred to as carbonaceous zeolite, extended the application of ion exchange to hydrogen cycle operation, allowing for the reduction of alkalinity as well as hardness. Soon, an anion exchange resin (a condensation product of polyamines and formaldehyde) was developed. The new anion resin was used with the hydrogen cycle cation resin in an attempt to demineralize (remove all dissolved salts from) water. However, early anion exchangers were unstable and could not remove such weakly ionized acids as silicic and carbonic acid.

In the middle 1940's, ion exchange resins were developed based on the copolymerization of styrene cross-linked with divinylbenzene. These resins were very stable and had much greater exchange capacities than their predecessors. The polystyrene–divinylbenzene-based anion exchanger could remove all anions, including silicic and carbonic acids. This innovation made the complete demineralization of water possible.

Polystyrene–divinylbenzene resins are still used in the majority of ion exchange applications. Although the basic resin components are the same, the resins have been modified in many ways to meet the requirements of specific applications and provide a longer resin life. One of the most significant changes has been the development of the macroreticular, or macroporous, resin structure.

Standard gelular resins, such as those shown in Figure 8-1, have a permeable membrane structure. This structure meets the chemical and physical requirements of most applications. However, in some applications the physical strength and chemical resistance required of the resin struc-

Figure 8-1. Microscopic view of gelular resin beads (20–50 mesh) of a sulfonated styrene–divinylbenzene strong acid cation exchanger. (Courtesy of Rohm and Haas Company.)

ture is beyond the capabilities of the typical gel structure. Macroreticular resins feature discrete pores within a highly cross-linked polystyrene–divinylbenzene matrix. These resins possess a higher physical strength than gels, as well as a greater resistance to thermal degradation and oxidizing agents. Macroreticular anion resins (Figure 8-2) are also more resistant to organic fouling due to their more porous structure. In addition to polystyrene–divinylbenzene resins (Figure 8-3), there are newer resins with an acrylic structure, which increases their resistance to organic fouling.

In addition to a plastic matrix, ion exchange resin contains ionizable functional groups. These functional groups consist of both positively charged cation elements and negatively charged

Figure 8-2. Microscopic view of a macroporous strong base anion resin. (Courtesy of the Dow Chemical Company.)

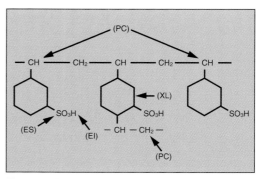

Figure 8-3. Chemical structural formula of sulfonic strong acid cation resin (Amberlite IR-120). (XL): cross-link; (PC): polymer chain; (ES): exchange site; (EI): exchangeable ion.

anion elements. However, only one of the ionic species is mobile. The other ionic group is attached to the bead structure. Figure 8-4 is a schematic illustration of a strong acid cation exchange resin bead, which has ionic sites consisting of immobile anionic (SO_3^-) radicals and mobile sodium cations (Na^+). Ion exchange occurs when raw water ions diffuse into the bead structure and exchange for the mobile portion of the functional

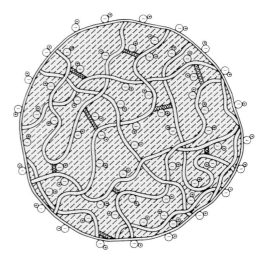

⊖ **FIXED NEGATIVELY CHARGED EXCHANGE SITE, i.e., SO_3^-**

⊕ **MOBILE, POSITIVELY CHARGED, EXCHANGEABLE CATION, i.e., Na^+**

〰 **POLYSTYRENE CHAIN**

XXXX **DIVINYLBENZENE CROSS-LINK**

///// **WATER OF HYDRATION**

Figure 8-4. Schematic of hydrated strong acid cation exchanger. (Courtesy of Rohm and Haas Company.)

group. Ions displaced from the bead diffuse back into the water solution.

CLASSIFICATIONS OF ION EXCHANGE RESINS

Ionizable groups attached to the resin bead determine the functional capability of the resin. Industrial water treatment resins are classified into four basic categories:

- Strong Acid Cation (SAC)
- Weak Acid Cation (WAC)
- Strong Base Anion (SBA)
- Weak Base Anion (WBA)

SAC resins can neutralize strong bases and convert neutral salts into their corresponding acids. SBA resins can neutralize strong acids and convert neutral salts into their corresponding bases. These resins are utilized in most softening and full demineralization applications. WAC and WBA resins are able to neutralize strong bases and acids, respectively. These resins are used for dealkalization, partial demineralization, or (in combination with strong resins) full demineralization.

SAC resins derive their functionality from sulfonic acid groups (HSO_3^-). When used in demineralization, SAC resins remove nearly all raw water cations, replacing them with hydrogen ions, as shown below:

$$\begin{bmatrix} Ca \\ Mg \\ 2Na \end{bmatrix} \cdot \begin{bmatrix} 2HCO_3 \\ SO_4 \\ 2Cl \end{bmatrix} + 2ZSO_3 \cdot H \rightleftharpoons 2ZSO_3 \cdot \begin{bmatrix} Ca \\ Mg \\ 2Na \end{bmatrix} + \begin{bmatrix} 2H_2CO_3 \\ H_2SO_4 \\ 2HCl \end{bmatrix}$$

The exchange reaction is reversible. When its capacity is exhausted, the resin can be regenerated with an excess of mineral acid.

Strong acid cation exchangers function well at all pH ranges. These resins have found a wide range of applications. For example, they are used in the sodium cycle (sodium as the mobile ion) for softening and in the hydrogen cycle for decationization.

Weak acid cation exchange resins derive their exchange activity from a carboxylic group (–COOH). When operated in the hydrogen form, WAC resins remove cations that are associated with alkalinity, producing carbonic acid as shown:

$$\begin{bmatrix} Ca \\ Mg \\ 2Na \end{bmatrix} \cdot (HCO_3)_2 + 2ZCOO \cdot H \rightleftharpoons 2ZCOO \cdot \begin{bmatrix} Ca \\ Mg \\ 2Na \end{bmatrix} + 2H_2CO_3$$

These reactions are also reversible and permit the return of the exhausted WAC resin to the regenerated form. WAC resins are not able to remove all of the cations in most water supplies. Their primary asset is their high regeneration efficiency in comparison with SAC resins. This high efficiency reduces the amount of acid required to regenerate the resin, thereby reducing the waste acid and minimizing disposal problems.

Weak acid cation resins are used primarily for softening and dealkalization of high-hardness, high-alkalinity waters, frequently in conjunction with SAC sodium cycle polishing systems. In full demineralization systems, the use of WAC and SAC resins in combination provides the economy of the more efficient WAC resin along with the full exchange capabilities of the SAC resin.

SBA resins derive their functionality from quaternary ammonium functional groups. Two types of quaternary ammonium groups, referred to as Type I and Type II, are used. Type I sites have three methyl groups:

$$(R \; - \; \underset{\underset{CH_3}{|}}{\overset{\overset{CH_3}{|}}{N}} \; - \; CH_3)^+$$

In a Type II resin one of the methyl groups is replaced with an ethanol group. The Type I resin has a greater stability than the Type II resin and is able to remove more of the weakly ionized acids. Type II resins provide a greater regeneration efficiency and a greater capacity for the same amount of regenerant chemical used.

When in the hydroxide form, SBA resins remove all commonly encountered anions as shown below:

$$\begin{bmatrix} H_2SO_4 \\ 2HCl \\ 2H_2SiO_3 \\ 2H_2CO_3 \end{bmatrix} + 2Z \cdot OH \rightleftharpoons 2Z \cdot \begin{bmatrix} SO_4 \\ 2Cl \\ 2HSiO_3 \\ 2HCO_3 \end{bmatrix} + 2H_2O$$

As with the cation resins, these reactions are reversible, allowing for the regeneration of the resin with a strong alkali, such as caustic soda, to return the resin to the hydroxide form.

Weak base resin functionality originates in primary ($R-NH_2$), secondary ($R-NHR'$), or tertiary ($R-NR'_2$) amine groups. WBA resins readily remove sulfuric, nitric, and hydrochloric acids, as represented by the following reaction:

$$\begin{bmatrix} H_2SO_4 \\ 2HCl \\ 2HNO_3 \end{bmatrix} + 2Z \cdot OH \rightleftharpoons 2Z \cdot \begin{bmatrix} SO_4 \\ 2Cl \\ NO_3 \end{bmatrix} + 2H_2O$$

WBA resins possess the same efficiency characteristic as WAC resins and can be regenerated with caustic soda, soda ash, or ammonia. WBA resins are more resistant than SBA resins to organics present in many water supplies. They can be used upstream of SBA resins for improved regeneration efficiency and protection of the SBA resin.

SODIUM ZEOLITE SOFTENING

Sodium zeolite softening is the most widely applied use of ion exchange. In zeolite softening, water containing scale-forming ions, such as calcium and magnesium, passes through a resin bed containing SAC resin in the sodium form. In the resin, the hardness ions are exchanged with the sodium, and the sodium diffuses into the bulk water solution. The hardness-free water, termed soft water, can then be used for low to medium pressure boiler feedwater, reverse osmosis system makeup, some chemical processes, and commercial applications, such as laundries.

Principles of Zeolite Softening

The removal of hardness from water by a zeolite softening process is described by the following reaction:

$$\begin{bmatrix} Ca \\ Mg \end{bmatrix} \cdot \begin{bmatrix} SO_4 \\ 2Cl \\ 2HCO_3 \end{bmatrix} + Na_2 \cdot Z \rightarrow Z \cdot \begin{bmatrix} Ca \\ Mg \end{bmatrix} + \begin{bmatrix} Na_2SO_4 \\ 2NaCl \\ 2NaHCO_3 \end{bmatrix}$$

Water from a properly operated zeolite softener is nearly free from detectable hardness. However, some small amounts of hardness, known as leakage, are present in the treated water. The level of hardness leakage is dependent on the hardness and sodium level in the influent water and the amount of salt used for regeneration.

Figure 8-5 is a typical profile of effluent hardness from a zeolite softener during a service cycle.

After final rinse, the softener produces a low, nearly constant level of hardness until the ion exchange resin nears exhaustion. At exhaustion, the effluent hardness increases sharply, and regeneration is required.

As illustrated by the softening reactions, SAC resin readily accepts calcium and magnesium ions in exchange for sodium ions. When exhausted resin is regenerated, a high concentration of sodium ions is applied to the resin to replace calcium and magnesium. The resin is treated with a 10% sodium chloride solution, and regeneration proceeds according to the following equation:

$$Z \cdot \begin{bmatrix} Ca \\ Mg \end{bmatrix} + \underset{\text{(concentrated)}}{2NaCl} \rightarrow Na_2 \cdot Z + \begin{bmatrix} Ca \\ Mg \end{bmatrix} \cdot Cl_2$$

During regeneration, a large excess of regenerant (approximately 3 times the amount of calcium and magnesium in the resin) is used. The eluted hardness is removed from the softening unit in the waste brine and by rinsing.

After regeneration, small residual amounts of hardness remain in the resin. If resin is allowed to sit in a stagnant vessel of water, some hardness will diffuse into the bulk water. Therefore, at the initiation of flow, the water effluent from a zeolite softener can contain hardness even if it has been regenerated recently. After a few minutes of flow, the hardness is rinsed from the softener, and the treated water is soft.

The duration of a service cycle depends on the rate of softener flow, the hardness level in the water, and the amount of salt used for regeneration. Table 8-1 shows the effect of regenerant level on the softening capacity of a gelular strong cation resin. Note that the capacity of the resin increases as the regenerant dosage increases, but the increase is not proportional. The regeneration is less efficient at the higher regenerant levels. Therefore, softener operating costs increase as the regenerant level increases. As shown by the data in Table 8-1, a 150% increase in regenerant salt provides only a 67% increase in operating capacity.

Table 8-1. Effect of regenerant salt level on strong acid cation resin softening capacity.

Salt (lb/ft³)	Capacity (gr/ft³)
6	18,000
8	20,000
10	24,000
15	30,000

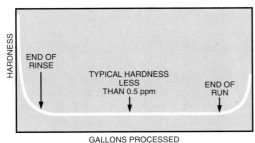

Figure 8-5. Typical sodium zeolite softener effluent profile.

Equipment

The equipment used for sodium zeolite softening consists of a softener exchange vessel, control valves and piping, and a system for brining, or regenerating, the resin. Usually, the softener tank is a vertical steel pressure vessel with dished heads as shown in Figure 8-6. Major features of the softening vessel include an inlet distribution system, free-board space, a regenerant distribution system, ion exchange resin, and a resin-retaining underdrain collection system.

The inlet distribution system is usually located at the top of the tank. The inlet system provides even distribution of influent water. This prevents

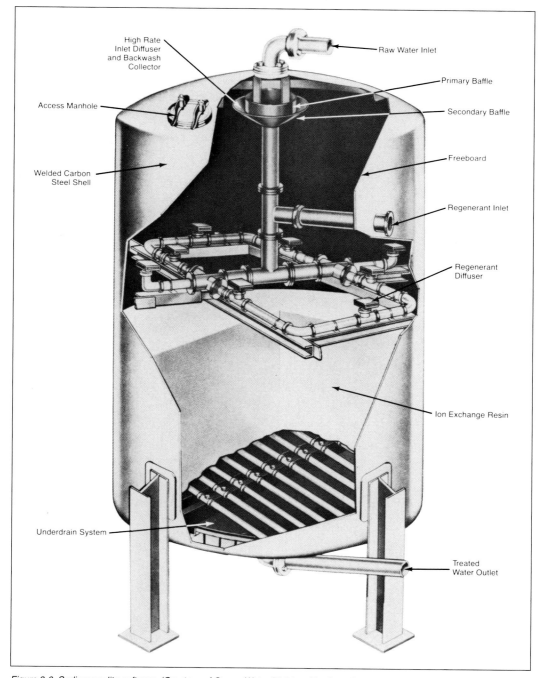

Figure 8-6. Sodium zeolite softener. (Courtesy of Graver Water Division, Ecodyne Corporation.)

the water from hollowing out flow channels in the resin bed, which would reduce system capacity and effluent quality. The inlet system also acts as a collector for backwash water.

The inlet distributor consists of a central header/hub with distributing laterals/radials or simple baffle plates, which direct the flow of water evenly over the resin bed. If water is not prevented from flowing directly onto the bed or tank walls, channeling will result.

The volume between the inlet distributor and the top of the resin bed is called the free-board space. The free-board allows for the expansion of the resin during the backwash portion of the regeneration without loss of resin. It should be a minimum of 50% of the resin volume (80% preferred).

The regenerant distributor is usually a header-lateral system that evenly distributes the regenerant brine during regeneration. The location of the distributor, 6 in. above the top of the resin bed, prevents the dilution of regenerant by water in the free-board space. It also reduces water and time requirements for displacement and fast rinse. The regenerant distributor should be secured to the tank structure to prevent breakage and subsequent channeling of the regenerant.

Water is softened by the bed of strong acid cation exchange resin in the sodium form. The quantity of resin required depends on the water flow, total hardness, and time desired between regeneration cycles. A minimum bed depth of 24 in. is recommended for all systems.

The underdrain system, located at the bottom of the vessel, retains ion exchange resin in the tank, evenly collects the service flow, and evenly distributes the backwash flow. Uneven collection of water in service or uneven distribution of the backwash water can result in channeling, resin fouling, or resin loss.

Although several underdrain designs are used, there are two primary types—subfill and resin-retaining. A subfill system consists of multiple layers of support media (such as graded gravel or anthracite) which support the resin, and a collection system incorporating drilled pipes or subfill strainers. As long as the support layers remain intact, the resin will remain in place. If the supporting media becomes disturbed, usually due to improper backwash, the resin can move through the disrupted layers and exit the vessel. A resin-retaining collector, such as a screened lateral or profile wire strainer, is more expensive

than a subfill system but protects against resin loss.

The main valve and piping system directs the flow of water and regenerant to the proper locations. The valve system consists of a valve nest or a single multiport valve. A valve nest includes six main valves: service inlet and outlet, backwash inlet and outlet, regenerant inlet, and regenerant/rinse drain. The valves may be operated manually, or automatically controlled by air, electrical impulse, or water pressure. In some systems, a single multiport valve is used in place of the valve nest. As the valve rotates through a series of fixed positions, ports in the valve direct flow in the same manner as a valve nest. Multiport valves can eliminate operational errors caused by opening of the incorrect valve but must be properly maintained to avoid leaks through the port seals.

The brining system consists of salt dissolving/brine measuring equipment, and dilution control equipment to provide the desired regenerant strength. The dissolving/measuring equipment is designed to provide the correct amount of concentrated brine (approximately 26% NaCl) for each regeneration, without allowing any undissolved salt into the resin. Most systems use a float-operated valve to control the fill and draw-down of the supply tank, thereby controlling the amount of salt used in the regeneration. Usually, the concentrated brine is removed from the tank by means of an eductor system, which also dilutes the brine to the optimum regenerant strength (8–10% NaCl). The brine can also be pumped from the concentrated salt tank and mixed with dilution water to provide the desired regenerant strength.

Softener Operation

A sodium zeolite softener operates through two basic cycles: the service cycle, which produces soft water for use, and the regeneration cycle, which restores resin capacity at exhaustion.

In the service cycle, water enters the softener through the inlet distribution system and flows through the bed. The hardness ions diffuse into the resin and exchange with sodium ions, which return to the bulk water. Soft water is collected in the underdrain system and discharged. Service water flow to the softener should be as constant as possible to prevent sudden surges and frequent on–off operation.

Due to resin requirements and vessel designs,

the softening operation is most efficient when a service flow rate between 6 and 12 gpm per square foot of resin surface area is maintained. Most equipment is designed to operate in this range, but some special designs utilize a deep resin bed to permit operation at 15–20 gpm/ft². Continuous operation above the manufacturer's suggested limits can lead to bed compaction, channeling, premature hardness breakthrough, and hardness leakage. Operating well below the manufacturer's recommended flow rates can also negatively affect softener performance. At low flow rates, the water is not sufficiently distributed, and the optimum resin–water contact cannot take place.

When a softener is exhausted, the resin must be regenerated. Monitoring of the effluent hardness reveals resin exhaustion. When hardness increases, the unit is exhausted. Automatic monitors provide a more constant indication of the condition of the softener than periodic operator sampling and testing, but require frequent maintenance to ensure accuracy. Many facilities regenerate softeners before exhaustion, based on a predetermined time period or number of gallons processed.

Most softening systems consist of more than one softener. They are often operated so that one softener is in regeneration or standby while the other units are in service. This ensures an uninterrupted flow of soft water. Prior to placing a standby softener into service, the unit should be rinsed to remove any hardness that has entered the water during the standing time.

Softener Regeneration

The regeneration cycle of a sodium zeolite softener consists of four steps: backwash, regeneration (brining), displacement (slow rinse), and fast rinse.

Backwash. During the service cycle, the downward flow of water causes suspended material to accumulate on the resin bed. Resin is an excellent filter and can trap particulate matter that has passed through upstream filtration equipment. The backwash step removes accumulated material and reclassifies the resin bed. In the backwash step, water flows from the underdrain distributor up through the resin bed and out the service distributor to waste. The upward flow lifts and expands the resin, allowing for removal of particulate material and resin fines and the classification of the resin. Resin classification brings the smaller beads to the top of the unit while the larger beads settle to the bottom. This

enhances the distribution of the regenerant chemical and service water.

Backwashing should continue for a minimum of 10 min or until effluent from the backwash outlet is clear. The backwash flow should be sufficient to expand the resin bed volume by 50% or more, depending on the available free-board. Insufficient backwash can lead to bed fouling and channeling. Excessive backwash flow rates result in the loss of resin. Backwash flow rates usually vary between 4–8 (ambient temperature) and 12–15 (hot service) gpm per square foot of bed area, but each manufacturer's recommendation should be followed. The ability of water to expand the resin is greatly affected by temperature. Less flow is required to expand the bed with cold water than with warm water. Resin bed expansion should be checked regularly and the flow rate adjusted as needed to maintain proper bed expansion.

Usually, the backwash water is filtered raw water. Water leaving the backwash outlet is unchanged in chemistry but can contain suspended solids. In order to conserve water, the backwash effluent can be returned to the clarifier or filter influent for treatment.

Regeneration (Brining). After backwash, regenerant brine is applied. The brine stream enters the unit through the regenerant distributor and flows down through the resin bed at a slow rate (usually between 0.5 and 1 gpm per square foot of resin). Brine flow is collected through the underdrain and sent to waste. The slow flow rate increases contact between the brine and resin. To achieve optimum efficiency from the brine, the solution strength should be 10% during brine introduction.

Displacement (Slow Rinse). Following the introduction of regenerant brine, a slow flow of water continues through the regenerant distribution system. This water flow displaces the regenerant through the bed at the desired flow rate. The displacement step completes the regeneration of the resin by ensuring proper contact of the regenerant with the bottom of the resin bed. The flow rate for the displacement water is usually the same rate used for the dilution of the concentrated brine. The duration of the displacement step should be sufficient to allow for approximately one resin bed volume of water to pass through the unit. This provides a "plug" of displacement water which gradually moves the brine completely through the bed.

Fast Rinse. After completion of the displacement rinse, water is introduced through the inlet distributor at a high flow rate. This rinse water removes the remaining brine as well as any residual hardness from the resin bed. The fast rinse flow rate is normally between 1.5 and 2 gpm per square foot of resin. Sometimes it is determined by the service rate for the softener.

Initially, the rinse effluent contains large amounts of hardness and sodium chloride. Usually, hardness is rinsed from the softener before excess sodium chloride. In many operations, the softener can be returned to service as soon as the hardness reaches a predetermined level, but some uses require rinsing until the effluent chlorides or conductivity are near influent levels. An effective fast rinse is important to ensure high effluent quality during the service run. If the softener has been in standby following a regeneration, a second fast rinse, known as a service rinse, can be used to remove any hardness that has entered the water during standby.

HOT ZEOLITE SOFTENING

Zeolite softeners can be used to remove residual hardness in the effluent from a hot process lime or lime–soda softener. The hot process effluent flows through filters and then through a bed of strong acid cation resin in the sodium form (Figure 8-7). The equipment and operation of a hot zeolite softener is identical to that of an ambient temperature softener, except that the

valves, piping, controllers, and instrumentation must be suitable for the high temperature (220–250 °F). Standard strong cation resin can be used at temperatures of up to 270 °F, but for a longer service life a premium gel or macroreticular resin is recommended. When operating a zeolite system following a hot process softener, it is important to design the system to eliminate flow surges in the hot lime unit. Common designs include the use of backwash water storage tanks in the hot lime unit and extended slow rinses for the zeolite in lieu of a standard fast rinse.

Applications and Advantages

Scale and deposit buildup in boilers and the formation of insoluble soap curds in washing operations have created a large demand for softened water. Because sodium zeolite softeners are able to satisfy this demand economically, they are widely used in the preparation of water for low and medium pressure boilers, laundries, and chemical processes. Sodium zeolite softening also offers the following advantages over other softening methods:

- treated water has a very low scaling tendency because zeolite softening reduces the hardness level of most water supplies to less than 2 ppm

- operation is simple and reliable; automatic and semiautomatic regeneration controls are available at a reasonable cost

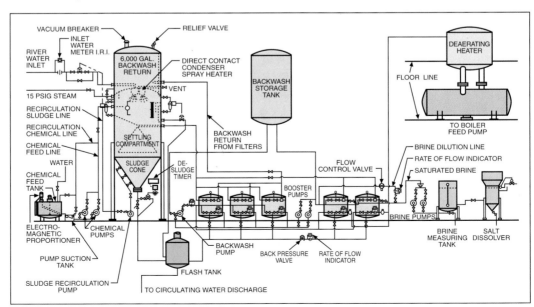

Figure 8-7. Combined hot lime/hot zeolite softening system. (Courtesy of Water Conditioning Division, Envirex, Inc.)

- salt is inexpensive and easy to handle
- no waste sludge is produced; usually, waste disposal is not a problem
- within certain limits, variations in water flow rate have little effect on treated water quality
- because efficient operation can be obtained in units of almost any size, sodium zeolite softeners are suitable for both large and small installations

Limitations

Although sodium zeolite softeners efficiently reduce the amount of dissolved hardness in a water supply, the total solids content, alkalinity, and silica in the water remain unaffected. A sodium zeolite softener is not a direct replacement for a hot lime–soda softener. Plants that have replaced their hot process softeners with only zeolite softeners have experienced problems with silica and alkalinity levels in their boilers.

Because the resin is such an efficient filter, sodium zeolite softeners do not function efficiently on turbid waters. Continued operation with an influent turbidity in excess of 1.0 JTU causes bed fouling, short service runs, and poor effluent quality. Most city and well waters are suitable, but many surface supplies must be clarified and filtered before use.

The resin can be fouled by heavy metal contaminants, such as iron and aluminum, which are not removed during the course of a normal regeneration. If excess iron or manganese is present in the water supply, the resin must be cleaned periodically. Whenever aluminum coagulants are used ahead of zeolite softeners, proper equipment operation and close control of clarifier pH are essential to good softener performance.

Strong oxidizing agents in the raw water attack and degrade the resin. Chlorine, present in most municipal supplies, is a strong oxidant and should be removed prior to zeolite softening by activated carbon filtration or reaction with sodium sulfite.

DEMINERALIZATION

Softening alone is insufficient for most high-pressure boiler feedwaters and for many process streams, especially those used in the manufacture of electronics equipment. In addition to the removal of hardness, these processes require removal of all dissolved solids, such as sodium, silica, alkalinity, and the mineral anions (Cl^-, SO_4^{2-}, NO_3^-).

Demineralization of water is the removal of essentially all inorganic salts by ion exchange. In this process, strong acid cation resin in the hydrogen form converts dissolved salts into their corresponding acids, and strong base anion resin in the hydroxide form removes these acids. Demineralization produces water similar in quality to distillation at a lower cost for most fresh waters.

Principles of Demineralization

A demineralizer system consists of one or more ion exchange resin columns, which include a strong acid cation unit and a strong base anion unit. The cation resin exchanges hydrogen for the raw water cations as shown by the following reactions:

$$\begin{bmatrix} Ca \\ Mg \\ 2Na \end{bmatrix} \cdot \begin{bmatrix} 2HCO_3 \\ SO_4 \\ 2Cl \\ 2NO_3 \end{bmatrix} + 2Z \cdot H \rightarrow 2Z \cdot \begin{bmatrix} Ca \\ Mg \\ 2Na \end{bmatrix} + \begin{bmatrix} 2H_2CO_3 \\ H_2SO_4 \\ 2HCl \\ 2HNO_3 \end{bmatrix}$$

A measure of the total concentration of the strong acids in the cation effluent is the free mineral acidity (FMA). In a typical service run, the FMA content is stable most of the time, as shown in Figure 8-8. If cation exchange were 100% efficient, the FMA from the exchanger would be equal to the theoretical mineral acidity (TMA) of the water. The FMA is usually slightly lower than the TMA because a small amount of sodium leaks through the cation exchanger. The amount of sodium leakage depends on the regenerant level, the flow rate, and the proportion of sodium to the other cations in the raw water. In general, sodium leakage increases as the ratio of sodium to total cations increases.

As a cation exchange unit nears exhaustion, FMA in the effluent drops sharply, indicating that the exchanger should be removed from service. At this time the resin should be regenerated with an acid solution, which returns the exchange sites

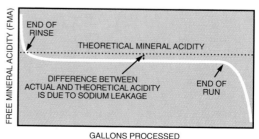

Figure 8-8. Typical effluent profile for strong acid cation exchanger.

to the hydrogen form. Sulfuric acid is normally used due to its affordable cost and its availability. However, improper use of sulfuric acid can cause irreversible fouling of the resin with calcium sulfate.

To prevent this occurrence, the sulfuric acid is usually applied at a high flow rate (1 gpm per square foot of resin) and an initial concentration of 2% or less. The acid concentration is gradually increased to 6–8% to complete regeneration.

Some installations use hydrochloric acid for regeneration. This necessitates the use of special materials of construction in the regenerant system. As with a sodium zeolite unit, an excess of regenerant (sulfuric or hydrochloric acid) is required—up to three times the theoretical dose.

To complete the demineralization process, water from the cation unit is passed through a strong base anion exchange resin in the hydroxide form. The resin exchanges hydrogen ions for both highly ionized mineral ions and the more weakly ionized carbonic and silicic acids, as shown below:

$$\begin{bmatrix} H_2SO_4 \\ 2HCl \\ 2H_2SiO_3 \\ 2H_2CO_3 \\ 2HNO_3 \end{bmatrix} + 2Z \cdot OH \rightleftharpoons 2Z \cdot \begin{bmatrix} SO_4 \\ 2Cl \\ 2HSiO_3 \\ 2HCO_3 \\ 2NO_3 \end{bmatrix} + 2H_2O$$

The above reactions indicate that demineralization completely removes the cations and anions from the water. In reality, because ion exchange reactions are equilibrium reactions, some leakage occurs. Most leakage from cation units is sodium. This sodium leakage is converted to sodium hydroxide in the anion units. Therefore, the effluent pH of a two bed cation–anion demineralizer system is slightly alkaline. The caustic produced in the anions causes a small amount of silica leakage. The extent of leakage from the anions depends on the chemistry of the water being processed and the regenerant dosage being used.

Demineralization using strong anion resins removes silica as well as other dissolved solids. Effluent silica and conductivity are important parameters to monitor during a demineralizer service run. Both silica and conductivity are low at the end of the fast rinse, as shown in Figure 8-9.

When silica breakthrough occurs at the end of a service run, the treated water silica level increases sharply. Often, the conductivity of the water decreases momentarily, then rises rapidly. This

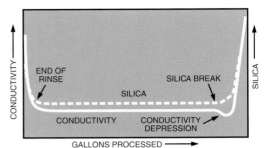

Figure 8-9. Conductivity/silica profile for strong base anion exchanger.

temporary drop in conductivity is easily explained. During the normal service run, most of the effluent conductivity is attributed to the small level of sodium hydroxide produced in the anion exchanger. When silica breakthrough occurs, the hydroxide is no longer available, and the sodium from the cation exchanger is converted to sodium silicate, which is much less conductive than sodium hydroxide. As anion resin exhaustion progresses, the more conductive mineral ions break through, causing a subsequent increase in conductivity.

When the end of a demineralizer run is detected, the unit must be removed from service immediately. If the demineralizer is allowed to remain in service past the breakpoint, the level of silica in the treated water can rise above that of the influent water, due to the concentrating of silica that takes place in the anion resin during the service run.

Strong base anion exchangers are regenerated with a 4% sodium hydroxide solution. As with cation regeneration, the relatively high concentration of hydroxide drives the regeneration reaction. To improve the removal of silica from the resin bed, the regenerant caustic is usually heated to 120 °F or to the temperature specified by the resin manufacturer. Silica removal is also enhanced by a resin bed preheat step before the introduction of warm caustic.

Equipment and Operation

The equipment used for cation–anion demineralization is similar to that used in zeolite softening. The primary difference is that the vessels, valves, and piping must be made of (or lined with) corrosion-resistant materials. Rubber and polyvinyl chloride (PVC) are commonly used for ion exchange vessel linings. The controls and regenerant systems for demineralizers are more complex, to allow for such enhancements as stepwise acid and warm caustic regenerations.

Demineralizers are similar in operation to zeolite softeners. The service flow rate guidelines for a demineralizer range from 6 to 10 gpm per square foot of resin. Flow rates of over 10 gpm per square foot of resin cause increased sodium and silica leakage with certain waters. Anion resin is much lighter than cation resin. Therefore, the backwash flow rates for anion exchange resins are much lower than those for cation resins, and anion resin expansion is affected by the temperature of the water more than cation resin expansion. The water used for each step of anion resin regeneration should be free from hardness, to prevent precipitation of hardness salts in the alkaline anion resin bed.

Continuous conductivity instruments and silica analyzers are commonly used to monitor anion effluent water quality and detect the need for regeneration. In some instances, conductivity probes are placed in the resin bed above the underdrain collectors to detect resin exhaustion before silica breakthrough into the treated water occurs.

Advantages and Limitations

Demineralizers can produce high-purity water for nearly every use. Demineralized water is widely used for high pressure boiler feedwater and for many process waters. The quality of water produced is comparable to distilled water, usually at a fraction of the cost. Demineralizers come in a wide variety of sizes. Systems range from laboratory columns that produce only a few gallons per hour to systems that produce thousands of gallons per minute.

Like other ion exchange systems, demineralizers require filtered water in order to function efficiently. Resin foulants and degrading agents, such as iron and chlorine, should be avoided or removed prior to demineralization. Anion resins are very susceptible to fouling and attack from the organic materials present in many surface water supplies. Some forms of silica, known as colloidal, or non-reactive, are not removed by a demineralizer. Hot, alkaline boiler water dissolves the colloidal material, forming simple silicates that are similar to those that enter the boiler in a soluble form. As such, they can form deposits on tube surfaces and volatilize into the steam.

DEALKALIZATION

Often, boiler or process operating conditions require the removal of hardness and the reduc-

tion of alkalinity but not the removal of the other solids. Zeolite softening does not reduce alkalinity, and demineralization is too costly. For these situations, a dealkalization process is used. Sodium zeolite/hydrogen zeolite (split stream) dealkalization, chloride–anion dealkalization, and weak acid cation dealkalization are the most frequently used processes.

Sodium Zeolite/Hydrogen Zeolite (Split Stream) Dealkalization

In a split stream dealkalizer, a portion of the raw water flows through a sodium zeolite softener. The remainder flows through a hydrogen-form strong acid cation unit (hydrogen zeolite). The effluent from the sodium zeolite is combined with the hydrogen zeolite effluent. The effluent from the hydrogen zeolite unit contains carbonic acid, produced from the raw water alkalinity, and free mineral acids. When the two streams are combined, free mineral acidity in the hydrogen zeolite effluent converts sodium carbonate and bicarbonate alkalinity in the sodium zeolite effluent to carbonic acid as shown below:

$$2NaHCO_3 \; + \; \begin{bmatrix} 2HCl \\ H_2SO_4 \\ 2HNO_3 \end{bmatrix} \; \rightarrow \; \begin{bmatrix} 2NaCl \\ Na_2SO_4 \\ 2NaNO_3 \end{bmatrix} \; + \; 2H_2CO_3$$

Carbonic acid is unstable in water. It forms carbon dioxide gas and water. The blended effluents are sent to a decarbonator or degasser, where the carbon dioxide is stripped from the water by a countercurrent stream of air. Figure 8-10 shows a typical split stream dealkalization system.

The desired level of blended water alkalinity can be maintained through control of the percentage of sodium zeolite and hydrogen zeolite water in the mixture. A higher percentage of sodium zeolite water results in higher alkalinity, and an increased percentage of hydrogen zeolite water reduces alkalinity.

In addition to reducing alkalinity, a split stream dealkalizer reduces the total dissolved solids of the water. This is important in high alkalinity waters, because the conductivity of these waters affects the process and can limit boiler cycles of concentration.

Sodium Zeolite/Chloride Anion Dealkalization

Strong base anion resin in the chloride form can be used to reduce the alkalinity of a water. Water flows through a zeolite softener and then an

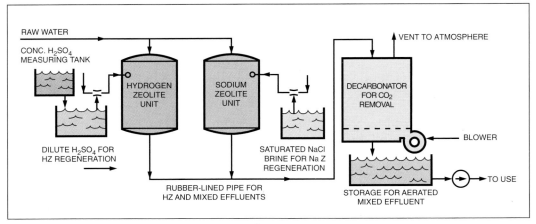

Figure 8-10. Sodium zeolite/hydrogen zeolite split stream softener.

anion unit, which replaces the carbonate, bicarbonate, sulfate, and nitrate ions with chloride ions as shown in these reactions:

$$\begin{bmatrix} Na_2SO_4 \\ 2NaHCO_3 \\ 2NaCO_3 \\ 2NaNO_3 \end{bmatrix} + 2Z \cdot Cl \longrightarrow 2NaCl + 2Z \cdot \begin{bmatrix} SO_4 \\ 2HCO_3 \\ 2NO_3 \end{bmatrix}$$

The chloride anion dealkalizer reduces alkalinity by approximately 90% but does not reduce total solids. When the resin nears exhaustion, treated water alkalinity increases rapidly, signaling the need for regeneration.

The zeolite softener is regenerated as previously described. In addition, the anion resin is also regenerated with a sodium chloride brine that returns the resin to the chloride form. Frequently, a small amount of caustic soda is added to the regenerant brine to enhance alkalinity removal.

Weak Acid Cation Dealkalization

Another method of dealkalization uses weak acid cation resins. Weak acid resins are similar in operation to strong acid cation resins, but only exchange for cations that are associated with alkalinity, as shown by these reactions:

$$\begin{bmatrix} Ca \\ Mg \\ 2Na \end{bmatrix} \cdot 2HCO_3 + 2Z \cdot H \longrightarrow 2Z \cdot \begin{bmatrix} Ca \\ Mg \\ 2Na \end{bmatrix} + 2H_2CO_3$$

where Z represents the resin. The carbonic acid (H_2CO_3) formed is removed by a decarbonator or degasser as in a split stream system.

The ideal influent for a weak acid cation system has a hardness level equal to the alkalinity (both expressed in ppm as $CaCO_3$). In waters that are higher in alkalinity than hardness, the alkalinity

is not removed to its lowest level. In waters containing more hardness than alkalinity, some hardness remains after treatment. Usually, these waters must be polished by a sodium zeolite softener to remove hardness. During the initial portion of a weak acid cation service run (the first 40–60%) some cations associated with mineral anions exchange, producing small amounts of mineral acids in the effluent. As the service cycle progresses, alkalinity appears in the effluent. When the alkalinity in the effluent exceeds 10% of the influent alkalinity, the unit is removed from service and regenerated with a 0.5% sulfuric acid solution. The concentration of regenerant acid should be kept below 0.5–0.7%, to prevent calcium sulfate precipitation in the resin. Weak acid cation resin exchange is very efficient. Therefore, the amount of acid required is virtually equal (chemically) to the amount of cations removed during the service cycle.

If the materials of construction for the downstream equipment or overall process cannot tolerate the mineral acidity present during the initial portions of the service cycle, a brine solution is passed through the regenerated weak acid resin prior to the final rinse. This solution removes the mineral acidity without a significant impact on the quality or length of the subsequent run.

Equipment used for a weak acid cation dealkalizer is similar to that used for a strong acid cation exchanger, with the exception of the resin. One variation of the standard design uses a layer of weak acid resin on top of strong acid cation resin. Because it is lighter, the weak acid resin remains on top. The layered resin system is regenerated with sulfuric acid and then with sodium chloride brine. The brine solution converts the strong acid resin to the sodium form. This resin then acts as a polishing softener.

Direct Acid Injection

In the process of direct acid injection and decarbonation, acid is used to convert alkalinity to carbonic acid. The carbonic acid dissociates to form carbon dioxide and water and the carbon dioxide is removed in a decarbonator. The use of an acid injection system should be approached with caution, because an acid overfeed or a breakdown in the pH control system can produce acidic feedwater, which corrodes the iron surfaces of feedwater systems and boilers. Proper pH monitoring and controlled caustic feed after decarbonation are required.

Advantages and Limitations of Dealkalization Systems

Ion exchange dealkalization systems produce hardness-free, low-alkalinity water at a reasonable cost, and with a high degree of reliability. They are well suited for processing feedwater for medium-pressure boilers, and for process water for the beverage industry. Split stream and weak acid cation systems also reduce the total dissolved solids. In addition to these advantages, the following disadvantages must be considered:

- dealkalizers do not remove all of the alkalinity and do not affect the silica content of a water

- dealkalizers require the same influent purity as other ion exchange processes; filtered water that is low in potential foulants must be used

- the water produced by a dealkalization system using a forced draft decarbonator becomes saturated with oxygen, so it is potentially corrosive

COUNTERFLOW AND MIXED BED DEIONIZATION

Due to increasing boiler operating pressures and the manufacture of products requiring contaminant-free water, there is a growing need for higher water quality than cation–anion demineralizers can produce. Therefore, it has become necessary to modify the standard demineralization process to increase the purity of the treated water. The most significant improvements in demineralized water purity have been produced by counterflow cation exchangers and mixed bed exchangers.

Counterflow Cation Exchangers

In a conventional demineralizer system, regenerant flow is in the same direction as the service flow, down through the resin bed. This scheme is known as co-current operation and is the basis for most ion exchange system designs. During the regeneration of a co-current unit, the contaminants are displaced through the resin bed during the regeneration. At the end of the regeneration, some ions, predominately sodium ions, remain in the bottom of the resin bed. Because the upper portion of the bed has been exposed to fresh regenerant, it is highly regenerated. As the water flows through the resin during service, cations are exchanged in the upper portion of the bed first, and then move down through the resin as the bed becomes exhausted. Sodium ions that remained in the bed during regeneration diffuse into the decationized water before it leaves the vessel. This sodium leakage enters the anion unit where anion exchange produces caustic, raising the pH and conductivity of the demineralized water.

In a counterflow regenerated cation exchanger, the regenerant flows in the opposite direction of the service flow. For example, if the service flow is downward through the bed, the regenerant acid flow is up through the bed. As a result, the most highly regenerated resin is located where the service water leaves the vessel. The highly regenerated resin removes the low level of contaminants that have escaped removal in the top of the bed. This results in higher water purity than co-current designs can produce. To maximize contact between the acid and resin and to keep the most highly regenerated resin from mixing with the rest of the bed, the resin bed must stay compressed during the regenerant introduction. This compression is usually achieved in one of two ways:

- a blocking flow of water or air is used

- the acid flow is split, and acid is introduced at both the top and the bottom of the resin bed (Figure 8-11)

Mixed Bed Exchangers

A mixed bed exchanger has both cation and anion resin mixed together in a single vessel. As water flows through the resin bed, the ion exchange process is repeated many times, "polishing" the water to a very high purity. During regeneration, the resin is separated into

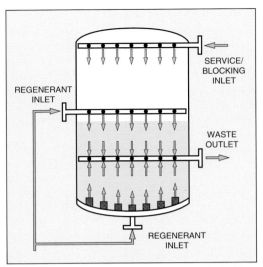

Figure 8-11. Counterflow cation profile showing dual acid flow blocking method.

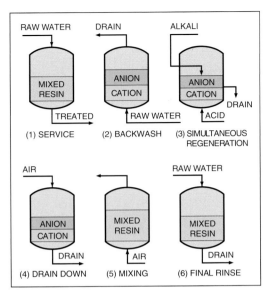

Figure 8-12. Significant steps in the regeneration sequence for a mixed bed exchanger.

distinct cation and anion fractions as shown in Figure 8-12. The resin is separated by backwashing, with the lighter anion resin settling on top of the cation resin. Regenerant acid is introduced through the bottom distributor, and caustic is introduced through distributors above the resin bed. The regenerant streams meet at the boundary between the cation and anion resin and discharge through a collector located at the resin interface. Following regenerant introduction and displacement rinse, air and water are used to mix the resins. Then the resins are rinsed, and the unit is ready for service.

Counterflow and mixed bed systems produce a purer water than conventional cation–anion demineralizers, but require more sophisticated equipment and have a higher initial cost. The more complicated regeneration sequences require closer operator attention than standard systems. This is especially true for a mixed bed unit.

OTHER DEMINERALIZATION PROCESSES

The standard cation–anion process has been modified in many systems to reduce the use of costly regenerants and the production of waste. Modifications include the use of decarbonators and degassers, weak acid and weak base resins, strong base anion caustic waste (to regenerate weak base anion exchangers), and reclamation of a portion of spent caustic for subsequent regeneration cycles. Several different approaches to demineralization using these processes are shown in Figure 8-13.

Decarbonators and Degassers

Decarbonators and degassers are economically beneficial to many demineralization systems, because they reduce the amount of caustic required for regeneration. Water from a cation exchanger is broken into small droplets by sprays and trays or packing in a decarbonator. The water then flows through a stream of air flowing in the opposite direction. Carbonic acid present in the cation effluent dissociates into carbon dioxide and water. The carbon dioxide is stripped from the water by the air, reducing the load to the anion exchangers. Typical forced draft decarbonators are capable of removing carbon dioxide down to 10–15 ppm. However, water effluent from a decarbonator is saturated with oxygen.

In a vacuum degasser, water droplets are introduced into a packed column that is operated under a vacuum. Carbon dioxide is removed from the water due to its decreased partial pressure in a vacuum. A vacuum degasser usually reduces carbon dioxide to less than 2 ppm and also removes most of the oxygen from the water. However, vacuum degassers are more expensive to purchase and operate than forced draft decarbonators.

Weak Acid and Weak Base Resins

Weak functionality resins have a much higher

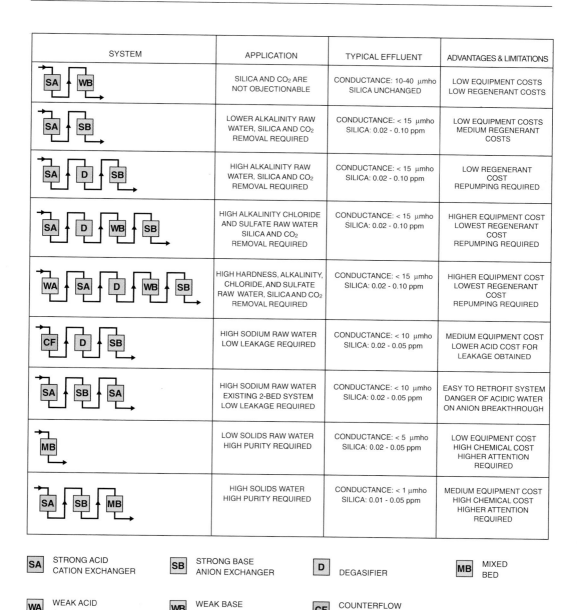

SYSTEM	APPLICATION	TYPICAL EFFLUENT	ADVANTAGES & LIMITATIONS
SA → WB	SILICA AND CO_2 ARE NOT OBJECTIONABLE	CONDUCTANCE: 10-40 µmho SILICA UNCHANGED	LOW EQUIPMENT COSTS LOW REGENERANT COSTS
SA → SB	LOWER ALKALINITY RAW WATER, SILICA AND CO_2 REMOVAL REQUIRED	CONDUCTANCE: < 15 µmho SILICA: 0.02 - 0.10 ppm	LOW EQUIPMENT COSTS MEDIUM REGENERANT COSTS
SA → D → SB	HIGH ALKALINITY RAW WATER, SILICA AND CO_2 REMOVAL REQUIRED	CONDUCTANCE: < 15 µmho SILICA: 0.02 - 0.10 ppm	LOW REGENERANT COST REPUMPING REQUIRED
SA → D → WB → SB	HIGH ALKALINITY CHLORIDE AND SULFATE RAW WATER SILICA AND CO_2 REMOVAL REQUIRED	CONDUCTANCE: < 15 µmho SILICA: 0.02 - 0.10 ppm	HIGHER EQUIPMENT COST LOWEST REGENERANT COST REPUMPING REQUIRED
WA → SA → D → WB → SB	HIGH HARDNESS, ALKALINITY, CHLORIDE, AND SULFATE RAW WATER, SILICA AND CO_2 REMOVAL REQUIRED	CONDUCTANCE: < 15 µmho SILICA: 0.02 - 0.10 ppm	HIGHER EQUIPMENT COST LOWEST REGENERANT COST REPUMPING REQUIRED
CF → D → SB	HIGH SODIUM RAW WATER LOW LEAKAGE REQUIRED	CONDUCTANCE: < 10 µmho SILICA: 0.02 - 0.05 ppm	MEDIUM EQUIPMENT COST LOWER ACID COST FOR LEAKAGE OBTAINED
SA → SB → SA	HIGH SODIUM RAW WATER EXISTING 2-BED SYSTEM LOW LEAKAGE REQUIRED	CONDUCTANCE: < 10 µmho SILICA: 0.02 - 0.05 ppm	EASY TO RETROFIT SYSTEM DANGER OF ACIDIC WATER ON ANION BREAKTHROUGH
MB	LOW SOLIDS RAW WATER HIGH PURITY REQUIRED	CONDUCTANCE: < 5 µmho SILICA: 0.02 - 0.05 ppm	LOW EQUIPMENT COST HIGH CHEMICAL COST HIGHER ATTENTION REQUIRED
SA → SB → MB	HIGH SOLIDS WATER HIGH PURITY REQUIRED	CONDUCTANCE: < 1 µmho SILICA: 0.01 - 0.05 ppm	MEDIUM EQUIPMENT COST HIGH CHEMICAL COST HIGHER ATTENTION REQUIRED

SA STRONG ACID CATION EXCHANGER	SB STRONG BASE ANION EXCHANGER	D DEGASIFIER	MB MIXED BED
WA WEAK ACID CATION EXCHANGER	WB WEAK BASE ANION EXCHANGER	CF COUNTERFLOW CATION	

Figure 8-13. Demineralizer systems.

regeneration efficiency than their strong functionality counterparts. Weak acid cation resins, as described in the dealkalization section, exchange with cations associated with alkalinity. Weak base resins exchange with the mineral acid anions (SO_4^{2-}, Cl^-, NO_3^-) in a strong acid solution. The regeneration efficiency of weak resins is virtually stoichiometric—the removal of 1 kgr of ions (as $CaCO_3$) requires only slightly more than 1 kgr of the regenerant ion (as $CaCO_3$). Strong

resins require three to four times the regenerant for the same contaminant removal.

Weak base resins are so efficient that it is common practice to regenerate a weak base exchanger with a portion of the "spent" caustic from regeneration of the strong base anion resin. The first fraction of the caustic from the strong base unit is sent to waste to prevent silica fouling of the weak base resin. The remaining caustic is used to regenerate the weak base resin. An additional feature of weak base resins is their ability to hold natural organic materials that foul strong base resins and release them during the regeneration cycle. Due to this ability, weak base resins are commonly used to protect strong base resins from harmful organic fouling.

Regenerant Reuse

Due to the high cost of caustic soda and the increasing problems of waste disposal, many demineralization systems are now equipped with a caustic reclaim feature. The reclaim system uses a portion of the spent caustic from the previous regeneration at the beginning of the next regeneration cycle. The reused caustic is followed by fresh caustic to complete the regeneration. The new caustic is then reclaimed for use in the next regeneration. Typically, sulfuric acid is not reclaimed, because it is lower in cost and calcium sulfate precipitation is a potential problem.

CONDENSATE POLISHING

Ion exchange uses are not limited to process and boiler water makeup. Ion exchange can be used to purify, or polish, returned condensate, removing corrosion products that could cause harmful deposits in boilers.

Typically, the contaminants in the condensate system are particulate iron and copper. Low levels of other contaminants may enter the system through condenser and pump seal leaks or carry-over of boiler water into the steam. Condensate polishers filter out the particulates and remove soluble contaminants by ion exchange.

Most paper mill condensate polishers operate at temperatures approaching 200 °F, precluding the use of anion resin. Cation resin, which is stable up to temperatures of over 270 °F, is used for deep bed condensate polishing in these applications. The resin is regenerated with sodium chloride brine, as in a zeolite softener. In situations where sodium leakage from the polisher adversely affects the boiler water internal chemical program or steam attemperating water purity, the resin can be regenerated with an ionized amine solution to prevent these problems.

The service flow rate for a deep bed polisher (20–50 gpm per square foot of resin surface area) is very high compared to that of a conventional softener. High flow rates are permissible because the level of soluble ions in the condensate can be usually very low. Particulate iron and copper are removed by filtration, while dissolved contaminants are reduced by exchange for the sodium or amine in the resin.

The deep bed cation resin condensate polisher is regenerated with 15 lb of sodium chloride per cubic foot of resin, in a manner similar to that used for conventional sodium zeolite regeneration. A solubilizing or reducing agent is often used to assist in the removal of iron. Sometimes, a supplemental backwash header is located just below the surface of the resin bed. This subsurface distributor, used prior to backwashing, introduces water to break up the crust that forms on the resin surface between regenerations.

An important consideration is the selection of a resin for condensate polishing. Because high pressure drops are generated by the high service flow rates and particulate loadings, and because many systems operate at high temperatures, considerable stress is imposed on the structure of the resin. A premium-grade gelular or macroreticular resin should be used in deep bed condensate polishing applications.

In systems requiring total dissolved solids and particulate removal, a mixed bed condensate polisher may be used. The temperature of the condensate should be below 140 °F, which is the maximum continuous operating temperature for the anion resin. Additionally, the flow through the unit is generally reduced to approximately 20 gpm/ft².

Ion exchange resins are also used as part of a precoat filtration system, as shown in Figure 8-14, for polishing condensate. The resin is crushed and mixed into a slurry, which is used to coat individual septums in a filter vessel. The powdered resin is a very fine filtering medium that traps particulate matter and removes some soluble contaminants by ion exchange. When the filter media becomes clogged, the precoat material is disposed of, and the septums are coated with a fresh slurry of powdered resin.

COMMON ION EXCHANGE SYSTEM PROBLEMS

As in any dynamic operating system incorporating electrical and mechanical equipment and chemical operations, problems do occur in ion exchange systems. The problems usually result in poor effluent quality, decreased service run lengths, or increased consumption of regenerant. To keep the ion exchange system operating efficiently and reliably, changes in water quality, run lengths, or regenerant consumption should be considered whenever problems are detected.

The cause–effect diagrams for short runs (Figure 8-15) and poor-quality effluent (Figure 8-16) show that there are many possible causes for

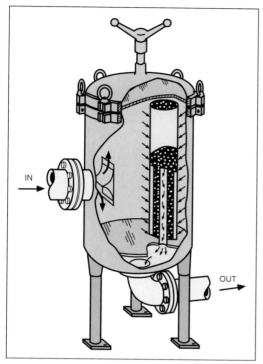

Figure 8-14. Powdered resin condensate polisher. (Courtesy of Graver Water Div., Ecodyne Corporation.)

reduced performance of a demineralization system. Some of the more common problems are discussed below.

Operational Problems

Changes in raw water quality have a significant impact on both the run length and the effluent quality produced by an ion exchange unit. Although most well waters have a consistent quality, most surface water compositions vary widely over time. A 10% increase in the hardness of the water to a sodium zeolite softener causes a 10% decrease in the service run length. An increase in the ratio of sodium to total cations causes increased sodium leakage from a demineralizer system. Regular chemical analysis of the influent water to ion exchangers should be performed to reveal such variations.

Other causes of ion exchange operational problems include:

- *Improper regenerations*, caused by incorrect regenerant flows, times, or concentrations. Manufacturer's recommendations should be followed when regenerating ion exchange resins.

- *Channeling*, resulting from either high or low flow rates, increased suspended solids loading or poor backwashing. This causes premature exhaustion even when much of the bed is in a regenerated state.

- *Resin fouling or degradation*, caused by poor-quality regenerant.

- *Failure to remove silica from the resin*, which can result from low regenerant caustic temperature. This can lead to increased silica leakage and short service runs.

- *Excess contaminants in the resin*, due to previous operation past exhaustion loads. Because the resin becomes loaded with more contaminants than a normal regeneration is designed to remove, a double regeneration is required following an extended service run.

Mechanical Problems

Typical mechanical problems associated with ion exchange systems include:

- *Leaking valves*, which cause poor quality effluent and prolonged rinses.

- *Broken or clogged distributor*, which leads to channeling.

- *Resin loss*, due to excessive backwashing or failure in the underdrain screening or support media.

- *Cation resin in the anion unit*, causing extended rinse times and sodium leakage into the demineralized water.

- *Instrumentation problems*, such as faulty totalizers or conductivity meters, which may indicate a problem when none exists, or may introduce poor quality water to service. Instrumentation in the demineralizer area should be checked regularly.

RESIN FOULING AND DEGRADATION

Resin can become fouled with contaminants that hinder the exchange process. Figure 8-17 shows a resin fouled with iron. The resin can also be attacked by chemicals that cause irreversible destruction. Some materials, such as natural organics (Figure 8-18), foul resins at first and then degrade the resin as time passes. This is the most common cause of fouling and degradation in ion exchange systems, and is discussed under "Organic Fouling," later in this chapter.

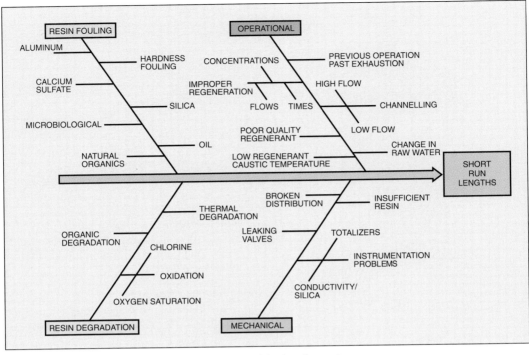

Figure 8-15. Cause–effect diagram for short runs in a two-bed demineralizer system.

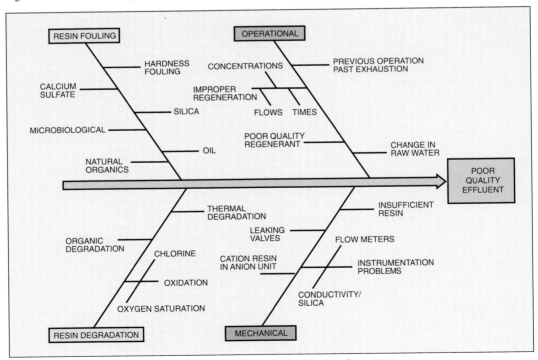

Figure 8-16. Cause–effect diagram for poor effluent quality in a two-bed demineralizer system.

Figure 8-17. Iron fouled resin.

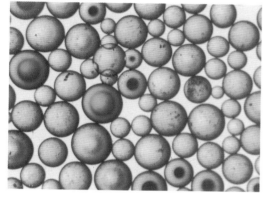

Figure 8-18. Anion resin fouled with organic material.

Causes of Resin Fouling

Iron and Manganese. Iron may exist in water as a ferrous or ferric inorganic salt or as a sequestered organic complex. Ferrous iron exchanges in resin, but ferric iron is insoluble and does not. Ferric iron coats cation resin, preventing exchange. An acid or a strong reducing agent must be used to remove this iron. Organically bound iron passes through a cation unit and fouls the anion resin. It must be removed along with the organic material. Manganese, present in some well waters, fouls a resin in the same manner as iron.

Aluminum. Aluminum is usually present as aluminum hydroxide, resulting from alum or sodium aluminate use in clarification or precipitation softening. Aluminum floc, if carried through filters, coats the resin in a sodium zeolite softener. It is removed by cleaning with either acid or caustic. Usually, aluminum is not a foulant in a demineralizer system, because it is removed from the resin during a normal regeneration.

Hardness Precipitates. Hardness precipitates carry through a filter from a precipitation softener or form after filtration by post-precipitation. These precipitates foul resins used for sodium zeolite softening. They are removed with acid.

Sulfate Precipitation. Calcium sulfate precipitation can occur in a strong acid cation unit operated in the hydrogen cycle. At the end of a service cycle, the top of the resin bed is rich in calcium. If sulfuric acid is used as the regenerant, and it is introduced at too high a concentration or too low a flow rate, precipitation of calcium sulfate occurs, fouling the resin. After calcium sulfate has formed, it is very difficult to redissolve; therefore, resin fouled by calcium sulfate is usually discarded. Mild cases of calcium sulfate fouling may be reversed with a prolonged soak in hydrochloric acid.

Barium sulfate is even less soluble than calcium sulfate. If a water source contains measurable amounts of barium, hydrochloric acid regeneration should be considered.

Oil Fouling. Oil coats resin, blocking the passage of ions to and from exchange sites. A surfactant can be used to remove oil. Care must be exercised to select a surfactant that does not foul resin. Oil-fouled anion resins should be cleaned with nonionic surfactants only.

Microbiological Fouling. Microbiological fouling can occur in resin beds, especially beds that are allowed to sit without service flow. Microbiological fouling can lead to severe plugging of the resin bed, and even mechanical damage due to an excessive pressure drop across the fouled resin. If microbiological fouling in standby units is a problem, a constant flow of recirculating water should be used to minimize the problem. Severe conditions may require the application of suitable sterilization agents and surfactants.

Silica Fouling. Silica fouling can occur in strong base anion resins if the regenerant temperature is too low, or in weak base resins if the effluent caustic from the SBA unit used to regenerate the weak base unit contains too much silica. At low pH levels, polymerization of the silica can occur in a weak base resin. It can also be a problem in an exhausted strong base anion resin. Silica fouling is removed by a prolonged soak in warm (120 °F) caustic soda.

Causes of Irreversible Resin Degradation

Oxidation. Oxidizing agents, such as chlorine, degrade both cation and anion resins. Oxidants attack the divinylbenzene cross-links in a cation resin, reducing the overall strength of the resin bead. As the attack continues, the cation resin begins to lose its spherical shape and rigidity, causing it to compact during service. This compaction increases the pressure drop across the resin bed and leads to channeling, which reduces the effective capacity of the unit.

In the case of raw water chlorine, the anion resin is not directly affected, because the chlorine is consumed by the cation resin. However, downstream strong base anion resins are fouled by certain degradation products from oxidized cation resin.

If chlorine is present in raw water, it should be removed prior to ion exchange with activated carbon filtration or sodium sulfite. Approximately 1.8 ppm of sodium sulfite is required to consume 1 ppm of chlorine.

Oxygen-saturated water, such as that found following forced draft decarbonation, accelerates the destruction of strong base exchange sites that occurs naturally over time. It also accelerates degradation due to organic fouling.

Thermal Degradation. Thermal degradation occurs if the anion resin becomes overheated during the service or regeneration cycle. This is especially true for acrylic resins, which have temperature limitations as low as 100 °F, and Type II strong base anion resins, which have a temperature limit of 105 °F when in the hydroxide form.

Organic Fouling

Organic fouling is the most common and expensive form of resin fouling and degradation. Usually, only low levels of organic materials are found in well waters. However, surface waters can contain hundreds of parts per million of natural and man-made organic matter. Natural organics are derived from decaying vegetation. They are aromatic and acidic in nature, and can complex heavy metals, such as iron. These contaminants include tannins, tannic acid, humic acid, and fulvic acid.

Initially, organics block the strong base sites on a resin. This blockage causes long final rinses and reduces salt splitting capacity. As the foulant continues to remain on the resin, it begins to degrade the strong base sites, reducing the salt splitting capacity of the resin. The functionality of the site changes from strong base to weak base, and finally to a nonactive site. Thus, a resin in the early stages of degradation exhibits high total capacity, but reduced salt splitting capacity. At

this stage, cleaning of the resin can still return some, but not all, of the lost operating capacity. A loss in salt splitting capacity reduces the ability of the resin to remove silica and carbonic acid.

Organic fouling of anion resin is evidenced by the color of the effluent from the anion unit during regeneration, which ranges from tea-colored to dark brown. During operation, the treated water has higher conductivity and a lower pH.

Prevention. The following methods are used, either alone or in combination, to reduce organic fouling:

- *Prechlorination and clarification.* Water is prechlorinated at the source, and then clarified with an organic removal aid.

- *Filtration through activated carbon.* It should be noted that a carbon filter has a finite capacity for removal of organic material and that the removal performance of the carbon should be monitored frequently.

- *Macroporous and weak base resin ahead of strong base resin.* The weak base or macroporous resin absorbs the organic material and is eluted during regeneration.

- *Specialty resins.* Acrylic and other specialty resins that are less susceptible to organic fouling have been developed.

Inspection and Cleaning. In addition to these preventive procedures, a program of regular inspection and cleaning of the ion exchange system helps to preserve the life of anion resin. Most cleaning procedures use one of the following:

- *Warm (120 °F) brine and caustic.* Mild oxidants or solubilizing agents can be added to improve the cleaning.

- *Hydrochloric acid.* When resins are also fouled with significant amounts of iron, hydrochloric acids are used.

- *Solutions of 0.25–0.5% sodium hypochlorite.* This procedure destroys the organic material but also significantly degrades the resin. Hypochlorite cleaning is considered a last resort.

It is important to clean an organically fouled resin before excessive permanent degradation of the strong base sites occurs. Cleaning after permanent degradation has occurred removes significant amounts of organic material but does not improve unit performance. The condition of the resin should be closely monitored to identify the optimum schedule for cleaning.

RESIN TESTING AND ANALYSIS

To track the condition of ion exchange resin and determine the best time for cleaning it, the resin should be periodically sampled and analyzed for physical stability, foulant levels, and the ability to perform the required ion exchange.

Samples should be representative of the entire resin bed. Therefore, samples should be collected at different levels within the bed, or a grain thief or hollow pipe should be used to obtain a "core" sample. During sampling, the inlet and regenerant distributor should be examined, and the condition of the top of the resin bed should be noted. Excessive hills or valleys in the resin bed are an indication of flow distribution problems.

The resin sample should be examined microscopically for signs of fouling and cracked or broken beads. It should also be tested for physical properties, such as density and moisture content (Figure 8-19). The level of organic and inorganic foulants in the resin should be determined and compared to known standards and the previous condition of the resin. Finally, the salt

Figure 8-19. Periodic sampling and evaluation of the resin is required to keep performance and efficiency at optimum levels.

splitting and total capacity should be measured on anion resin samples to evaluate the rate of degradation or organic fouling.

MEMBRANE SYSTEMS

Since the 1940's, ion exchange resins have been used to remove dissolved salts from water. These resins exchange ions in the water for ions on the resin exchange sites and hold them until released by a regeneration solution (see Chapter 8 for a more detailed discussion). Many ion exchange processes exist for a variety of industrial water and wastewater applications. The ion exchange process consumes large quantities of regeneration chemicals, such as brine, acid, and caustic—materials that can present significant handling and disposal problems.

In recent years, membrane processes have been used increasingly for the production of "pure" waters from fresh water and seawater. Membrane processes are also being applied in process and wastewater systems.

Although typically thought to be expensive and relatively experimental, membrane technology is advancing quickly—becoming less expensive, improving performance, and extending life expectancy.

MEMBRANE PROCESSES

Common membrane processes include ultrafiltration (UF), reverse osmosis (RO), electrodialysis (ED), and electrodialysis reversal (EDR). These processes (with the exception of UF) remove most ions; RO and UF systems also provide efficient removal of nonionized organics and particulates. Because UF membrane porosity is too large for ion rejection, the UF process is used to remove contaminants, such as oil and grease, and suspended solids.

Reverse Osmosis

Osmosis is the flow of solvent through a semipermeable membrane, from a dilute solution to a concentrated solution. This flow results from the driving force created by the difference in pressure between the two solutions. Osmotic pressure is the pressure that must be added to the concentrated solution side in order to stop the solvent flow through the membrane. Reverse osmosis is the process of reversing the flow, forcing water through a membrane from a concentrated solution to a dilute solution to produce pure water. Figure 9-1 illustrates the processes of osmosis and reverse osmosis.

Reverse osmosis is created when sufficient pressure is applied to the concentrated solution to overcome the osmotic pressure. This pressure is provided by feedwater pumps. Concentrated contaminants (brine) are removed from the high-pressure side of the RO membrane, and pure water (permeate) is removed from the low-pressure side. Figure 9-2 is a simplified schematic of an RO process. Membrane modules may be staged in various design configurations, producing the highest-quality permeate with the least amount of waste. An example of a multistage RO configuration is shown in Figure 9-3.

Typically, 95% of dissolved salts are removed from the brine. All particulates are removed. However, due to their molecular porosity, RO membranes do not remove dissolved gases, such as Cl_2, CO_2, and O_2.

RO Membranes. The two most common RO membranes used in industrial water treatment are cellulose acetate (CA) and polyamide (PA) composite. Currently, most membranes are spiral wound; however, hollow fiber configurations are available. In the spiral wound configuration, a flat sheet membrane and spacers are wound around the permeate collection tube to produce flow channels for permeate and feedwater. This design maximizes flow while minimizing the membrane module size.

Hollow fiber systems are bundles of tiny, hair-like membrane tubes. Ions are rejected when the

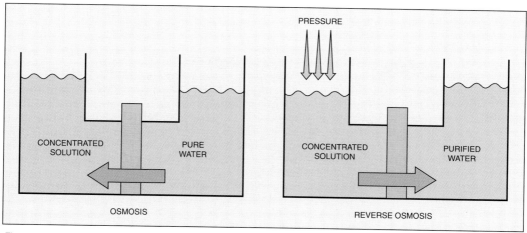

Figure 9-1. In the osmosis process, water flows through a membrane from the dilute solution side to the more concentrated solution side. In reverse osmosis, applied pressure causes water to flow from the concentrated solution to the dilute solution.

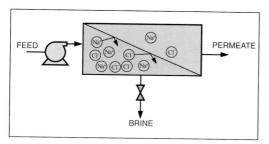

Figure 9-2. A reverse osmosis system converts a feed stream into a purified stream (permeate) and a concentrated stream (brine).

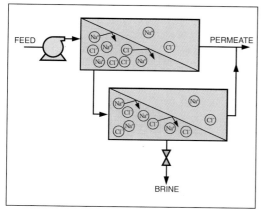

Figure 9-3. A multistage reverse osmosis system configured to reduce the quantity of waste brine.

feedwater permeates the walls of these tubes, and permeate is collected through the hollow center of the fibers. Concentrated brine is produced on the outside of the fibers contained by the module housing.

Figure 9-4 shows the construction and flow patterns in a spiral wound membrane configuration. Figure 9-5 shows the construction and flow patterns in a hollow fiber membrane system.

Electrodialysis

Electrodialysis (ED) processes transfer ions of dissolved salts across membranes, leaving purified water behind. Ion movement is induced by direct current electrical fields. A negative electrode (cathode) attracts cations, and a positive electrode (anode) attracts anions. Systems are compartmentalized in stacks by alternating cation and anion transfer membranes. Alternating compartments carry concentrated brine and purified permeate. Typically, 40–60% of dissolved ions are removed or rejected. Further improvement in water quality is obtained by staging (operation of stacks in series). ED processes do not remove particulate contaminants or weakly ionized contaminants, such as silica. Figure 9-6 is a simplified schematic of an ED process.

Electrodialysis Reversal

Electrodialysis reversal (EDR) processes operate on the same principles as ED; however, EDR operation reverses system polarity (typically 3–4 times per hour). This reversal stops the buildup of concentrated solutions on the membrane and thereby reduces the accumulation of inorganic and organic deposition on the membrane surface. EDR systems are similar to ED systems, designed with adequate chamber area to collect both product water and brine. EDR produces water of the same purity as ED.

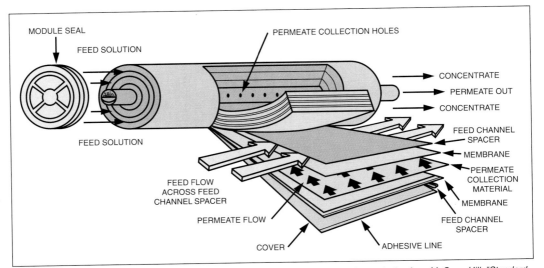

Figure 9-4. *Spiral wound reverse osmosis modules are widely used. (Reprinted with permission from McGraw-Hill, "Standard Handbook of Environmental Engineering.")*

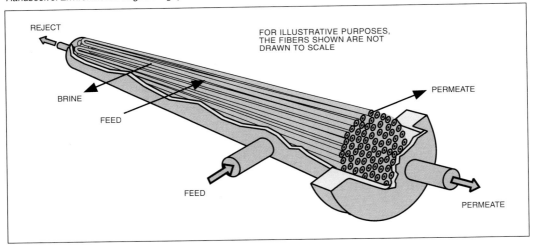

Figure 9-5. *Hollow fiber reverse osmosis modules.*

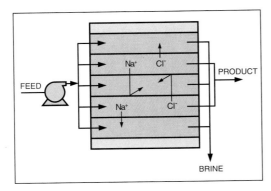

Figure 9-6. *In electrodialysis and electrodialysis reversal, ions pass through alternating cation and anion transfer membranes.*

Ultrafiltration

In many process and wastewater applications, removal of dissolved ions is not required but efficient removal of colloidal inorganic or organic molecules is. Ultrafiltration (UF) membrane configurations and system designs are similar to those used in the single-stage RO process. Because the large molecules removed by UF exhibit negligible osmotic pressure, operating pressures are usually much lower than in RO systems. Figure 9-7 illustrates the performance of ultrafiltration membranes. Typical applications include removal of oil and grease and recovery of valuable contaminants in process waste streams.

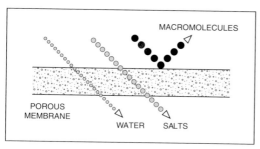

Figure 9-7. Ultrafiltration membranes pass inorganic ions but reject large organic molecules and colloidal particles.

PRETREATMENT

Processes that rely on microporous membranes must be protected from fouling. Membrane fouling causes a loss of water production (flux), reduced permeate quality, and increased trans-membrane pressure drop.

Membrane fouling is typically caused by precipitation of inorganic salts, particulates of metal oxides, colloidal silt, and the accumulation or growth of microbiological organisms on the membrane surface. These fouling problems can lead to serious damage and necessitate more frequent replacement of membranes.

SOLIDS REMOVAL

Membrane feedwater should be relatively free from colloidal particulates. The most common particulates encountered in industrial membrane systems are silt, iron oxides, and manganese oxides.

Silt Density Index (SDI) testing should be used to confirm sufficient water quality for the specific membrane system employed. SDI evaluates the potential of feedwater to foul a 0.45 µm filter. Unacceptable SDI measurements can be produced even when water quality is relatively high by most industrial water treatment standards. Where pretreatment is inadequate or

ineffective, chemical dispersants may be used to permit operation at higher-than-recommended SDI values. RO systems are highly susceptible to particulate fouling, ED and EDR systems are more forgiving, and UF systems are designed to handle "dirty" waters.

SCALE CONTROL

Membrane processes produce a concentration gradient of dissolved salts approaching the membrane surfaces. The concentration at the membrane may exceed the solubility limits of certain species. Calcium carbonate ($CaCO_3$) and calcium sulfate ($CaSO_4$) are typical precipitates formed. Silica, barium, and strontium salts are also frequently identified in membrane deposits. Because of their low solubility, very low levels of feedwater barium or strontium can cause membrane fouling.

Various saturation indexes, such as the Stiff-Davis and Langelier, should be maintained below precipitating values in the brine (through pH control or deposit control agents) to prevent calcium carbonate fouling. Other precipitates may be controlled by the proper application of deposit control agents.

MICROBIOLOGICAL FOULING

Cellulose acetate membranes can be degraded by microbiological activity. Proper maintenance of chlorine residuals can prevent microbiological attack of these membranes.

Polyacrylamide membranes are resistant to microbiological degradation; however, they are susceptible to chemical oxidation. Therefore, chlorination is not an acceptable treatment. If inoculation occurs, microbiological fouling can become a problem. Nonoxidizing antimicrobials and biodispersants should be used if serious microbiological fouling potential exists.

Boiler Water Systems

Water reclamation and reuse are increasingly important and practical ways to preserve fresh water resources.

BOILER FEEDWATER DEAERATION

The dissolved gases normally present in water cause many corrosion problems. For instance, oxygen in water produces pitting that is particularly severe because of its localized nature (see Figure 10-1). Carbon dioxide corrosion is frequently encountered in condensate systems and less commonly in water distribution systems. Water containing ammonia, particularly in the presence of oxygen, readily attacks copper and copper-bearing alloys. The resulting corrosion leads to deposits on boiler heat transfer surfaces and reduces efficiency and reliability.

In order to meet industrial standards for both oxygen content and the allowable metal oxide levels in feedwater, nearly complete oxygen removal is required. This can be accomplished only by efficient mechanical deaeration supplemented by an effective and properly controlled chemical oxygen scavenger.

Several principles apply to the mechanical deaeration of feedwater:

1. the solubility of any gas in a liquid is directly proportional to the partial pressure of the gas at the liquid surface

2. the solubility of a gas in a liquid decreases with increasing liquid temperature (see Figure 10-2)

3. efficiency of removal is increased when the liquid and gas are thoroughly mixed

The solubility of a gas in a liquid is expressed by Henry's Law:

$$C_{total} = kP$$

where

C_{total} = total concentration of the gas in solution

P = partial pressure of the gas above solution

k = a proportionality constant known as Henry's Law Constant

For example, 8 ppm of oxygen can be dissolved in water when the partial pressure of oxygen is

0.2 atmosphere; only 4 ppm of oxygen can be dissolved in water if the partial pressure of oxygen is reduced to 0.1 atmosphere.

As is evident from Henry's Law, a dissolved gas can be removed from water by a reduction of the partial pressure of that gas in the atmosphere contacting the liquid. This can be accomplished in either of two ways:

- a vacuum is applied to the system and the unwanted gas is vented

- a new gas is introduced into the system while the unwanted gas is vented

Vacuum deaeration has been used successfully in water distribution systems. However, pressure deaeration (with steam as the purge gas) is

Figure 10-1. Oxygen-pitted boiler feedwater pipe.

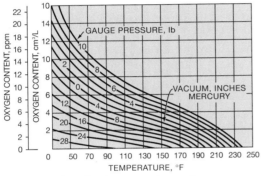

Figure 10-2. Solubility of oxygen vs. temperature.

normally used to prepare boiler feedwater. Steam is chosen as the purge gas for several reasons:

■ it is readily available

■ it heats the water and reduces the solubility of oxygen

■ it does not contaminate the water

■ only a small quantity of steam must be vented, because most of the steam used to scrub the water is condensed and becomes a part of the deaerated water

In order to deaerate the boiler feedwater, water is sprayed into a steam atmosphere. This heats the water to within a few degrees of the temperature of the saturated steam. Because the solubility of oxygen in water is very low under these conditions, 97 to 98% of the oxygen in the incoming water is released to the steam and is purged from the system by venting. Although the remaining oxygen is not soluble under equilibrium conditions, it is not readily released to the steam. Therefore,

water leaving the heating section of the deaerator must be scrubbed vigorously with steam to maximize removal.

EQUIPMENT

The purpose of a deaerator is to reduce dissolved gases, particularly oxygen, to a low level and improve a plant's thermal efficiency by raising the water temperature. In addition, deaerators provide feedwater storage and proper suction conditions for boiler feedwater pumps.

Pressure deaerators, or deaerating heaters, can be classified under two major categories: tray-type and spray-type (see Figure 10-3). Tray-type deaerators are also referred to as "spray-tray" type, because the water is initially introduced by spray valves or nozzles. The spray type is also referred to as the "spray-scrubber" type because a separate scrubbing section is used to provide additional steam–water contact after spraying.

The tray-type deaerating heater, shown in Figures 10-4 and 10-5, consists of a shell, spray nozzles to distribute and spray the water, a direct-contact vent condenser, tray stacks, and protective interchamber walls. Although the shell is constructed of low carbon steel, more corrosion-resistant stainless steels are used for the spray nozzles, vent condenser, trays, and interchamber walls.

The operation of this deaerator is illustrated in Figure 10-5. Incoming water is sprayed into a steam atmosphere, where it is heated to within a few degrees of the saturation temperature of the steam. Most of the noncondensable gases (principally oxygen and free carbon dioxide) are

Figure 10-3. Horizontal spray-tray type deaerating heater with storage tank.

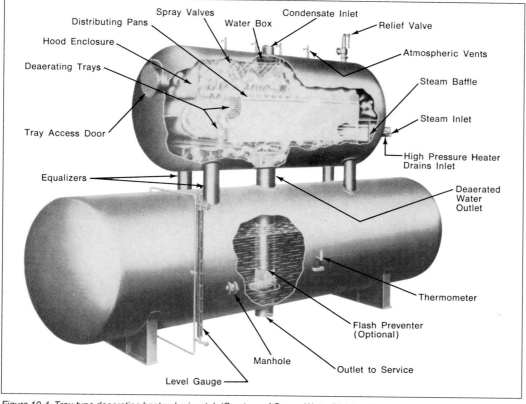

Figure 10-4. Tray-type deaerating heater, horizontal. (Courtesy of Graver Water Division, Ecodyne Corporation.)

released to the steam as the water is sprayed into the unit. Seals prevent the recontamination of tray stack water by gases from the spray section.

In the tray section, water cascades from tray to tray, breaking into fine droplets or films which intimately contact the incoming steam. The steam heats the water to the saturation temperature of the steam and removes all but the very last traces of oxygen. Deaerated water falls to the storage space below, where a steam blanket protects it from recontamination.

The steam, entering the deaerators through ports in the tray compartment, flows down through the tray stack parallel to the water flow. A very small amount of steam condenses in this section as the water temperature rises to the saturation temperature of the steam. The remainder of the steam scrubs the cascading water.

Upon leaving the tray compartment, the steam flows upward between the shell and the interchamber walls to the spray section. Most of the steam is condensed and becomes a part of the deaerated water. A small portion of the steam, which contains the noncondensable gases released

from the water, is vented to the atmosphere. It is essential that sufficient venting is provided at all times or deaeration will be incomplete.

As mentioned, most tray and spray-type deaerators use spring-loaded spray nozzles, which evenly distribute the inlet water (see Figure 10-6). Newer spray valves are designed to provide a uniform spray pattern under varying load conditions for efficient steam–water contact. The valve is designed to provide atomization of the inlet water into small droplets to improve heat transfer and to provide efficient scrubbing of the inlet water oxygen.

Steam flow through the tray stack may be cross-flow, counter-current, or co-current to the water. The deaerated water is usually stored in a separate tank, as illustrated in Figure 10-4.

The spray-type deaerating heater consists of a shell, spring-loaded inlet spray valves, a direct-contact vent condenser, and a steam scrubber for final deaeration. The inlet spray valves and direct contact vent condenser section are stainless steel; the shell and steam scrubber may be low carbon steel.

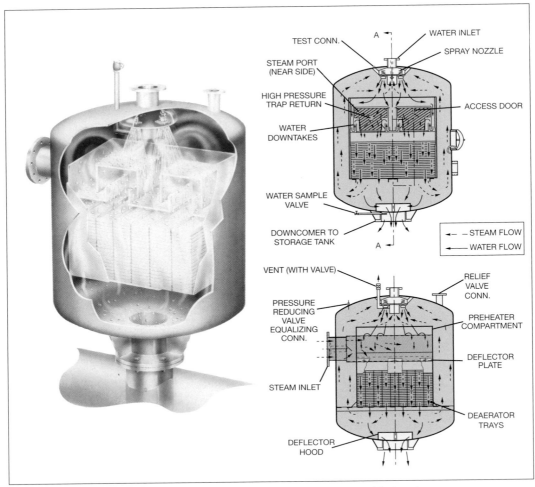

Figure 10-5. Tray-type deaerating heater, vertical. (Courtesy of Cochrane Environmental Systems.)

Figure 10-6. Deaerator spray nozzle/valve. (Courtesy of Graver Water Division, Ecodyne Corporation.)

The incoming water is sprayed into a steam atmosphere and heated to within a few degrees of the saturation temperature of the steam. Most of the noncondensable gases are released to the steam, and the heated water falls to a water seal and drains to the lowest section of the steam scrubber.

The water is scrubbed by a large volume of steam and heated to the saturation temperature prevailing at that point. The intimate steam–water contact achieved in the scrubber efficiently strips the water of dissolved gases. As the steam–water mixture rises in the scrubber, the deaerated water is a few degrees above the saturation temperature, due to a slight pressure loss. A small amount of flashing results, which aids in the release of dissolved gases. The deaerated water overflows from the steam scrubber to the storage section below.

Steam enters the deaerator through a chest on

the side and flows to the steam scrubber. Because the volume of steam is large compared to the volume of water, thorough scrubbing is achieved. The steam, after flowing through the scrubber, passes up into the spray heater section to heat the incoming water. Most of the steam condenses in the spray section to become a part of the deaerated water. A small portion of the steam is vented to the atmosphere to remove the noncondensable gases.

In the jet-atomizing segment of spray-type deaerators, the incoming water is sprayed into a steam atmosphere. Here, the water is heated sufficiently to release the majority of the noncondensable gases. The water is then delivered into a high-velocity steam jet. It impinges on a baffle and is atomized into fine droplets. The high-velocity steam heats the water to its saturation temperature and scrubs all but the last traces of oxygen from the fine water droplets.

Other types of deaerating equipment, less common in industrial plants, include film-type and bubbling device type deaerators.

In film-type deaerators, the water flows along a surface, such as Raschig rings, in a thin film counter-current to steam flow. Oxygen is removed along the film surface. Water is collected at evenly spaced intervals.

In a bubbling device deaerator, oxygen is removed following preheating of the water, through intimate contact of steam and water moving over perforated plates.

Deaerating Condensers

In power generating stations, main turbine condensers have air ejectors to remove dissolved gases. Sometimes the pressure deaerator is omitted from the feedwater cycle. However, there is a danger of air leaking into the system, both during start-up/shutdown and while the condensers are operating at low loads. This may necessitate steam blanketing and increased chemical deaeration.

Vacuum Deaeration

Vacuum deaeration is used at temperatures below the atmospheric boiling point to reduce the corrosion rate in water distribution systems. A vacuum is applied to the system to bring the water to its saturation temperature. Spray nozzles break the water into small particles to facilitate gas removal and vent the exhaust gases.

Incoming water enters through spray nozzles and falls through a column packed with Raschig rings or other synthetic packings. In this way,

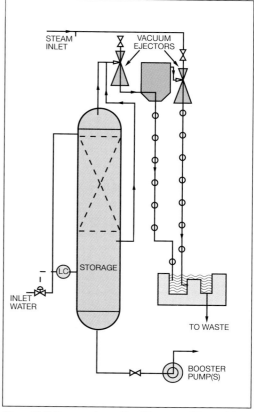

Figure 10-7. Vacuum degassifier. (Courtesy of Graver Water Division, Ecodyne Corporation.)

water is reduced to thin films and droplets, which promote the release of dissolved gases. The released gases and water vapor are removed through the vacuum, which is maintained by steam jet eductors or vacuum pumps, depending on the size of the system. Vacuum deaerators remove oxygen less efficiently than pressure units. A typical vacuum deaerator is shown in Figure 10-7.

IMPORTANT CONSIDERATIONS

Inlet water to the deaerators should be largely free from suspended solids, which can clog spray valves and ports of the inlet distributor and the deaerator trays. In addition, spray valves, ports, and deaerator trays may plug with scale which forms when the water being deaerated has high levels of hardness and alkalinity.

Pressure deaerators reduce oxygen to very low levels. Yet even trace amounts of oxygen may cause corrosion damage to a system. Therefore, good operating practice requires supplemental removal of oxygen by means of a chemical oxygen scavenger such as sodium sulfite or hydrazine, or other materials, such as organic, volatile oxygen scavengers.

Although deaeration removes free carbon dioxide, it removes only small amounts of combined carbon dioxide. The majority of the combined carbon dioxide is released with the steam in the boiler and subsequently dissolves in the condensate, frequently causing corrosion problems. These problems can be controlled through the use of volatile neutralizing amines, filming amines, and metal oxide conditioners.

MONITORING PERFORMANCE

Pressure deaerators, used to prepare boiler feedwater, produce deaerated water which is very low in dissolved oxygen and free carbon dioxide. Vendors usually guarantee less than 0.005 cm³/L (7 ppb) of oxygen.

Vacuum deaerators, used to protect water distribution lines, are not designed to deaerate as thoroughly as pressure deaerators. Usually, they reduce the oxygen content to about 0.25 to 0.50 cm³/L (330 to 650 ppb).

In order to ensure maximum oxygen removal, spot or continuous monitoring of dissolved oxygen in the effluent of the deaerator is essential. Continuous monitoring with an on-line oxygen meter is normally recommended (see Figure 10-8). For performance testing of the deaerator, the feed of the chemical oxygen scavenger is stopped for a brief period of time.

It is good practice to check the operation of the unit regularly. Care should be taken to ensure that the unit is not operated beyond its capacity. The system should also be checked for water hammer and thermal stress, which can be caused by the introduction of cold condensate. Thorough off-line inspection should be performed as often as possible and should include the following:

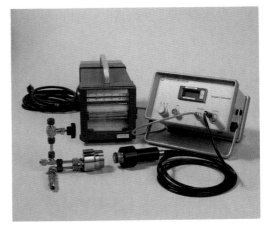

Figure 10-8. Orbisphere continuous oxygen meter.

- inlet water regulating valves and controls for storage tank level control

- high and low alarms for storage tank levels

- overflow valve and controller for prevention of high water level

- steam pressure reducing valves to maintain required minimum deaerator pressure

- safety relief valves

- temperature and pressure gauges for proper monitoring of makeup water, deaerator, and storage tank

- steam vent for removal of gases and vent condenser for integrity

- steam inlet baffles for integrity

- inlet spray valves for deposits and operation

- trays for proper position

- weld areas for damage (particularly cracking)

PREBOILER AND BOILER CORROSION CONTROL

Corrosion is one of the main causes of reduced reliability in steam generating systems. It is estimated that problems due to boiler system corrosion cost industry billions of dollars per year.

Many corrosion problems occur in the hottest areas of the boiler—the water wall, screen, and superheater tubes. Other common problem areas include deaerators, feedwater heaters, and economizers.

Methods of corrosion control vary depending upon the type of corrosion encountered. The most common causes of corrosion are dissolved gases (primarily oxygen and carbon dioxide), under-deposit attack, low pH, and attack of areas weakened by mechanical stress, leading to stress and fatigue cracking.

These conditions may be controlled through the following procedures:

■ maintenance of proper pH and alkalinity levels

■ control of oxygen and boiler feedwater contamination

■ reduction of mechanical stresses

■ operation within design specifications, especially for temperature and pressure

■ proper precautions during start-up and shutdown

■ effective monitoring and control

CORROSION TENDENCIES OF BOILER SYSTEM COMPONENTS

Most industrial boiler and feedwater systems are constructed of carbon steel. Many have copper alloy and/or stainless steel feedwater heaters and condensers. Some have stainless steel superheater elements.

Proper treatment of boiler feedwater effectively protects against corrosion of feedwater heaters, economizers, and deaerators. The ASME Consensus for Industrial Boilers (see Chapter 13) specifies maximum levels of contaminants for corrosion and deposition control in boiler systems.

The consensus is that feedwater oxygen, iron, and copper content should be very low (e.g., less than 7 ppb oxygen, 20 ppb iron, and 15 ppb copper for a 900 psig boiler) and that pH should be maintained between 8.5 and 9.5 for system corrosion protection.

In order to minimize boiler system corrosion, an understanding of the operational requirements for all critical system components is necessary.

Feedwater Heaters

Boiler feedwater heaters are designed to improve boiler efficiency by extracting heat from streams such as boiler water blowdown and turbine extraction or excess exhaust steam. Feedwater heaters are generally classified as low-pressure (ahead of the deaerator), high-pressure (after the deaerator), or deaerating heaters.

Regardless of feedwater heater design, the major problems are similar for all types. The primary problems are corrosion, due to oxygen and improper pH, and erosion from the tube side or the shell side. Due to the temperature increase across the heater, incoming metal oxides are deposited in the heater and then released during changes in steam load and chemical balances. Stress cracking of welded components can also be a problem. Erosion is common in the shell side, due to high-velocity steam impingement on tubes and baffles.

Corrosion can be minimized through proper design (to minimize erosion), periodic cleaning, control of oxygen, proper pH control, and the use of high-quality feedwater (to promote passivation of metal surfaces).

Deaerators

Deaerators are used to heat feedwater and reduce oxygen and other dissolved gases to acceptable levels. Corrosion fatigue at or near welds is a major problem in deaerators. Most corrosion fatigue cracking has been reported to be the result of mechanical factors, such as manufacturing procedures, poor welds, and lack of stress-relieved welds. Operational problems such as water/steam hammer can also be a factor.

Effective corrosion control requires the following practices:

- regular monitoring of operation

- minimization of stresses during start-up

- maintenance of stable temperature and pressure levels

- control of dissolved oxygen and pH in the feedwater

- regular out-of-service inspection using established nondestructive techniques

Other forms of corrosive attack in deaerators include stress corrosion cracking of the stainless steel tray chamber, inlet spray valve spring cracking, corrosion of vent condensers due to oxygen pitting, and erosion of the impingement baffles near the steam inlet connection.

Economizers

Economizer corrosion control involves procedures similar to those employed for protecting feedwater heaters.

Economizers help to improve boiler efficiency by extracting heat from flue gases discharged from the fireside of a boiler. Economizers can be classified as nonsteaming or steaming. In a steaming economizer, 5–20% of the incoming feedwater becomes steam. Steaming economizers are particularly sensitive to deposition from feedwater contaminants and resultant under-deposit corrosion. Erosion at tube bends is also a problem in steaming economizers.

Oxygen pitting, caused by the presence of oxygen and temperature increase, is a major problem in economizers; therefore, it is necessary to maintain essentially oxygen-free water in these units. The inlet is subject to severe pitting, because it is often the first area after the deaerator to be exposed to increased heat. Whenever possible, tubes in this area should be inspected closely for evidence of corrosion.

Economizer heat transfer surfaces are subject to corrosion product buildup and deposition of incoming metal oxides. These deposits can slough off during operational load and chemical changes.

Corrosion can also occur on the gas side of the economizer due to contaminants in the flue gas, forming low-pH compounds. Generally, economizers are arranged for downward flow of gas and upward flow of water. Tubes that form the heating surface may be smooth or provided with extended surfaces.

Superheaters

Superheater corrosion problems are caused by a number of mechanical and chemical conditions. One major problem is the oxidation of superheater metal due to high gas temperatures, usually occurring during transition periods, such as start-up and shutdown. Deposits due to carryover can contribute to the problem. Resulting failures usually occur in the bottom loops—the hottest areas of the superheater tubes.

Oxygen pitting, particularly in the pendant loop area, is another major corrosion problem in superheaters. It is caused when water is exposed to oxygen during downtime. Close temperature control helps to minimize this problem. In addition, a nitrogen blanket and chemical oxygen scavenger can be used to maintain oxygen-free conditions during downtime.

Low-Pressure Steam and Hot Water Heating Systems

Hot water boilers heat and circulate water at approximately 200 °F. Steam heating boilers are used to generate steam at low pressures, such as 15 psig. Generally, these two basic heating systems are treated as closed systems, because makeup requirements are usually very low.

High-temperature hot water boilers operate at pressures of up to 500 psig, although the usual range is 35–350 psig. System pressure must be maintained above the saturation pressure of the heated water to maintain a liquid state. The most common way to do this is to pressurize the system with nitrogen. Normally, the makeup is of good quality (e.g., deionized or sodium zeolite softened water). Chemical treatment consists of sodium sulfite (to scavenge the oxygen), pH adjustment, and a synthetic polymer dispersant to control possible iron deposition.

The major problem in low-pressure heating systems is corrosion caused by dissolved oxygen and low pH. These systems are usually treated with an inhibitor (such as molybdate or nitrite) or with an oxygen scavenger (such as sodium sulfite), along with a synthetic polymer for deposit control. Sufficient treatment must be fed to water added to make up for system losses, which usually occur as a result of circulating pump leakage. Generally, 200–400 ppm P-alkalinity is maintained in the water for effective control of pH. Inhibitor requirements vary depending on the system.

Electric boilers are also used for heating. There are two basic types of electric boilers: resistance and electrode. Resistance boilers generate heat by means of a coiled heating element. High-quality makeup water is necessary, and sodium sulfite is usually added to remove all traces of dissolved oxygen. Synthetic polymers have been used for deposit control. Due to the high heat transfer rate at the resistance coil, a treatment that precipitates hardness should not be used.

Electrode boilers operate at high or low voltage and may employ submerged or water-jet electrodes. High-purity makeup water is required. Depending on the type of system, sodium sulfite is normally used for oxygen control and pH adjustment. Some systems are designed with copper alloys, so chemical addition must be of the correct type, and pH control must be in the range suitable for copper protection.

TYPES OF CORROSION

Corrosion control techniques vary according to the type of corrosion encountered. Major methods of corrosion control include maintenance of the proper pH, control of oxygen, control of deposits, and reduction of stresses through design and operational practices.

Galvanic Corrosion

Galvanic corrosion occurs when a metal or alloy is electrically coupled to a different metal or alloy.

The most common type of galvanic corrosion in a boiler system is caused by the contact of dissimilar metals, such as iron and copper. These differential cells can also be formed when deposits are present. Galvanic corrosion can occur at welds due to stresses in heat-affected zones or the use of different alloys in the welds. Anything that results in a difference in electrical potential at discrete surface locations can cause a galvanic reaction. Causes include:

- scratches in a metal surface

- differential stresses in a metal

- differences in temperature

- conductive deposits

A general illustration of a corrosion cell for iron in the presence of oxygen is shown in Figure 11-1. Pitting of boiler tube banks has been encountered due to metallic copper deposits. Such deposits may form during acid cleaning procedures if the procedures do not completely compensate for the amount of copper oxides in the deposits or if a copper removal step is not included. Dissolved copper may be plated out on freshly cleaned surfaces, establishing anodic corrosion areas and forming pits, which are very similar to oxygen pits in form and appearance. This process is illustrated by the following reactions involving hydrochloric acid as the cleaning solvent.

Magnetite is dissolved and yields an acid solution containing both ferrous (Fe^{2+}) and ferric (Fe^{3+}) chlorides (ferric chlorides are very corrosive to steel and copper):

$$\underset{\text{magnetite}}{Fe_3O_4} + \underset{\substack{\text{hydrochloric}\\\text{acid}}}{8HCl} \rightarrow \underset{\substack{\text{ferrous}\\\text{chloride}}}{FeCl_2} + \underset{\substack{\text{ferric}\\\text{chloride}}}{2FeCl_3} + \underset{\text{water}}{4H_2O}$$

Metallic or elemental copper in boiler deposits is dissolved in the hydrochloric acid solution by the following reaction:

$$\underset{\substack{\text{ferric}\\\text{chloride}}}{FeCl_3} + \underset{\text{copper}}{Cu} \rightarrow \underset{\substack{\text{cuprous}\\\text{chloride}}}{CuCl} + \underset{\substack{\text{ferrous}\\\text{chloride}}}{FeCl_2}$$

Once cuprous chloride is in solution, it is immediately redeposited as metallic copper on the steel surface according to the following reaction:

$$\underset{\substack{\text{cuprous}\\\text{chloride}}}{2CuCl} + \underset{\text{iron}}{Fe} \rightarrow \underset{\substack{\text{ferrous}\\\text{chloride}}}{FeCl_2} + \underset{\text{copper}}{2Cu^0}$$

Thus, hydrochloric acid cleaning can cause galvanic corrosion unless the copper is prevented from plating on the steel surface. A complexing agent is added to prevent the copper from

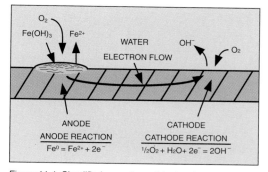

Figure 11-1. Simplified corrosion cell for iron in water.

redepositing. The following chemical reaction results:

$$FeCl_3 + Cu + \text{Complexing} \rightarrow FeCl_2 + CuCl$$

ferric   copper   Agent   ferrous   cuprous
chloride   chloride   chloride
                                             complex

This can take place as a separate step or during acid cleaning. Both iron and the copper are removed from the boiler, and the boiler surfaces can then be passivated.

In most cases, the copper is localized in certain tube banks and causes random pitting. When deposits contain large quantities of copper oxide or metallic copper, special precautions are required to prevent the plating out of copper during cleaning operations.

Caustic Corrosion

Concentration of caustic (NaOH) can occur either as a result of steam blanketing (which allows salts to concentrate on boiler metal surfaces) or by localized boiling beneath porous deposits on tube surfaces.

Caustic corrosion (gouging) occurs when caustic is concentrated and dissolves the protective magnetite (Fe_3O_4) layer. Iron, in contact with the boiler water, forms magnetite and the protective layer is continuously restored. However, as long as a high caustic concentration exists, the magnetite is constantly dissolved, causing a loss of base metal and eventual failure (see Figure 11-2).

Steam blanketing is a condition that occurs when a steam layer forms between the boiler water and the tube wall. Under this condition, insufficient water reaches the tube surface for efficient heat transfer. The water that does reach the overheated boiler wall is rapidly vaporized, leaving behind a concentrated caustic solution, which is corrosive.

Porous metal oxide deposits also permit the development of high boiler water concentrations. Water flows into the deposit and heat applied to the tube causes the water to evaporate, leaving a very concentrated solution. Again, corrosion may occur.

Caustic attack creates irregular patterns, often referred to as gouges. Deposition may or may not be found in the affected area.

Boiler feedwater systems using demineralized or evaporated makeup or pure condensate may be protected from caustic attack through coordinated phosphate/pH control. Phosphate buffers the boiler water, reducing the chance of large pH changes due to the development of high caustic concentrations. Excess caustic combines with disodium phosphate and forms trisodium phosphate. Sufficient disodium phosphate must be available to combine with all of the free caustic in order to form trisodium phosphate.

Disodium phosphate neutralizes caustic by the following reaction:

$$Na_2HPO_4 + NaOH \rightarrow Na_3PO_4 + H_2O$$

disodium   sodium   trisodium   water
phosphate   hydroxide   phosphate

This results in the prevention of caustic buildup beneath deposits or within a crevice where leakage is occurring. Caustic corrosion (and caustic embrittlement, discussed later) does not occur, because high caustic concentrations do not develop (see Figure 11-3).

Figure 11-4 shows the phosphate/pH relationship recommended to control boiler corrosion. Different forms of phosphate consume or add

Figure 11-2. Boiler system tube shows high-pH caustic gouging.

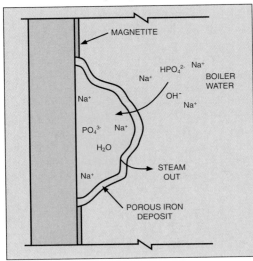

Figure 11-3. Caustic under-deposit corrosion can be controlled through a coordinated phosphate/pH program.

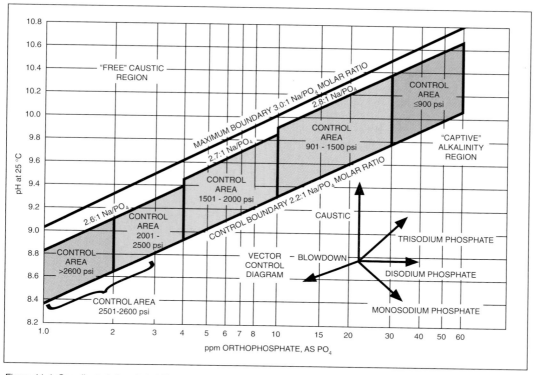

Figure 11-4. Coordinated phosphate/pH program controls free caustic and resultant corrosion.

caustic as the phosphate shifts to the proper form. For example, addition of monosodium phosphate consumes caustic as it reacts with caustic to form disodium phosphate in the boiler water according to the following reaction:

$$NaH_2PO_4 \;+\; NaOH \;\rightarrow\; Na_2HPO_4 \;+\; H_2O$$
monosodium phosphate   sodium hydroxide   disodium phosphate   water

Conversely, addition of trisodium phosphate adds caustic, increasing boiler water pH:

$$Na_3PO_4 \;+\; H_2O \;\rightarrow\; Na_2HPO_4 \;+\; NaOH$$
trisodium phosphate   water   disodium phosphate   sodium hydroxide

Control is achieved through feed of the proper type of phosphate to either raise or lower the pH while maintaining the proper phosphate level. Increasing blowdown lowers both phosphate and pH. Therefore, various combinations and feed rates of phosphate, blowdown adjustment, and caustic addition are used to maintain proper phosphate/pH levels.

Elevated temperatures at the boiler tube wall or deposits can result in some precipitation of phosphate. This effect, termed "phosphate hideout," usually occurs when loads increase. When the load is reduced, phosphate reappears.

Clean boiler water surfaces reduce potential concentration sites for caustic. Deposit control treatment programs, such as those based on chelants and synthetic polymers, can help provide clean surfaces.

Where steam blanketing is occurring, corrosion can take place even without the presence of caustic, due to the steam/magnetite reaction and the dissolution of magnetite. In such cases, operational changes or design modifications may be necessary to eliminate the cause of the problem.

Acidic Corrosion

Low makeup or feedwater pH can cause serious acid attack on metal surfaces in the preboiler and boiler system. Even if the original makeup or feedwater pH is not low, feedwater can become acidic from contamination of the system. Common causes include the following:

■ improper operation or control of demineralizer cation units

■ process contamination of condensate (e.g., sugar contamination in food processing plants)

■ cooling water contamination from condensers

Acid corrosion can also be caused by chemical cleaning operations. Overheating of the cleaning solution can cause breakdown of the inhibitor used, excessive exposure of metal to cleaning agent, and high cleaning agent concentration. Failure to neutralize acid solvents completely before start-up has also caused problems.

In a boiler and feedwater system, acidic attack can take the form of general thinning, or it can be localized at areas of high stress such as drum baffles, "U" bolts, acorn nuts, and tube ends.

Hydrogen Embrittlement

Hydrogen embrittlement is rarely encountered in industrial plants. The problem usually occurs only in units operating at or above 1,500 psi.

Hydrogen embrittlement of mild steel boiler tubing occurs in high-pressure boilers when atomic hydrogen forms at the boiler tube surface as a result of corrosion. Hydrogen permeates the tube metal, where it can react with iron carbides to form methane gas, or with other hydrogen atoms to form hydrogen gas. These gases evolve predominantly along grain boundaries of the metal. The resulting increase in pressure leads to metal failure.

The initial surface corrosion that produces hydrogen usually occurs beneath a hard, dense scale. Acidic contamination or localized low-pH excursions are normally required to generate atomic hydrogen. In high-purity systems, raw water in-leakage (e.g., condenser leakage) lowers boiler water pH when magnesium hydroxide precipitates, resulting in corrosion, formation of atomic hydrogen, and initiation of hydrogen attack.

Coordinated phosphate/pH control can be used to minimize the decrease in boiler water pH that results from condenser leakage. Maintenance of clean surfaces and the use of proper procedures for acid cleaning also reduce the potential for hydrogen attack.

Oxygen Attack

Without proper mechanical and chemical deaeration, oxygen in the feedwater will enter the boiler. Much is flashed off with the steam; the remainder can attack boiler metal. The point of attack varies with boiler design and feedwater distribution. Pitting is frequently visible in the feedwater distribution holes, at the steam drum waterline, and in downcomer tubes.

Oxygen is highly corrosive when present in hot water. Even small concentrations can cause serious problems. Because pits can penetrate deep into the metal, oxygen corrosion can result in rapid failure of feedwater lines, economizers, boiler tubes, and condensate lines. Additionally, iron oxide generated by the corrosion can produce iron deposits in the boiler.

Oxygen corrosion may be highly localized or may cover an extensive area. It is identified by well defined pits or a very pockmarked surface. The pits vary in shape, but are characterized by sharp edges at the surface. Active oxygen pits are distinguished by a reddish brown oxide cap (tubercle). Removal of this cap exposes black iron oxide within the pit (see Figure 11-5).

Oxygen attack is an electrochemical process that can be described by the following reactions:

Anode:

$$Fe \longrightarrow Fe^{2+} + 2e^-$$

Cathode:

$$\frac{1}{2}O_2 + H_2O + 2e^- \longrightarrow 2OH^-$$

Overall:

$$Fe + \frac{1}{2}O_2 + H_2O \longrightarrow Fe(OH)_2$$

The influence of temperature is particularly important in feedwater heaters and economizers. A temperature rise provides enough additional energy to accelerate reactions at the metal surfaces, resulting in rapid and severe corrosion.

At 60 °F and atmospheric pressure, the solubility of oxygen in water is approximately 8 ppm. Efficient mechanical deaeration reduces dissolved oxygen to 7 ppb or less. For complete protection from oxygen corrosion, a chemical scavenger is required following mechanical deaeration.

Major sources of oxygen in an operating system include poor deaerator operation, in-leakage of air on the suction side of pumps, the breathing

Figure 11-5. Oxygen pitting of a boiler feedwater pipe.

action of receiving tanks, and leakage of un-deaerated water used for pump seals.

The acceptable dissolved oxygen level for any system depends on many factors, such as feed-water temperature, pH, flow rate, dissolved solids content, and the metallurgy and physical condition of the system. Based on experience in thousands of systems, 3–10 ppb of feedwater oxygen is not significantly damaging to economizers. This is reflected in industry guidelines.

■ the ASME consensus is less than 7 ppb (ASME recommends chemical scavenging to "essentially zero" ppb)

■ TAPPI engineering guidelines are less than 7 ppb

■ EPRI fossil plant guidelines are less than 5 ppb dissolved oxygen

MECHANICAL CONDITIONS AFFECTING CORROSION

Many corrosion problems are the result of mechanical and operational problems. The fol-lowing practices help to minimize these corro-sion problems:

■ selection of corrosion-resistant metals

■ reduction of mechanical stress where possible (e.g., use of proper welding procedures and stress-relieving welds)

■ minimization of thermal and mechanical stresses during operation

■ operation within design load specifications, without over-firing, along with proper start-up and shutdown procedures

■ maintenance of clean systems, including the use of high-purity feedwater, effective and closely controlled chemical treatment, and acid cleaning when required

Stress Corrosion Cracking

Stress corrosion cracking occurs from the com-bined action of corrosion and stress. The corrosion may be initiated by improper chemical cleaning, high dissolved oxygen levels, pH excursions in the boiler water, the presence of free hydroxide, and high levels of chlorides. Stresses are either residual in the metal or caused by thermal excursions. Rapid start-up or shutdown can cause or further aggravate stresses.

Tube failures occur near stressed areas such as welds, supports, or cold worked areas.

Caustic Embrittlement

Caustic embrittlement (caustic stress corrosion cracking), or intercrystalline cracking, has long been recognized as a serious form of boiler metal failure. Because chemical attack of the metal is normally undetectable, failure occurs suddenly—often with catastrophic results.

For caustic embrittlement to occur, three conditions must exist:

■ the boiler metal must have a high level of stress

■ a mechanism for the concentration of boiler water must be present

■ the boiler water must have embrittlement-producing characteristics

Where boiler tubes fail as a result of caustic embrittlement, circumferential cracking can be seen. In other components, cracks follow the lines of greatest stress. A microscopic examina-tion of a properly prepared section of embrittled metal shows a characteristic pattern, with crack-ing progressing along defined paths or grain boundaries in the crystal structure of the metal (see Figure 11-6). The cracks do not penetrate the crystals themselves, but travel between them; therefore, the term "intercrystalline cracking" is used.

Good engineering practice dictates that the boiler water be evaluated for embrittling charac-teristics. An embrittlement detector (described in Chapter 14) is used for this purpose.

If a boiler water possesses embrittling charac-teristics, steps must be taken to prevent attack of the boiler metal. Sodium nitrate is a standard treatment for inhibiting embrittlement in lower-pressure boiler systems. The inhibition of em-brittlement requires a definite ratio of nitrate to the caustic alkalinity present in the boiler water.

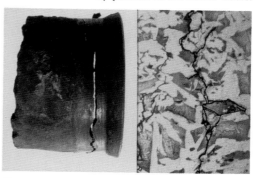

Figure 11-6. Caustic stress corrosion cracking (embrittlement) of a boiler tube. Photomicrograph shows intercrystalline cracking.

In higher-pressure boiler systems, where de-mineralized makeup water is used, embrittling characteristics in boiler water can be prevented by the use of coordinated phosphate/pH treatment control, described previously under "Caustic Corrosion." This method prevents high concentrations of free sodium hydroxide from forming in the boiler, eliminating embrittling tendencies.

Fatigue Cracking

Fatigue cracking (due to repeated cyclic stress) can lead to metal failure. The metal failure occurs at the point of the highest concentration of cyclic stress. Examples of this type of failure include cracks in boiler components at support brackets or rolled in tubes when a boiler undergoes thermal fatigue due to repeated start-ups and shutdowns.

Thermal fatigue occurs in horizontal tube runs as a result of steam blanketing and in water wall tubes due to frequent, prolonged lower header blowdown.

Corrosion fatigue failure results from cyclic stressing of a metal in a corrosive environment. This condition causes more rapid failure than that caused by either cyclic stressing or corrosion alone. In boilers, corrosion fatigue cracking can result from continued breakdown of the protective magnetite film due to cyclic stress.

Corrosion fatigue cracking occurs in deaerators near the welds and heat-affected zones. Proper operation, close monitoring, and detailed out-of-service inspections (in accordance with published recommendations) minimize problems in deaerators.

Steam Side Burning

Steam side burning is a chemical reaction between steam and the tube metal. It is caused by excessive heat input or poor circulation, resulting in insufficient flow to cool the tubes. Under such conditions, an insulating superheated steam film develops. Once the tube metal temperature has reached 750 °F in boiler tubes or 950–1000 °F in superheater tubes (assuming low alloy steel construction), the rate of oxidation increases dramatically; this oxidation occurs repeatedly and consumes the base metal. The problem is most frequently encountered in superheaters and in horizontal generating tubes heated from the top.

Erosion

Erosion usually occurs due to excessive velocities. Where two-phase flow (steam and water) exists, failures due to erosion are caused by the impact of the fluid against a surface. Equipment vulnerable to erosion includes turbine blades, low-pressure steam piping, and heat exchangers that are subjected to wet steam. Feedwater and condensate piping subjected to high-velocity water flow are also susceptible to this type of attack. Damage normally occurs where flow changes direction.

METALLIC OXIDES IN BOILER SYSTEMS

Iron and copper surfaces are subject to corrosion, resulting in the formation of metal oxides. This condition can be controlled through careful selection of metals and maintenance of proper operating conditions.

Iron Oxide Formation

Iron oxides present in operating boilers can be classified into two major types. The first and most important is the 0.0002–0.0007 in. (0.2–0.7 mil) thick magnetite formed by the reaction of iron and water in an oxygen-free environment. This magnetite forms a protective barrier against further corrosion.

Magnetite forms on boiler system metal surfaces from the following overall reaction:

$$\underset{\text{iron}}{3Fe} + \underset{\text{water}}{4H_2O} \rightarrow \underset{\text{magnetite}}{Fe_3O_4} + \underset{\text{hydrogen}}{4H_2\uparrow}$$

The magnetite, which provides a protective barrier against further corrosion, consists of two layers. The inner layer is relatively thick, compact, and continuous. The outer layer is thinner, porous, and loose in structure. Both of these layers continue to grow due to water diffusion (through the porous outer layer) and lattice diffusion (through the inner layer). As long as the magnetite layers are left undisturbed, their growth rate rapidly diminishes.

The second type of iron oxide in a boiler is the corrosion products, which may enter the boiler system with the feedwater. These are frequently termed "migratory" oxides, because they are not usually generated in the boiler. The oxides form an outer layer over the metal surface. This layer is very porous and easily penetrated by water and ionic species.

Iron can enter the boiler as soluble ferrous ions and insoluble ferrous and ferric hydroxides or oxides. Oxygen-free, alkaline boiler water converts iron to magnetite, Fe_3O_4. Migratory magnetite deposits on the protective layer and is normally gray to black in color.

Copper Oxide Formation

A truly passive oxide film does not form on copper or its alloys. In water, the predominant

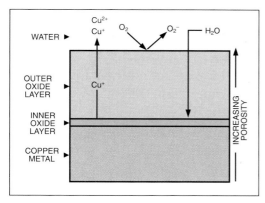

Figure 11-7. Model of oxide layers on copper shows thickness of outer oxide layer.

copper corrosion product is cuprous oxide (Cu_2O). A typical corrosion reaction follows:

$$8Cu + O_2 + 2H_2O \rightarrow 4Cu_2O + 2H_2$$
copper oxygen water cuprous oxide hydrogen

As shown in Figure 11-7, the oxide that develops on the copper surfaces is comprised of two layers. The inner layer is very thin, adherent, nonporous, and comprised mostly of cupric oxide (CuO). The outer layer is thick, adherent, porous and comprised mainly of cuprous oxide (Cu_2O). The outer layer is formed by breakup of the inner layer. At a certain thickness of the outer layer, an equilibrium exists at which the oxide continually forms and is released into the water.

Maintenance of the proper pH, elimination of oxygen, and application of metal-conditioning agents can minimize the amount of copper alloy corrosion.

Metal Passivation

The establishment of protective metal oxide layers through the use of reducing agents (such as hydrazine, hydroquinone, and other oxygen scavengers) is known as metal passivation or metal conditioning. Although "metal passivation" refers to the direct reaction of the compound with the metal oxide and "metal conditioning" more broadly refers to the promotion of a protective surface, the two terms are frequently used interchangeably.

The reaction of hydrazine and hydroquinone, which leads to the passivation of iron-based metals, proceeds according to the following reactions:

$$N_2H_4 + 6Fe_2O_3 \rightarrow 4Fe_3O_4 + 2H_2O + N_2$$
hydrazine hematite magnetite water nitrogen

$$C_6H_4(OH)_2 + 3Fe_2O_3 \rightarrow 2Fe_3O_4 + C_6H_4O_2 + H_2O$$
hydroquinone hematite magnetite benzoquinone water

Similar reactions occur with copper-based metals:

$$N_2H_4 + 4CuO \rightarrow 2Cu_2O + 2H_2O + N_2$$
hydrazine cupric oxide cuprous oxide water nitrogen

$$C_6H_6O_2 + 2CuO \rightarrow Cu_2O + C_6H_4O_2 + H_2O$$
hydroquinone cupric oxide cuprous oxide benzoquinone water

Magnetite and cuprous oxide form protective films on the metal surface. Because these oxides are formed under reducing conditions, removal of the dissolved oxygen from boiler feedwater and condensate promotes their formation. The effective application of oxygen scavengers indirectly leads to passivated metal surfaces and less metal oxide transport to the boiler whether or not the scavenger reacts directly with the metal surface.

A significant reduction in feedwater oxygen and metal oxides can occur with proper application of oxygen scavengers (see Figure 11-8).

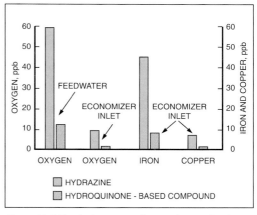

Figure 11-8. Feedwater oxygen, iron, and copper levels show dramatic reduction when hydroquinone-based materials are used instead of hydrazine (data taken during start-ups and excursions).

CORROSION CONTROL FACTORS

Steel and Steel Alloys

Protection of steel in a boiler system depends on temperature, pH, and oxygen content. Generally, higher temperatures, high or low pH levels, and higher oxygen concentrations increase steel corrosion rates.

Mechanical and operational factors, such as velocities, metal stresses, and severity of service can strongly influence corrosion rates. Systems vary in corrosion tendencies and should be evaluated individually.

Copper and Copper Alloys

Many factors influence the corrosion rate of copper alloys:

- temperature
- pH
- oxygen concentration
- amine concentration
- ammonia concentration
- flow rate

The impact of each of these factors varies depending on characteristics of each system. Temperature dependence results from faster reaction times and greater solubility of copper oxides at elevated temperatures. Maximum temperatures specified for various alloys range from 200 to 300 °F.

Methods of minimizing copper and copper alloy corrosion include:

- replacement with a more resistant metal
- elimination of oxygen
- maintenance of high-purity water conditions
- operation at the proper pH level
- reduction of water velocities
- application of materials which passivate the metal surfaces

pH Control

Maintenance of proper pH throughout the boiler feedwater, boiler, and condensate systems is essential for corrosion control. Most low-pressure boiler system operators monitor boiler water alkalinity because it correlates very closely with pH, while most feedwater, condensate, and high-pressure boiler water requires direct monitoring of pH. Control of pH is important for the following reasons:

- corrosion rates of metals used in boiler systems are sensitive to variations in pH
- low pH or insufficient alkalinity can result in corrosive acidic attack
- high pH or excess alkalinity can result in caustic gouging/cracking and foaming, with resultant carryover
- speed of oxygen scavenging reactions is highly dependent on pH levels

The pH or alkalinity level maintained in a boiler system depends on many factors, such as sys-

tem pressure, system metals, feedwater quality, and type of chemical treatment applied.

The corrosion rate of carbon steel at feedwater temperatures approaches a minimum value in the pH range of 9.2–9.6 (see Figure 11-9). It is important to monitor the feedwater system for corrosion by means of iron and copper testing. For systems with sodium zeolite or hot lime softened makeup, pH adjustment may not be necessary. In systems that use deionized water makeup, small amounts of caustic soda or neutralizing amines, such as morpholine and cyclohexylamine, can be used.

In the boiler, either high or low pH increases the corrosion rates of mild steel (see Figure 11-10). The pH or alkalinity that is maintained

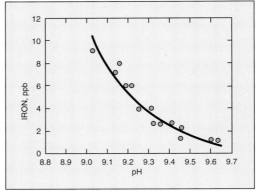

Figure 11-9. Iron corrosion product release from carbon steel in boiler feedwater.

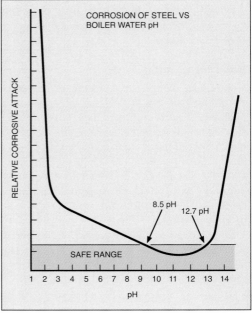

Figure 11-10. High or low boiler water pH corrodes boiler steel.

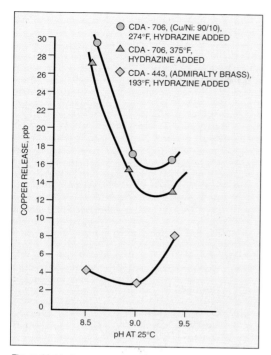

Figure 11-11. Average copper release as a function of pH shows optimum pH in range of 8.8 to 9.2 for different copper-based alloys. (Courtesy of the Electric Power Research Institute.)

depends on the pressure, makeup water characteristics, chemical treatment, and other factors specific to the system.

The best pH for protection of copper alloys is somewhat lower than the optimum level for carbon steel. For systems that contain both metals, the condensate and feedwater pH is often maintained between 8.8 and 9.2 for corrosion protection of both metals. The optimum pH varies from system to system and depends on many factors, including the alloy used (see Figure 11-11).

To elevate pH, neutralizing amines should be used instead of ammonia, which (especially in the presence of oxygen) accelerates copper alloy corrosion rates. Also, amines form protective films on copper oxide surfaces that inhibit corrosion.

Oxygen Control

Chemical Oxygen Scavengers. The oxygen scavengers most commonly used in boiler systems are sodium sulfite, sodium bisulfite, hydrazine, catalyzed versions of the sulfites and hydrazine, and organic oxygen scavengers, such as hydroquinone and ascorbate.

It is of critical importance to select and properly use the best chemical oxygen scavenger for a given system. Major factors that determine the

best oxygen scavenger for a particular application include reaction speed, residence time in the system, operating temperature and pressure, and feedwater pH. Interferences with the scavenger/oxygen reaction, decomposition products, and reactions with metals in the system are also important factors. Other contributing factors include the use of feedwater for attemperation, the presence of economizers in the system, and the end use of the steam. Chemical oxygen scavengers should be fed to allow ample time for the scavenger/oxygen reaction to occur. The deaerator storage system and the feedwater storage tank are commonly used feed points.

In boilers operating below 1,000 psig, sodium sulfite and a concentrated liquid solution of catalyzed sodium bisulfite are the most commonly used materials for chemical deaeration due to low cost and ease of handling and testing. The oxygen scavenging property of sodium sulfite is illustrated by the following reaction:

$$2Na_2SO_3 \; + \; O_2 \; \rightarrow \; 2Na_2SO_4$$
$$\text{sodium sulfite} \qquad \text{oxygen} \qquad \text{sodium sulfate}$$

Theoretically, 7.88 ppm of chemically pure sodium sulfite is required to remove 1.0 ppm of dissolved oxygen. However, due to the use of technical grades of sodium sulfite, combined with handling and blowdown losses during normal plant operation, approximately 10 lb of sodium sulfite per pound of oxygen is usually required. The concentration of excess sulfite maintained in the feedwater or boiler water also affects the sulfite requirement.

Sodium sulfite must be fed continuously for maximum oxygen removal. Usually, the most suitable point of application is the drop leg between the deaerator and the storage compartment. Where hot process softeners are followed by hot zeolite units, an additional feed is recommended at the filter effluent of the hot process units (prior to the zeolite softeners) to protect the ion exchange resin and softener shells.

As with any oxygen scavenging reaction, many factors affect the speed of the sulfite–oxygen reaction. These factors include temperature, pH, initial concentration of oxygen scavenger, initial concentration of dissolved oxygen, and catalytic or inhibiting effects. The most important factor is temperature. As temperature increases, reaction time decreases; in general, every 18 °F increase in temperature doubles reaction speed. At temperatures of 212 °F and above, the reaction is rapid. Overfeed of sodium sulfite also increases reaction rate. The reaction proceeds most rapidly at pH values in the range of 8.5–10.0.

Certain materials catalyze the oxygen–sulfite reaction. The most effective catalysts are the heavy metal cations with valences of two or more. Iron, copper, cobalt, nickel, and manganese are among the more effective catalysts.

Figure 11-12 compares the removal of oxygen using commercial sodium sulfite and a catalyzed sodium sulfite. After 25 seconds of contact, catalyzed sodium sulfite removed the oxygen completely. Uncatalyzed sodium sulfite removed less than 50% of the oxygen in this same time period. In a boiler feedwater system, this could result in severe corrosive attack.

The following operational conditions necessitate the use of catalyzed sodium sulfite:

■ low feedwater temperature

■ incomplete mechanical deaeration

■ rapid reaction required to prevent pitting in the system

■ short residence time

■ use of economizers

High feedwater sulfite residuals and pH values above 8.5 should be maintained in the feedwater to help protect the economizer from oxygen attack.

Some natural waters contain materials that can inhibit the oxygen/sulfite reaction. For example, trace organic materials in a surface supply used for makeup water can reduce speed of scavenger/oxygen reaction time. The same problem can occur where contaminated condensate is used as a portion of the boiler feedwater. The organic materials complex metals (natural or formulated catalysts) and prevent them from increasing the rate of reaction.

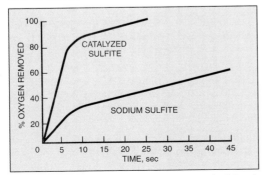

Figure 11-12. Comparison of the reaction rates of catalyzed sulfite and sodium sulfite with dissolved oxygen.

Sodium sulfite must be fed where it will not contaminate feedwater to be used for attemporation or desuperheating. This prevents the addition of solids to the steam.

At operating pressures of 1,000 psig and higher, hydrazine or organic oxygen scavengers are normally used in place of sulfite. In these applications, the increased dissolved solids contributed by sodium sulfate (the product of the sodium sulfite–oxygen reaction) can become a significant problem. Also, sulfite decomposes in high-pressure boilers to form sulfur dioxide (SO_2) and hydrogen sulfide (H_2S). Both of these gases can cause corrosion in the return condensate system and have been reported to contribute to stress corrosion cracking in turbines. Hydrazine has been used for years as an oxygen scavenger in high-pressure systems and other systems in which sulfite materials cannot be used. Hydrazine is a reducing agent that removes dissolved oxygen by the following reaction:

$$\underset{\text{hydrazine}}{N_2H_4} \quad + \quad \underset{\text{oxygen}}{O_2} \quad \rightarrow \quad \underset{\text{water}}{2H_2O} \quad + \quad \underset{\text{nitrogen}}{N_2}$$

Because the products of this reaction are water and nitrogen, the reaction adds no solids to the boiler water. The decomposition products of hydrazine are ammonia and nitrogen. Decomposition begins at approximately 400 °F and is rapid at 600 °F. The alkaline ammonia does not attack steel. However, if enough ammonia and oxygen are present together, copper alloy corrosion increases. Close control of the hydrazine feed rate can limit the concentration of ammonia in the steam and minimize the danger of attack on copper-bearing alloys. The ammonia also neutralizes carbon dioxide and reduces the return line corrosion caused by carbon dioxide.

Hydrazine is a toxic material and must be handled with extreme care. Because the material is a suspected carcinogen, federally published guidelines must be followed for handling and reporting. Because pure hydrazine has a low flash point, a 35% solution with a flash point of greater than 200 °F is usually used. Theoretically, 1.0 ppm of hydrazine is required to react with 1.0 ppm of dissolved oxygen. However, in practice 1.5–2.0 parts of hydrazine are required per part of oxygen.

The factors that influence the reaction time of sodium sulfite also apply to other oxygen scavengers. Figure 11-13 shows rate of reaction as a function of temperature and hydrazine con-

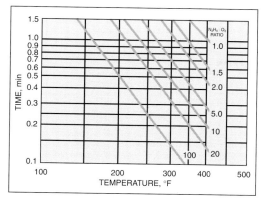

Figure 11-13. Time/temperature relationship for 90% oxygen removal by hydrazine at a pH of 9.5.

centration. The reaction is also dependent upon pH (the optimum pH range is 9.0–10.0).

In addition to its reaction with oxygen, hydrazine can also aid in the formation of magnetite and cuprous oxide (a more protective form of copper oxide), as shown in the following reactions:

$$N_2H_4 \; + \; 6Fe_2O_3 \; \rightarrow \; 4Fe_3O_4 \; + \; N_2 \; + \; 2H_2O$$
hydrazine   hematite   magnetite   nitrogen   water

and

$$N_2H_4 \; + \; 4CuO \; \rightarrow \; 2Cu_2O \; + \; N_2 \; + \; 2H_2O$$
hydrazine   cupric   cuprous   nitrogen   water
          oxide   oxide

Because hydrazine and organic scavengers add no solids to the steam, feedwater containing these materials is generally satisfactory for use as attemperating or desuperheating water.

The major limiting factors of hydrazine use are its slow reaction time (particularly at low temperatures), ammonia formation, effects on copper-bearing alloys, and handling problems.

Organic Oxygen Scavengers. Several organic compounds are used to remove dissolved oxygen from boiler feedwater and condensate. Among the most commonly used compounds are hydroquinone and ascorbate. These materials are less toxic than hydrazine and can be handled more safely. As with other oxygen scavengers, temperature, pH, initial dissolved oxygen concentration, catalytic effects, and scavenger concentration affect the rate of reaction with dissolved oxygen. When fed to the feedwater in excess of oxygen demand or when fed directly to the condensate, some organic oxygen scavengers carry forward to protect steam and condensate systems.

Hydroquinone is unique in its ability to react quickly with dissolved oxygen, even at ambient temperature. As a result of this property, in addition to its effectiveness in operating systems, hydroquinone is particularly effective for use in boiler storage and during system start-ups and shutdowns. It is also used widely in condensate systems.

Hydroquinone reacts with dissolved oxygen as shown in the following reactions:

$$C_6H_4(OH)_2 \; + \; \tfrac{1}{2}O_2 \; \rightarrow \; C_6H_4O_2 \; + \; H_2O$$
hydroquinone   oxygen   benzoquinone   water

Benzoquinone reacts further with oxygen to form polyquinones:

$$C_6H_4O_2 \; + \; O_2 \; \rightarrow \; \text{polyquinones}$$
benzoquinone   oxygen

These reactions are not reversible under the alkaline conditions found in boiler feedwater and condensate systems. In fact, further oxidation and thermal degradation (in higher-pressure systems) leads to the final product of carbon dioxide. Intermediate products are low molecular weight organic compounds, such as acetates.

Oxygen Level Monitoring. Oxygen monitoring provides the most effective means of controlling oxygen scavenger feed rates. Usually, a slight excess of scavenger is fed. Feedwater and boiler water residuals provide an indication of excess scavenger feed and verify chemical treatment feed rates. It is also necessary to test for iron and copper oxides in order to assess the effectiveness of the treatment program. Proper precautions must be taken in sampling for metal oxides to ensure representative samples.

Due to volatility and decomposition, measurement of boiler residuals is not a reliable means of control. The amount of chemical fed should be recorded and compared with oxygen levels in the feedwater to provide a check on the control of dissolved oxygen in the system. With sodium sulfite, a drop in the chemical residual in the boiler water or a need to increase chemical feed may indicate a problem. Measures must be taken to determine the cause so that the problem can be corrected.

Sulfite residual limits are a function of boiler operating pressure. For most low- and medium-pressure systems, sulfite residuals should be in excess of 20 ppm. Hydrazine control is usually based on a feedwater excess of 0.05–0.1 ppm. For different organic scavengers, residuals and tests vary.

MONITORING AND TESTING

Effective corrosion control monitoring is essential to ensure boiler reliability. A well planned monitoring program should include the following:

- proper sampling and monitoring at critical points in the system
- completely representative sampling
- use of correct test procedures
- checking of test results against established limits
- a plan of action to be carried out promptly when test results are not within established limits
- a contingency plan for major upset conditions
- a quality improvement system and assessment of results based on testing and inspections

Monitoring Techniques

Appropriate monitoring techniques vary with different systems. Testing should be performed at least once per shift. Testing frequency may have to be increased for some systems where control is difficult, or during periods of more variable operating conditions. All monitoring data, whether spot sampling or continuous, should be recorded.

Boiler feedwater hardness, iron, copper, oxygen, and pH should be measured. Both iron and copper, as well as oxygen, can be measured on a daily basis. It is recommended that, when possible, a continuous oxygen meter be installed in the feedwater system to detect oxygen intrusions. Iron and copper, in particular, should be measured with care due to possible problems of sample contamination.

If a continuous oxygen meter is not installed, periodic testing with spot sampling ampoules should be used to evaluate deaerator performance and potential for oxygen contamination from pump seal water and other sources.

For the boiler water, the following tests should be performed:

- phosphate (if used)
- P-alkalinity or pH
- sulfite (if used)
- conductivity

Sampling

It is critical to obtain representative samples in order to monitor conditions in the boiler feedwater system properly. Sample lines, continuously flowing at the proper velocity and volume, are required. Generally, a velocity of 5–6 ft/sec and a flow of 800–1000 mL/min are satisfactory. The use of long sample lines should be avoided. Iron and copper sampling should be approached with extreme care because of the difficulty of obtaining representative samples and properly interpreting results. Trends, rather than individual samples, should be used to assess results. Copper sampling requires special precautions, such as acidification of the stream. Composite sampling, rather than spot sampling, can also be a valuable tool to determine average concentrations in a system.

Oxygen sampling should be performed as close to the line as possible, because long residence time in sampling lines can allow the oxygen scavenger to further react and reduce oxygen readings. Also, if in-leakage occurs, falsely high data may be obtained. Sampling for oxygen should also be done at both the effluent of the deaerator and effluent of the boiler feedwater pump, to verify that oxygen ingress is not occurring.

Results and Action Required

All inspections of equipment should be thorough and well documented.

Conditions noted must be compared to data from previous inspections. Analytical results and procedures must be evaluated to ensure that quality standards are maintained and that steps are taken for continual improvement. Cause-and-effect diagrams (see Figure 11-14) can be used either to verify that all potential causes of problems are reviewed, or to troubleshoot a particular corrosion-related problem.

CORROSION PROTECTION DURING DOWNTIME AND STORAGE

Oxygen corrosion in boiler feedwater systems can occur during start-up and shutdown and while the boiler system is on standby or in storage, if proper procedures are not followed. Systems must be stored properly to prevent corrosion damage, which can occur in a matter of hours in the absence of proper lay-up procedures. Both the water/steam side and the fireside are subject to downtime corrosion and must be protected.

Off-line boiler corrosion is usually caused by oxygen in-leakage. Low pH causes further corrosion. Low pH can result when oxygen reacts with iron to form hydroferric acid. This corrosion product, an acidic form of iron, forms at water–air interfaces.

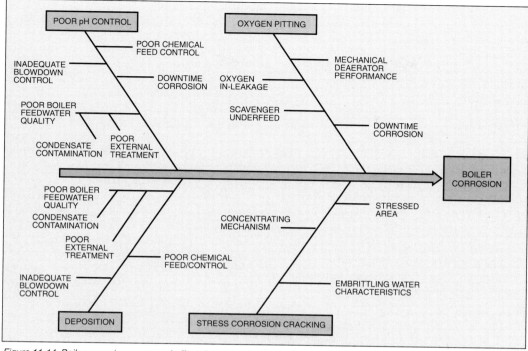

Figure 11-14. Boiler corrosion cause-and-effect diagram shows major types and causes of corrosion.

Corrosion also occurs in boiler feedwater and condensate systems. Corrosion products generated both in the preboiler section and the boiler may deposit on critical heat transfer surfaces of the boiler during operation and increase the potential for localized corrosion or overheating.

The degree and speed of surface corrosion depend on the condition of the metal. If a boiler contains a light surface coating of boiler sludge, surfaces are less likely to be attacked because they are not fully exposed to oxygen-laden water. Experience has indicated that with the improved cleanliness of internal boiler surfaces, more attention must be given to protection from oxygen attack during storage. Boilers that are idle even for short time periods (e.g., weekends) are susceptible to attack.

Boilers that use undeaerated water during start-up and during their removal from service can be severely damaged. The damage takes the form of oxygen pitting scattered at random over the metal surfaces. Damage due to these practices may not be noticed for many years after installation of the unit.

The choice of storage methods depends on the length of downtime expected and the boiler complexity. If the boiler is to be out of service for a month or more, dry storage may be preferable.

Wet storage is usually suitable for shorter downtime periods or if the unit may be required to go on-line quickly. Large boilers with complex circuits are difficult to dry, so they should be stored by one of the wet storage methods.

Dry Storage

For dry storage, the boiler is drained, cleaned, and dried completely. All horizontal and non-drainable boiler and superheater tubes must be blown dry with compressed gas. Particular care should be taken to purge water from long horizontal tubes, especially if they have bowed slightly.

Heat is applied to optimize drying. After drying, the unit is closed to minimize air circulation. Heaters should be installed as needed to maintain the temperature of all surfaces above the dew point.

Immediately after surfaces are dried, one of the three following desiccants is spread on watertight wood or corrosion-resistant trays:

■ *quicklime*—used at a rate of 6 lb/100 ft³ of boiler volume

■ *silica gel*—used at a rate of 17 lb/100 ft³ of boiler volume

■ *activated alumina*—used at a rate of 27 lb/100 ft³ of boiler volume

The trays are placed in each drum of a water tube boiler, or on the top flues of a fire-tube unit. All manholes, handholes, vents, and connections are blanked and tightly closed. The boiler should be opened every month for inspection of the desiccant. If necessary, the desiccant should be renewed.

Wet Storage

For wet storage, the unit is inspected, cleaned if necessary, and filled to the normal water level with deaerated feedwater.

Sodium sulfite, hydrazine, hydroquinone, or another scavenger is added to control dissolved oxygen, according to the following requirements:

- *Sodium sulfite*. 3 lb of sodium sulfite and 3 lb of caustic soda should be added per 1000 gal of water contained in the boiler (minimum 400 ppm P-alkalinity as $CaCO_3$ and 200 ppm sulfite as SO_3).

- *Hydrazine*. 5 lb of a 35% solution of hydrazine and 0.1 lb of ammonia or 2–3 lb of a 40% solution of neutralizing amine can be added per 1000 gal (minimum 200 ppm hydrazine and 10.0 pH). Due to the handling problems of hydrazine, organic oxygen scavengers are normally recommended.

- *Hydroquinone*. Hydroquinone-based materials are added to achieve approximately 200 ppm as hydroquinone in previously passivated on-line systems. In new systems, or those considered to have a poorly formed magnetite film, the minimum feed rate is 400 ppm as hydroquinone. pH should be maintained at 10.0.

No matter which treatment is used, pH or alkalinity adjustment to minimum levels is required.

After chemical addition, with vents open, heat is applied to boil the water for approximately 1 hr. The boiler must be checked for proper concentration of chemicals, and adjustments made as soon as possible.

If the boiler is equipped with a nondrainable superheater, the superheater is filled with high-quality condensate or demineralized water and treated with a volatile oxygen scavenger and pH control agent. The normal method of filling nondrainable superheaters is by back-filling and discharging into the boiler. After the superheater is filled, the boiler should be filled completely with deaerated feedwater. Morpholine, cyclohexylamine, or similar amines are used to maintain the proper pH.

If the superheater is drainable or if the boiler does not have a superheater, the boiler is allowed to cool slightly after firing. Then, before a vacuum is created, the unit is filled completely with deaerated feedwater.

A surge tank (such as a 55-gal drum) containing a solution of treatment chemicals or a nitrogen tank at 5 psig pressure is connected to the steam drum vent to compensate for volumetric changes due to temperature variations.

The drain between the nonreturn valve and main steam stop valve is left open wide. All other drains and vents are closed tightly.

The boiler water should be tested weekly with treatment added as necessary to maintain treatment levels. When chemicals are added, they should be mixed by one of the following methods:

- circulate the boiler water with an external pump
- reduce the water level to the normal operating level and steam the boiler for a short time

If the steaming method is used, the boiler should subsequently be filled completely, in keeping with the above recommendations.

Although no other treatment is required, standard levels of the chemical treatment used when the boiler is operating can be present.

Boilers can be protected with nitrogen or another inert gas. A slightly positive nitrogen (or other inert gas) pressure should be maintained after the boiler has been filled to the operating level with deaerated feedwater.

Storage of Feedwater Heaters and Deaerators

The tube side of a feedwater heater is treated in the same way the boiler is treated during storage. The shell side can be steam blanketed or flooded with treated condensate.

All steel systems can use the same chemical concentrations recommended for wet storage. Copper alloy systems can be treated with half the amount of oxygen scavenger, with pH controlled to 9.5.

Deaerators are usually steam or nitrogen blanketed; however, they can be flooded with a lay-up solution as recommended for wet lay-up of boilers. If the wet method is used, the deaerator should be pressurized with 5 psig of nitrogen to prevent oxygen ingress.

Cascading Blowdown

For effective yet simple boiler storage, clean, warm, continuous blowdown can be distributed into a convenient bottom connection on an idle boiler. Excess water is allowed to overflow to an appropriate disposal site through open vents. This method decreases the potential for oxygen ingress and ensures that properly treated water enters the boiler. This method should not be used for boilers equipped with nondrainable superheaters.

Cold Weather Storage

In cold weather, precautions must be taken to prevent freezing. Auxiliary heat, light firing of the boiler, cascade lay-up, or dry storage may be employed to prevent freezing problems. Sometimes, a 50/50 water and ethylene glycol mixture is used for freeze protection. However, this method requires that the boiler be drained, flushed, and filled with fresh feedwater prior to start-up.

Disposal of Lay-up Solutions

The disposal of lay-up chemicals must be in compliance with applicable federal, state, and local regulations.

Fireside Storage

When boilers are removed from the line for extended periods of time, fireside areas must also be protected against corrosion.

Fireside deposits, particularly in the convection, economizer, and air heater sections, are hygroscopic in nature. When metal surface temperatures drop below the dew point, condensation occurs, and if acidic hygroscopic deposits are present, corrosion can result.

The fireside areas (particularly the convection, economizer, and air heater sections) should be cleaned prior to storage.

High-pressure alkaline water is an effective means of cleaning the fireside areas. Before alkaline water is used for this purpose, a rinse should be made with fresh water of neutral pH to prevent the formation of hydroxide gels in the deposits (these deposits can be very difficult to remove).

Following chemical cleaning with a water solution, the fireside should be dried by warm air or a small fire. If the boiler is to be completely closed up, silica gel or lime can be used to absorb any water of condensation. As an alternative, metal surfaces can be sprayed or wiped with a light oil.

If the fireside is to be left open, the metal surfaces must be maintained above the dew point by circulation of warm air.

BOILER DEPOSITS: OCCURRENCE AND CONTROL

Deposition is a major problem in the operation of steam generating equipment. The accumulation of material on boiler surfaces can cause over-heating and/or corrosion. Both of these conditions frequently result in unscheduled downtime.

Boiler feedwater pretreatment systems have advanced to such an extent that it is now possible to provide boilers with ultrapure water. However, this degree of purification requires the use of elaborate pretreatment systems. The capital expenditures for such pretreatment equipment trains can be considerable and are often not justified when balanced against the capability of internal treatment.

The need to provide boilers with high-quality feedwater is a natural result of the advances made in boiler performance. The ratio of heating surface to evaporation has decreased. Consequently, heat transfer rates through radiant water wall tubes have increased—occasionally in excess of 200,000 Btu/ft^2/hr. The tolerance for deposition is very low in these systems.

The quality of feedwater required is dependent on boiler operating pressure, design, heat transfer rates, and steam use. Most boiler systems have sodium zeolite softened or demineralized make-up water. Feedwater hardness usually ranges from 0.01 to 2.0 ppm, but even water of this purity does not provide deposit-free operation. Therefore, good internal boiler water treatment programs are necessary.

DEPOSITS

Common feedwater contaminants that can form boiler deposits include calcium, magnesium, iron, copper, aluminum, silica, and (to a lesser extent) silt and oil. Most deposits can be classified as one of two types (Figure 12-1):

- scale that crystallized directly onto tube surfaces

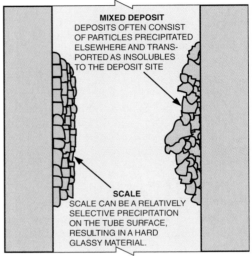

Figure 12-1. Boiler tube deposits can be crystalline or amorphous.

- sludge deposits that precipitated elsewhere and were transported to the metal surface by the flowing water

Scale is formed by salts that have limited solubility but are not totally insoluble in boiler water. These salts reach the deposit site in a soluble form and precipitate when concentrated by evaporation. The precipitates formed usually have a fairly homogeneous composition and crystal structure.

High heat transfer rates cause high evaporation rates, which concentrate the remaining water in the area of evaporation. A number of different scale-forming compounds can precipitate from the concentrated water. The nature of the scale formed depends on the chemical composition of the concentrated water. Normal deposit constituents are calcium, magnesium, silica, aluminum, iron, and (in some cases) sodium.

The exact combinations in which they exist vary from boiler to boiler, and from location to location within a boiler (Table 12-1). Scale may form as calcium silicate in one boiler and as sodium iron silicate in another.

Compared to some other precipitation reactions, such as the formation of calcium phosphate, the crystallization of scale is a slow process. As a result, the crystals formed are well defined, and a hard, dense, and highly insulating material is formed on the tube metal. Some forms of scale are so tenacious that they resist any type of removal—mechanical or chemical.

Sludge is the accumulation of solids that precipitate in the bulk boiler water or enter the boiler as suspended solids. Sludge deposits can be hard, dense, and tenacious. When exposed to high heat levels (e.g., when a boiler is drained hot), sludge deposits are often baked in place. Sludge deposits hardened in this way can be as troublesome as scale.

Once deposition starts, particles present in the circulating water can become bound to the deposit. Intraparticle binding does not need to occur between every particle in a deposit mass. Some nonbound particles can be captured in a network of bound particles.

Binding is often a function of surface charge and loss of water of hydration. Iron oxide, which

Table 12-1. Crystalline scale constituents identified by X-ray diffraction.

Name	Formula
Acmite	$Na_2O \cdot Fe_2O_3 \cdot 4SiO_2$
Analcite	$Na_2O \cdot Al_2O_3 \cdot 4SiO_2 \cdot 2H_2O$
Anhydrite	$CaSO_4$
Aragonite	$CaCO_3$
Brucite	$Mg(OH)_2$
Calcite	$CaCO_3$
Cancrinite	$4Na_2O \cdot CaO \cdot 4Al_2O_3 \cdot 2CO_2 \cdot 9SiO_2 \cdot 3H_2O$
Hematite	Fe_2O_3
Hydroxyapatite	$Ca_{10}(OH)_2(PO_4)_6$
Magnetite	Fe_3O_4
Noselite	$4Na_2O \cdot 3Al_2O_3 \cdot 6SiO_2 \cdot SO_4$
Pectolite	$Na_2O \cdot 4CaO \cdot 6SiO_2 \cdot H_2O$
Quartz	SiO_2
Serpentine	$3MgO \cdot 2SiO_2 \cdot 2H_2O$
Thenardite	Na_2SO_4
Wallastonite	$CaSiO_3$
Xonotlite	$5CaO \cdot 5SiO_2 \cdot H_2O$

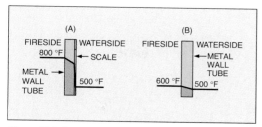

Figure 12-2. Deposition reduces heat transfer from boiler tube to boiler water, increasing the tube metal temperature. Tube metal overheating and failure can occur.

exists in many hydrated and oxide forms, is particularly prone to bonding. Some silicates will do the same, and many oil contaminants are notorious deposit binders, due to polymerization and degradation reactions.

In addition to causing material damage by insulating the heat transfer path from the boiler flame to the water (Figure 12-2), deposits restrict boiler water circulation. They roughen the tube surface and increase the drag coefficient in the boiler circuit. Reduced circulation in a generating tube contributes to accelerated deposition, overheating, and premature steam–water separation.

BOILER CIRCULATION

Figures 12-3 and 12-4 illustrate the process of boiler circulation. The left legs of the U-tubes represent downcomers and are filled with relatively cool water. The right legs represent generating tubes and are heated. The heat generates steam bubbles, and convection currents create circulation. As more heat is applied, more steam is generated and the circulation rate increases.

If deposits form (Figure 12-4), the roughened surface and partially restricted opening resist flow, reducing circulation. At a constant heat input the same amount of steam is generated, so the steam–water ratio in the generating tube is increased. The water in the tube becomes more concentrated, increasing the potential for deposition of boiler water salts.

In extreme cases, deposition becomes heavy enough to reduce circulation to a point at which premature steam–water separation occurs. When this happens in a furnace tube, failure due to overheating is rapid. When deposits are light they may not cause tube failures, but they reduce any safety margin in the boiler design.

Up to the point of premature steam–water separation, the circulation rate of a boiler is increased with increased heat input. Often, as

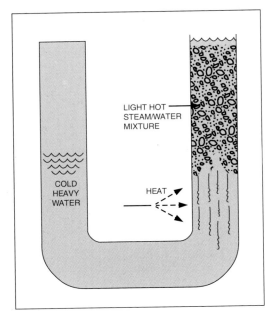

Figure 12-3. U-tube illustrates water circulation and steam generation in a clean circuit.

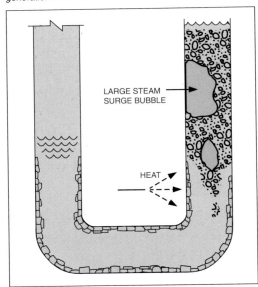

Figure 12-4. U-tube illustrates water circulation and steam generation with deposits.

illustrated in Figure 12-5, the inflection point (A) is above the nominal boiler rating. When the circuit is dirty, the inflection point of the circulation-to-heat input curve moves to the left, and the overall water circulation is reduced. This is represented by the lower broken line.

Circulation and deposition are closely related. The deposition of particles is a function of water sweep as well as surface charge (Figure 12-6). If the surface charge on a particle is relatively neutral in its tendency to cause the particle either to adhere to the tube wall or to remain suspended, an adequate water sweep will keep it off the tube. If the circulation through a circuit is not adequate to provide sufficient water sweep, the neutral particle may adhere to the tube. In cases of extremely low circulation, total evaporation can occur and normally soluble sodium salts deposit.

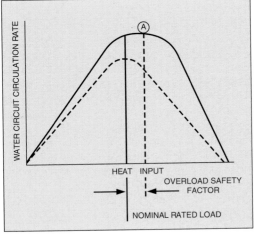

Figure 12-5. Circulation as a function of heat input in a boiler circuit.

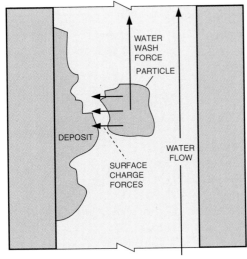

Figure 12-6. Opposing forces act on water-carried particles. Surface charges may attract particles to a deposit. Water flow "sweeps" the particle along.

CHEMICAL TREATMENT

Sodium carbonate treatment was the original method of controlling calcium sulfate scale. Today's methods are based on the use of phosphates and chelants. The former is a precipitating program, the latter a solubilizing program.

Carbonate Control

Before the acceptance of phosphate treatment in the 1930's, calcium sulfate scaling was a major boiler problem. Sodium carbonate treatment was used to precipitate calcium as calcium carbonate to prevent the formation of calcium sulfate. The driving force for the formation of calcium carbonate was the maintenance of a high concentration of carbonate ion in the boiler water. Even where this was accomplished, major scaling by calcium carbonate was common. As boiler pressures and heat transfer rates slowly rose, the calcium carbonate scale became unacceptable, as it led to tube overheating and failure.

Phosphate Control

Calcium phosphate is virtually insoluble in boiler water. Even small levels of phosphate can be maintained to ensure the precipitation of calcium phosphate in the bulk boiler water—away from heating surfaces. Therefore, the introduction of phosphate treatment eliminated the problem of calcium carbonate scale. When calcium phosphate is formed in boiler water of sufficient alkalinity (pH 11.0–12.0), a particle with a relatively nonadherent surface charge is produced. This does not prevent the development of deposit accumulations over time, but the deposits can be controlled reasonably well by blowdown.

In a phosphate precipitation treatment program, the magnesium portion of the hardness contamination is precipitated preferentially as magnesium silicate. If silica is not present, the magnesium will precipitate as magnesium hydroxide. If insufficient boiler water alkalinity is being maintained, magnesium can combine with phosphate. Magnesium phosphate has a surface charge that can cause it to adhere to tube surfaces and then collect other solids. For this reason, alkalinity is an important part of a phosphate precipitation program.

The magnesium silicate formed in a precipitating program is not particularly adherent. However, it contributes to deposit buildup on a par with other contaminants. Analyses of typical boiler deposits show that magnesium silicate is present in roughly the same ratio to calcium phosphate as magnesium is to calcium in boiler feedwater.

Phosphate/Polymer Control

Phosphate treatment results are improved by organic supplements. Naturally occurring organics such as lignins, tannins, and starches were the first supplements used. The organics were added to promote the formation of a fluid sludge that would settle in the mud drum. Bottom blowdown from the mud drum removed the sludge.

There have been many advances in organic treatments (Figure 12-7). Synthetic polymers are now used widely, and the emphasis is on dispersion of particles rather than fluid sludge formation. Although this mechanism is quite complex, polymers alter the surface area and the surface charge to mass ratio of typical boiler solids. With proper polymer selection and application, the surface charge on the particle can be favorably altered (Figure 12-8).

Many synthetic polymers are used in phosphate precipitation programs. Most are effective in dispersing magnesium silicate and magnesium

Figure 12-7. Experimental boilers are used to evaluate chemical treatment programs under rigorous conditions.

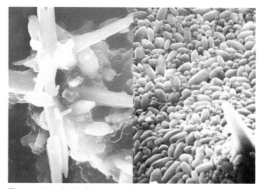

Figure 12-8. (Left) Scanning electron photomicrograph (4000X magnification) of calcium phosphate–magnesium silicate crystals formed in boiler water not treated with dispersant. (Right) With a sulfonated polymer, crystal growth is controlled.

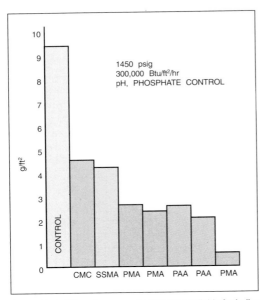

Figure 12-9. Although many polymers are available for boiler water treatment application, performance levels vary.

hydroxide as well as calcium phosphate. The polymers are usually low in molecular weight and have numerous active sites. Some polymers are used specifically for hardness salts or for iron; some are effective for a broad spectrum of ions. Figure 12-9 shows the relative performance of different polymers used for boiler water treatment.

Chelant Control

Chelants are the prime additives in a solubilizing boiler water treatment program. Chelants have the ability to complex many cations (hardness and heavy metals under boiler water conditions).

They accomplish this by locking metals into a soluble organic ring structure. The chelated cations do not deposit in the boiler. When applied with a dispersant, chelants produce clean waterside surfaces.

Suppliers and users of chelants have learned a great deal about their successful application since their introduction as a boiler feedwater treatment method in the early 1960's. Chelants were heralded as "miracle treatment" additives. However, as with any material, the greatest challenge was to understand the proper application.

Chelants are weak organic acids that are injected into boiler feedwater in the neutralized sodium salt form. The water hydrolyzes the chelant, producing an organic anion. The degree of hydrolysis is a function of pH; full hydrolysis requires a relatively high pH.

The anionic chelant has reactive sites that attract coordination sites on cations (hardness and heavy metal contaminants). Coordination sites are areas on the ion that are receptive to chemical bonding. For example, iron has six coordination sites, as does EDTA (ethylenediaminetetraacetic acid). Iron ions entering the boiler (e.g., as contamination from the condensate system) combine with EDTA. All coordination sites on the iron ion are used by the EDTA, and a stable metal chelate is formed (Figure 12-10).

NTA (nitrilotriacetic acid), another chelant applied to boiler feedwater, has four coordination sites and does not form as stable a complex as EDTA. With NTA, the unused coordination sites on the cation are susceptible to reactions with competing anions.

Table 12-2. Phosphate/polymer performance can be maintained at high heat transfer rates through selection of the appropriate polymer.

Treatment Type	Boiler Treatment Concentration (ppm)	Heat Transfer Rate (Btu/ft²/hr)	Operating Pressure (psig)	% Scale Reduction
Synthetic polymer A	10	185,000	300	44
Synthetic polymer B	10	185,000	300	93
Synthetic polymer C	10	185,000	300	94
Synthetic polymer B	5	185,000	300	56
Synthetic polymer C	5	185,000	300	94
Synthetic polymer B	10	185,000	900	64
Synthetic polymer C	10	185,000	900	92
Synthetic polymer B	10	300,000	900	44
Synthetic polymer C	10	300,000	900	86
Synthetic polymer B	10	300,000	1200	30
Synthetic polymer C	10	240,000	1200	90
Synthetic polymer C	10	300,000	1200	83

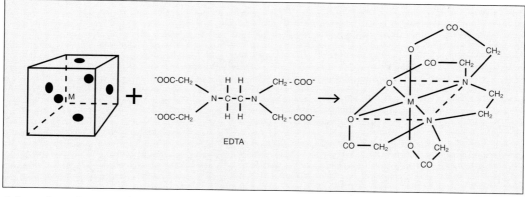

Cation	Competing Anion	Chelant	Remarks	Cation	Competing Anion	Chelant	Remarks
Ca^{2+}	CO_3^{2-}	EDTA/NTA	no problem	Fe^{3+}	$OH^- > 5$ ppm	NTA	deposition
Ca^{2+}	PO_4^{3-}	NTA	deposition	Mg^{2+}	OH^-	NTA/EDTA	no problem
Ca^{2+}	$PO_4^{3-} > 10$ ppm	EDTA	some deposition	Mg^{2+}	$SiO_2 > 40$ ppm	NTA	deposition
Fe^{2+}	OH^-	EDTA	no problem	Mg^{2+}	$SiO_2 > 300$ ppm	EDTA	some deposition
Fe^{2+}	$OH^- > 5$ ppm	NTA	deposition	Al^{3+}	$OH^- < 30$	EDTA/NTA	Al-chelant complex
Fe^{3+}	$OH^- > 5$ ppm	EDTA	some deposition	Al^{3+}	$OH^- > 30$	EDTA/NTA	no deposition[a]

[a]Aluminate is soluble at high pH levels

Figure 12-10. Most metals (represented at left) have six reactive coordination sites. EDTA can effectively tie into each coordination site and produce a stable complex.

Chelants combine with cations that form deposits, such as calcium, magnesium, iron, and copper. The metal chelate formed is water-soluble. When the chelate is stable, precipitation does not occur. Although there are many substances having chelating properties, EDTA and NTA are, to date, the most suitable chelants for boiler feedwater treatment.

The logarithm of the equilibrium constant for the chelant–metal ion reaction, frequently called the Stability Constant (K_s), can be used to assess the chemical stability of the complex formed. For the calcium–EDTA reaction:

$$K_s = \log \frac{(Ca–EDTA)^{2-}}{(Ca)^{2+}(EDTA)^{4-}} = 10.59$$

Table 12-3 lists stability constants for EDTA and NTA with common feedwater contaminants.

Table 12-3. Stability constants provide a measure of chemical stability of the chelant–metal ion complexes.

Metal Ion	EDTA	NTA
Ca^{2+}	10.59	6.41
Mg^{2+}	8.69	5.41
Fe^{2+}	14.33	8.82
Fe^{3+}	25.1	15.9

The effectiveness of a chelant program is limited by the concentration of the competing anions. With the exception of phosphate, the competing anion limitations on EDTA chelation are not usually severe. Alkalinity and silica, in addition to phosphate, are restricting considerations in the use of NTA.

Chelant/Polymer Control

Iron oxide is of particular concern in today's boiler water treatment programs. Deposition from low (less than 1.0 ppm) hardness boiler feedwater is eliminated with chelant programs and can be reduced by up to 95% by a good polymer/phosphate treatment program. Iron oxide is an increasingly significant contributor to boiler deposits because of the virtual elimination of hardness deposits in many systems and because the high heat transfer rates of many boilers encourage iron deposition.

Chelants with high stability values, such as EDTA, can complex iron deposits. However, this ability is limited by competition with hydrate ions. Experience has shown that relying on EDTA or other chelants alone is not the most satisfactory method for iron control.

At normal chelant feed rates, limited chelation of incoming particulate iron occurs. This is usually enough to solubilize some condensate iron contamination. The chelation of magnetite (the oxide formed under boiler conditions—a mix of Fe_2O_3 and FeO) is possible because the chelant combines with the ferrous (FeO) portion of the magnetite.

Overfeed (high levels) of chelant can remove large quantities of iron oxide. However, this is undesirable because high excess chelant cannot distinguish between the iron oxide that forms the protective magnetite coating and iron oxide that forms deposits.

A chelant/polymer combination is an effective approach to controlling iron oxide. Adequate chelant is fed to complex hardness and soluble iron, with a slight excess to solubilize iron contamination. Polymers are then added to condition and disperse any remaining iron oxide contamination (Figure 12-11).

A chelant/polymer program can produce clean waterside surfaces, contributing to much more reliable boiler operation (Figure 12-12). Out-of-service boiler cleaning schedules can be extended and, in some cases, eliminated. This depends on operational control and feedwater quality.

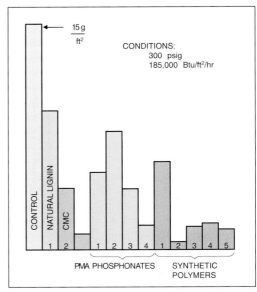

Figure 12-11. Chelant/polymer can provide a high degree of iron deposit protection, provided that the proper polymer is used. Even the members of the same family of polymers, such as polymethacrylate (PMA), can vary greatly in performance.

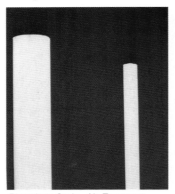
Phosphate Cycle—No Treatment

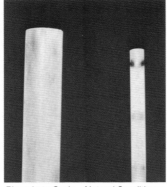
Phosphate Cycle—Natural Conditioner

Phosphate Cycle—Lignin Conditioner

Phosphate Cycle—Polymeric Dispersants

Phosphate Cycle—Chelant and Polymer Dispersant Blend

Chelation Cycle—Chelant and Polymer Dispersant Blend

Figure 12-12. Chelant/polymer provides the most deposit-free mode of internal treatment control. Test conditions: 600 psig; 60,000 (large probe) +180,000 (smaller probe) Btu/ft²/hr feedwater, makeup constant.

Chelants with high complexing stabilities are "forgiving" treatments—they can remove deposits that form when feedwater quality or treatment control periodically deviates from standard.

Boilers with moderate deposition in the forms of calcium carbonate and calcium phosphate can be cleaned effectively through an in-service chelant cleanup program. In-service chelant cleanup programs should be controlled and not attempted on a heavily deposited boiler or applied at too fast a pace. Chelants can cause large accumulations of deposit to slough off in a short period of time. These accumulations can plug headers or redeposit in critical circulation areas, such as furnace wall tubes.

In a chelant cleanup program, sufficient chelant is added to solubilize incoming feedwater hardness and iron. This is followed by a recommended excess chelant feed. Regular inspections (usually every 90 days) are highly recommended so that the progress of the treatment may be monitored.

The polymer level in the boiler should also be increased above the normal concentration. This confines particles to the bulk water as much as possible until they settle in the mud drum. An increased number of mud drum "blows" should be performed to remove the particles from the boiler.

In-service chelant cleanup programs are not advisable when deposit analyses reveal that major constituents are composed of silicates, iron oxide, or any scale that appears to be hard, tightly bound, or lacking in porosity. Because such scales are not successfully removed in most instances, an in-service chelant cleanup cannot be justified in these situations.

Phosphate/Chelant/Polymer Combinations

Combinations of polymer, phosphate, and chelant are commonly used to produce results comparable to chelant/polymer treatment in low- to medium-pressure boilers. Boiler cleanliness is improved over phosphate treatment, and the presence of phosphate provides an easy means of testing to confirm the presence of treatment in the boiler water.

Polymer-Only Treatment

Polymer-only treatment programs are also used with a degree of success. In this treatment, the polymer is usually used as a weak chelant to complex the feedwater hardness. These treatments are most successful when feedwater hardness is consistently very low.

High-Pressure Boiler Water Treatment

High-pressure boilers usually have areas of high heat flux and feedwater, composed of demineralized makeup water and a high percentage of condensate returns. Because of these conditions, high-pressure boilers are prone to caustic attack. Low-pressure boilers that use demineralized water and condensate as feedwater are also susceptible to caustic attack.

There are several means by which boiler water can become highly concentrated. One of the most common is iron oxide deposition on radiant wall tubes. Iron oxide deposits are often quite porous and act as miniature boilers. Water is drawn into the iron oxide deposit. Heat applied to the deposit from the tube wall generates steam, which passes out through the deposit. More water enters the deposit, taking the place of the steam. This cycle is repeated and the water beneath the deposit is concentrated to extremely high levels. It is possible to have 100,000 ppm of caustic beneath the deposit while the bulk water contains only about 5–10 ppm of caustic (Figure 12-13).

Steam generating units supplied with demineralized or evaporated makeup water or pure condensate may be protected from caustic corrosion by a treatment known by the general term "coordinated phosphate/pH control." Phosphate is a pH buffer in this program and limits the localized concentration of caustic. A detailed discussion of this treatment is included in Chapter 11.

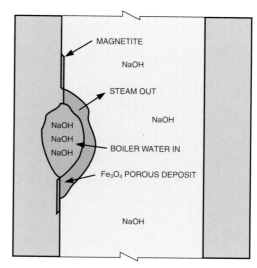

Figure 12-13. *Porous deposits provide conditions that promote high concentrations of boiler water solids, such as sodium hydroxide (NaOH).*

If deposits are minimized, the areas where caustic can be concentrated is reduced. To minimize the iron deposition in high-pressure (1000–1750 psig) boilers, specific polymers have been designed to disperse the iron and keep it in the bulk water.

As with phosphate precipitation and chelant control programs, the use of these polymers with coordinated phosphate/pH treatment improves deposit control. Figure 12-14 illustrates the effectiveness of dispersants in controlling iron oxide deposition. Test conditions were 1500 psig (590 °F), 240,000 Btu/ft²/hr heat flux, and coordinated phosphate/pH program water chemistry. A comparison of the untreated heat transfer surface (shown at left) with the polymer dispersant treated conditions (shown at right) provides a graphic illustration of the value of dispersants in preventing steam generator deposition. The ability to reduce iron oxide accumulations is an important requirement in the treatment of boiler systems operating at high pressures and with high-purity feedwater.

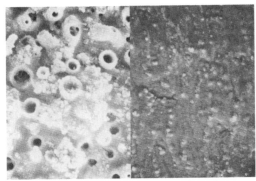

Figure 12-14. Experimental boiler heat transfer surfaces (800X magnifications) exposed to feedwater iron contamination. Heavy iron oxide deposition occurred (left) when no polymer was used. A virtually clean surface was achieved with an iron-specific polymer program (right).

Supercritical boilers use all-volatile treatments, generally consisting of ammonia and hydrazine. Because of the extreme potential for deposit formation and steam contamination, no solids can be tolerated in supercritical once-through boiler water, including treatment solids.

BOILER BLOWDOWN CONTROL

Boiler blowdown is the removal of water from a boiler. Its purpose is to control boiler water parameters within prescribed limits to minimize scale, corrosion, carryover, and other specific problems. Blowdown is also used to remove suspended solids present in the system. These solids are caused by feedwater contamination, by internal chemical treatment precipitates, or by exceeding the solubility limits of otherwise soluble salts.

In effect, some of the boiler water is removed (blown down) and replaced with feedwater. The percentage of boiler blowdown is as follows:

$$\frac{\text{quantity blowdown water}}{\text{quantity feedwater}} \times 100 = \% \text{ blowdown}$$

The blowdown can range from less than 1% when an extremely high-quality feedwater is available to greater than 20% in a critical system with poor-quality feedwater. In plants with sodium zeolite softened makeup water, the percentage is commonly determined by means of a chloride test. In higher-pressure boilers, a soluble, inert material may be added to the boiler water as a tracer to determine the percentage of blowdown. The formula for calculating blowdown percentage using chloride and its derivation are shown in Table 13-1.

LIMITING FACTORS AFFECTING BLOWDOWN

The primary purpose of blowdown is to maintain the solids content of boiler water within certain limits. This may be required for specific reasons, such as contamination of the boiler water. In this case, a high blowdown rate is required to eliminate the contaminants as rapidly as possible.

The blowdown rate required for a particular boiler depends on the boiler design, the operating conditions, and the feedwater contaminant levels.

Table 13-1. Algebraic proof of blowdown formula.

Let

x	=	quantity feedwater
y	=	quantity blowdown water
a	=	chloride concentration in feedwater
b	=	chloride concentration in boiler water
k	=	percent blowdown

By definition of percent blowdown

$$k = \frac{100y}{x}$$

Because the total chlorides entering the boiler must equal total chlorides leaving boiler,

$$xa = yb$$

Multiplying both sides by $\dfrac{100}{xb}$ gives:

$$\frac{100a}{b} = \frac{100y}{x}$$

Because by definition $\dfrac{100y}{x} = k$, then $k = \dfrac{100a}{b}$ or

$$\frac{\text{Cl}^- \text{ in feedwater}}{\text{Cl}^- \text{ in boiler water}} \times 100 = \% \text{ blowdown}$$

In many systems, the blowdown rate is determined according to total dissolved solids. In other systems, alkalinity, silica, or suspended solids levels determine the required blowdown rate.

For many years, boiler blowdown rates were established to limit boiler water contaminants to levels set by the American Boiler Manufacturers' Association (ABMA) in its Standard Guarantee of Steam Purity. These standards were used even though they were of a general nature and not applicable to each individual case. Today, the ASME "Consensus on Operating Practices for the Control of Feedwater and Boiler Water Quality in Modern Industrial Boilers," shown in Table 13-2, is frequently used for establishing blowdown rates.

Table 13-2. Suggested water quality limits.[a]

	Drum Operating Pressure,[b] MPa (psig)							
	0–2.07 (0–300)	2.08–3.10 (301–450)	3.11–4.14 (451–600)	4.15–5.17 (601–750)	5.18–6.21 (751–900)	6.22–6.89 (901–1000)	6.90–10.34 (1001–1500)	10.35–13.79 (1501–2000)
FEEDWATER[h]								
Dissolved oxygen (mg/L O_2) measured before oxygen scavenger addition[i]	<0.040	<0.040	<0.007	<0.007	<0.007	<0.007	<0.007	<0.007
Total iron (mg/L Fe)	≤0.100	≤0.050	≤0.030	≤0.025	≤0.020	≤0.020	≤0.010	≤0.010
Total copper (mg/L Cu)	≤0.050	≤0.025	≤0.020	≤0.020	≤0.015	≤0.015	≤0.010	≤0.010
Total hardness (mg/L $CaCO_3$)	≤0.300	≤0.300	≤0.200	≤0.200	≤0.100	≤0.050	—not detectable—	
pH range at 25 °C	7.5–10.0	7.5–10.0	7.5–10.0	7.5–10.0	7.5–10.0	8.5–9.5	9.0–9.6	9.0–9.6
Chemicals for preboiler system protection			use only volatile alkaline materials					
Nonvolatile TOC (mg/L C)[g]	<1	<1	<0.5	<0.5	<0.5	—as low as possible, <0.2—		
Oily matter (mg/L)	<1	<1	<0.5	<0.5	<0.5	—as low as possible, <0.2—		
BOILER WATER								
Silica (mg/L SiO_2)	≤150	≤90	≤40	≤30	≤20	≤8	≤2	<1
Total alkalinity (mg/L $CaCO_3$)	<350[d]	<300[d]	<250[d]	<200[d]	<150[d]	<100[d]	—not specified[e]—	
Free hydroxide alkalinity (mg/L $CaCO_3$)[c]			—not specified—				—not detectable[e]—	
Specific conductance (μS/cm) (μmho/cm) at 25 °C without neutralization	<3500[f]	<3000[f]	<2500[f]	<2000[f]	<1500[f]	<1000[f]	≤150	≤100

[a] Source: ASME Research Committee on Steam and Water in Thermal Power Systems. Boiler type: industrial water tube, high duty, primary fuel fired, drum type; makeup water percentage: up to 100% of feedwater; conditions: includes superheater, turbine drives, or process restriction on steam purity; saturated steam purity target.

[b] With local heat fluxes >473.2 kW/m² (>150,000 Btu/hr/ft²), use values for the next higher pressure range.

[c] Minimum level of OH⁻ alkalinity in boilers below 6.21 MPa (900 psig) must be specified individually with regard to silica solubility and other components of internal treatment.

[d] Maximum total alkalinity consistent with acceptable steam purity. If necessary, override conductance as blowdown control parameter. If makeup is demineralized water at 4.14 MPa (600 psig) to 6.89 MPa (1000 psig), boiler water alkalinity and conductance should be that in table for 6.90–10.34 MPa (1001–1500 psig) range.

[e] Refers to free sodium or potassium hydroxide alkalinity. Some small variable amount of total alkalinity will be present and measurable with the assumed congruent or coordinated phosphate-pH control or volatile treatment employed at these high pressure ranges.

[f] Maximum values often not achievable without exceeding suggested maximum total alkalinity values, especially in boilers below 6.21 MPa (900 psig) with >20% makeup of water whose total alkalinity is >20% of TDS naturally or after pretreatment by lime–soda or sodium cycle ion exchange softening. Actual permissible conductance values to achieve any desired steam purity must be established for each case by careful steam purity measurement. Relationship between conductance and steam purity is affected by too many variables to allow its reduction to a simple list of tabulated values.

[g] Nonvolatile TOC is that organic carbon not intentionally added as part of the water treatment regime.

[h] Boilers below 6.21 MPa (900 psig) with large furnaces, large steam release space, and internal chelant, polymer, and/or antifoam treatment can sometimes tolerate higher levels of feedwater impurities than those in the table and still achieve adequate deposition control and steam purity. Removal of these impurities by external pretreatment is always a more positive solution. Alternatives must be evaluated as to practicality and economics in each individual case.

[i] Values in table assume existence of a deaerator.

[j] No values given because steam purity achievable depends on many variables, including boiler water total alkalinity and specific conductance as well as design of boiler, steam drum internals, and operating conditions (see footnote f). Because boilers in this category require a relatively high degree of steam purity, other operating parameters must be set as low as necessary to achieve this high purity for protection of the superheaters and turbines and/or to prevent process contamination.

This consensus applies to deposition control as well as steam quality. Good engineering judgment must be used in all cases. Because each specific boiler system is different, control limits may be different as well. There are many mechanical factors that can affect the blowdown control limits, including boiler design, rating, water level, load characteristics, and type of fuel.

In some cases, the blowdown control limits for a particular system may be determined by operating experience, equipment inspections, or steam purity testing rather than ASME or ABMA water quality criteria. In certain cases, it is possible to exceed standard total solids (or conductivity), silica, or alkalinity limits. Antifoam agents have been applied successfully to allow higher-than-normal solids limits, as shown in Figure 13-1. Chelating and effective dispersant programs also may allow certain water criteria to be exceeded.

The maximum levels possible for each specific system can be determined only from experience. The effect of water characteristics on steam quality can be verified with steam purity testing. However, the effects on internal conditions must be determined from the results observed during the turnaround for the specific unit.

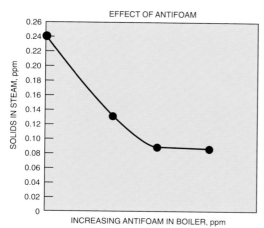

Figure 13-1. Effect of antifoam concentration on steam purity.

Certain boilers may require lower than normal blowdown levels due to unusual boiler design or operating criteria or an exceptionally pure feedwater requirement. In some plants, boiler blowdown limits are lower than necessary due to a conservative operating philosophy.

MANUAL BLOWDOWN

Intermittent manual blowdown is designed to remove suspended solids, including any sludge formed in the boiler water. The manual blowdown take-off is usually located in the bottom of the lowest boiler drum, where any sludge formed would tend to settle.

Properly controlled intermittent manual blowdown removes suspended solids, allowing satisfactory boiler operation. Most industrial boiler systems contain both a manual intermittent blowdown and a continuous blowdown system. In practice, the manual blowdown valves are opened periodically in accordance with an operating schedule. To optimize suspended solids removal and operating economy, frequent short blows are preferred to infrequent lengthy blows. Very little sludge is formed in systems using boiler feedwater of exceptionally high quality. The manual blowdown can be less frequent in these systems than in those using feedwater that is contaminated with hardness or iron. The water treatment consultant can recommend an appropriate manual blowdown schedule.

Blowdown valves on the water wall headers of a boiler should be operated in strict accordance with the manufacturer's recommendations. Usually,

due to possible circulation problems, water wall headers are not blown down while the unit is steaming. Blowdown normally takes place when the unit is taken out of service or banked. The water level should be watched closely during periods of manual blowdown.

CONTINUOUS BLOWDOWN

Continuous blowdown, as the term implies, is the continuous removal of water from the boiler. It offers many advantages not provided by the use of bottom blowdown alone. For instance, water may be removed from the location of the highest dissolved solids in the boiler water. As a result, proper boiler water quality can be maintained at all times. Also, a maximum of dissolved solids may be removed with minimal loss of water and heat from the boiler.

Another major benefit of continuous blowdown is the recovery of a large amount of its heat content through the use of blowdown flash tanks and heat exchangers. Control valve settings must be adjusted regularly to increase or decrease the blowdown according to control test results and to maintain close control of boiler water concentrations at all times.

When continuous blowdown is used, manual blowdown is usually limited to approximately one short blow per shift to remove suspended solids which may have settled out near the manual blowdown connection.

ENERGY CONSERVATION

Several factors can contribute to reduced energy consumption on the water side of steam generation equipment.

Scale Reduction

Heat transfer is inhibited by scale formation on internal surfaces. Scale reduction through proper pretreatment and internal chemical treatment results in cleaner internal surfaces for more efficient heat transfer and resultant energy savings.

Boiler Water Blowdown Reduction

A reduction in boiler water blowdown can result in significant fuel and water savings.

In some installations, boiler water solids are lower than the maximum level permissible. Through improved control methods, including automatic boiler blowdown equipment, boiler water blowdown can be reduced to maintain the solids close to but not above the maximum level permissible.

The rate of blowdown required depends on feedwater characteristics, load on the boiler, and mechanical limitations. Variations in these factors will change the amount of blowdown required, causing a need for frequent adjustments to the manually operated continuous blowdown system. Even frequent manual adjustment may be inadequate to meet the changes in operating conditions. Table 13-3 illustrates the savings possible with automatic boiler blowdown control.

Blowdown rate is often the most poorly controlled variable of an internal treatment program. Conductivity limits for manually controlled boiler blowdown are usually quite wide, with the lower limits below 70% of the maximum safe value. This is often necessary with manual control because a narrow range cannot be maintained safely.

In plants with sodium zeolite softened makeup water, automatic control systems can maintain boiler water conductivity within 5% of the setpoint. Plant operating records have verified that

with manual adjustment, continuous blowdown is within this 5% range no more than 20% of the time. In general, the average plant saves approximately 20% of boiler blowdown when changing from manually adjusted continuous blowdown to automatically controlled continuous blowdown. This reduction is gained without risk of scale or carryover due to high boiler water solids.

In some instances, an increase in feedwater quality permits a significant reduction in the blowdown rate at the existing maximum allowable solids level. This can be accomplished through reuse of additional condensate as feedwater, or through improvement of external treatment methods for higher makeup water quality.

Any reduction in blowdown contributes to water and fuel savings, as illustrated in Table 13-4. When uniform concentrations are maintained at or near maximum permissible levels in the boiler water, savings result in several areas, including makeup water demand, cost of processing water, cost of blowdown water waste treatment, fuel

Table 13-3: Example of savings with installation of automatic blowdown equipment (basis: one day).

Evaporation:	2,400,000 lb/day
Boiler pressure:	600 psig
Manual blowdown:	183,423 lb/day (7.1%)
Automatic blowdown:	145,069 lb/day (5.7%)
Blowdown reduction:	38,354 lb/day
Feedwater temperature:	240 °F
Makeup water temperature:	60 °F
Heat of liquid at 600 psig:	475 Btu/lb
Heat of liquid at 60 °F:	− 28 Btu/lb
Heat required:	447 Btu/lb
(blowdown reduction)	× 38,354 lb/day
Heat reduction	17,144,238 Btu/day
Fuel (gas):	1,040 Btu/ft³
(@ 80% boiler efficiency)	× .80
Available fuel heat:	832 Btu/ft³
(heat reduction)	17,144,238 Btu/day
	÷ 832 Btu/ft³
Fuel reduction:	20,606 ft³/day
Fuel savings @ $4.00/1,000 ft³:	$82.42
Labor reduction:	.5 hr
Daily Labor savings @ $30.00/hr:	$15.00
Water reduction:	4,598 gal/day
Daily water savings	
@ $.80/1,000 gal:	$3.68
Total daily savings:	$101.10
	× 365 days/yr
Total annual savings:	$36,902

Table 13-4. Example of fuel savings possible through reduction in blowdown (basis: one day).

Evaporation (steam):	2,000,000 lb
Present blowdown:	128,000 lb/day (6%)
Reduced blowdown:	− 41,000 lb/day (2%)
Reduction in blowdown:	87,000 lb/day
Feedwater (steam plus blowdown):	2,041,000 lb
Boiler Pressure:	200 psig
Feedwater Temperature:	215 °F
Makeup Water Temperature:	60 °F
Fuel (oil):	145,000 Btu/gal
@ 80% boiler efficiency	× .80
Available fuel heat:	116,00 Btu/gal
Heat of Liquid at Boiler Pressure:	362 Btu/lb
Heat of Liquid at 60°F:	− 28 Btu/lb
Heat Required:	334 Btu/lb
(reduction in blowdown)	87,000 lb/day
	× 334 Btu/lb
Total Heat Saving:	29,058,000 Btu/day
	÷ 116,000 Btu/gal
Fuel Savings:	250 gal
(@ $0.80/gal)	× .80
Daily savings:	$200
	× 365 days/yr
Annual savings:	$72,000

consumption, and chemical treatment require-ments. These savings are noticeably greater where makeup water quality is poor, where heat recovery equipment is nonexistent or inefficient, and where operating conditions are frequently changed.

Heat Recovery

Heat recovery is used frequently to reduce energy losses that result from boiler water blowdown. Figure 13-2 illustrates a typical boiler blowdown heat recovery system using a flash tank and heat exchanger.

Installation of heat recovery equipment is valuable only when energy from the flash tank or the blowdown water can be recovered and utilized. When an excess supply of exhaust or low-pressure steam is already available, there is little justification for installing heat recovery equipment.

If economically justified, boiler water blowdown can be used to heat process streams. In most cases, boiler water blowdown heat recovery systems use flash steam from the flash tank for deaeration. The blowdown from the flash tank is passed through an exchanger and used to preheat boiler makeup water. With the use of an efficient heat exchange unit, the only heat loss is the terminal temperature difference between the incoming makeup water and the blowdown water to the sewer. This difference usually amounts to 10–20 °F (5–10 °C).

Table 13-5 provides a typical calculation for determining the fuel savings achieved in a heat recovery system using a low-pressure flash tank

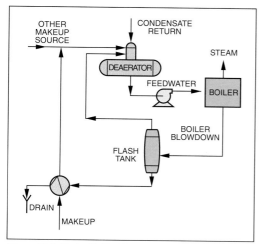

Figure 13-2. Typical boiler blowdown heat recovery system using a flash tank and heat exchanger.

Table 13-5. Example of fuel savings possible through use of heat recovery on continuous blowdown (basis: one day).

Evaporation (steam):	5,000,000 lb
Blowdown:	+ 263,000 lb/day (5.0%)
Feedwater (steam + blowdown):	5,263,000 lb
Boiler Pressure:	600 psig
Feedwater Temperature (live steam used):	240 °F
Makeup Water Temperature:	60 °F
Fuel (oil) volume:	145,000 Btu/gal
(@ 75% boiler efficiency)	× 0.75
Available fuel heat:	108,750 Btu/gal

Employing a flash tank at 5 psig, the quantity of steam available may be calculated from the formula:

$$\% \text{ flashed steam} = \frac{H_b - H_f}{V_f} \times 100$$

where

H_b: heat of liquid at boiler pressure	475 Btu/lb
H_f: heat of liquid at flash pressure	− 196 Btu/lb
	279 Btu/lb
V_f: latent heat of vaporization at flash press	÷ 960 Btu/lb
	× 100
% flashed steam:	29.1
(blowdown)	263,000 lb
(@29.1% of flashed steam)	× .291
Flashed steam available at 5 psig:	76,500 lb
Total heat of flashed steam at 5 psig:	1,156 Btu/lb
(Heat of makeup water at 60 °F)	− 28 Btu/lb
Heat available in flashed steam	1,128 Btu/lb
(flashed steam available)	× 76,500 lb
Heat savings in flashed steam	86,292,000 Btu
Heat of liquid at 5 psig:	196 Btu/lb
Heat of liquid at 80°F[a]:	− 48 Btu/lb
Heat recovery:	148 Btu/lb
(blowdown)	263,000 lb
(blowdown not flashed)	× 0.709
(heat recovery)	× 148 Btu/lb
Heat savings from heat exchanger:	27,597,000 Btu
(heat savings in flashed steam)	+ 86,292,000 Btu
Total heat savings:	113,889,000 Btu
(available fuel heat)	÷ 108,750 Btu/gal
Fuel savings:	1,047 gal
(@ $.80/gal)	× .80
Daily savings:	$837.60
	× 365 days/yr
Annual Savings	$305,724

[a] The drain water from the flash tank is passed through the heat exchanger and then to the sewer. It is assumed that the temperature of the water leaving the exchanger is 20 °F above the incoming makeup water, or 80 °F.

This chart is used to calculate the percent of boiler water discharged by a continuous blowdown system that can be flashed into steam at a reduced pressure and is recoverable as low pressure steam for heating or process.

Example: A boiler operates at a pressure of 450 psig. Continuous blowdown amounts to 10,000 lb/hr. What percentage of blowdown water can be recovered as flashed steam at 10 psig?

Solution: Locate 450 psig on left axis. Follow horizontally toward the right to the intersection with 10 psig "flash" curve (point A). Drop vertically downward to the bottom axis and read 24.5%. (24.5% of 10,000 lb/hr blowdown = 2450 lb/hr of flash steam at 10 psig pressure.)

These curves have been prepared from the formula:

$$\% \text{ flashed steam} = \frac{H_b - H_f}{V_f} \times 100$$

where

H_b = heat of liquid at boiler pressure, Btu/lb
H_f = heat of liquid at flash pressure, Btu/lb
V_f = latent heat of vaporization at flash pressure, Btu/lb

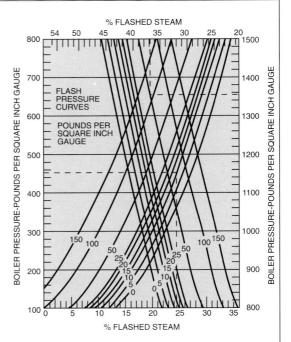

NOTE: For boiler pressures between 100 and 800 psig, use "flash" pressure curves slanting from lower left to upper right corner and the bottom axis. For boiler pressures above 800 psig, use "flash" pressure curves slanting from lower right to upper left corner and top axis.

Figure 13-3. Flash steam recoverable from continuous blowdown systems.

and heat exchanger. Figure 13-3 can be used to determine the quantity of flash steam recoverable from a flash tank.

EQUIPMENT EMPLOYED

Manual Blowdown

Equipment for manual blowdown, considered a part of the boiler and installed with the unit, usually consists of a take-off line, a quick-opening valve, and a shut-off valve. The take-off line is always located in the lowest part of the lowest boiler drum, where the greatest concentration of suspended solids should form.

Several types of water-tube boilers have more than one blowdown connection. They permit blowdown from both ends of the mud drum. Blowdown connections are installed on headers for draining and for removal of suspended solids which may accumulate and restrict circulation. The boiler manufacturer usually specifies certain restrictions on the blowdown of water wall headers. These restrictions should be followed closely.

Continuous Blowdown

Usually, continuous blowdown equipment is installed by the boiler manufacturer. The exact location of the continuous blowdown take-off line depends primarily on the water circulation pattern. Its position must ensure the removal of the most concentrated water. The line must also be located so that boiler feedwater or chemical feed solution does not flow directly into it. The size of the lines and control valves depends on the quantity of blowdown required.

Figure 13-4 illustrates a typical location in a steam drum for a continuous blowdown connection. In most units, the take-off line is several inches below the low water level. In other designs, the take-off is close to the bottom of the steam drum.

Automatic Blowdown

An automatic blowdown control system continuously monitors the boiler water, adjusts the rate of blowdown, and maintains the specific conductance of the boiler water at the desired level. The basic components of an automatic blowdown control system include a measurement

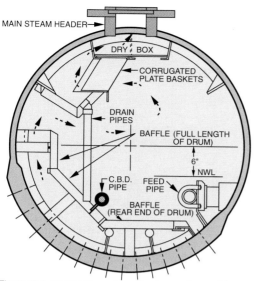

Figure 13-4. Typical steam drum showing continuous blow-down location. (Courtesy of Combustion Engineering, Inc.)

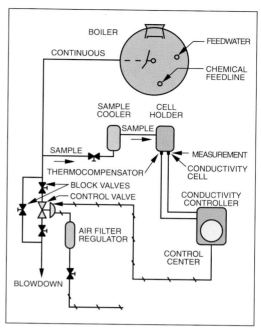

Figure 13-5. Modulating automatic boiler blowdown equipment.

assembly, a control center, and a modulating blowdown control valve. A typical modulating automatic boiler blowdown control system is shown in Figure 13-5.

BLOWDOWN CONTROL

If an economical blowdown rate is to be maintained, suitable boiler water tests must be run frequently to check concentrations in the boiler water. When sodium zeolite softened makeup is used, the need for boiler blowdown is usually determined by measurement of the boiler water conductivity, which provides an indirect measure of the boiler water dissolved solids.

Other boiler water constituents such as chlorides, sodium, and silica are also used as a means of controlling blowdown. The alkalinity test has been used as a supplementary blowdown control for systems in which boiler water alkalinity can be particularly high.

Total Solids

From a technical standpoint, gravimetric measurement provides a satisfactory way to determine boiler water total solids; however, this method is rarely used because the analysis is time-consuming and is too difficult for routine control. Also, a comparison of the total solids content of boiler water with the total solids content of feedwater does not necessarily provide an accurate measure of the feedwater concentration within the boiler, because of the following:

■ the boiler water samples may not show a representative suspended solids content due

to settling or deposit formation

■ internal treatment can add various solids to the boiler water

■ breakdown of bicarbonates and carbonates can liberate carbon dioxide gas and lower the total solids in the boiler water

Dissolved Solids

The specific conductance of boiler water provides an indirect measure of dissolved solids and can usually be used for blowdown control. However, establishing the rate of blowdown on the basis of the relative specific conductance of feedwater and boiler water does not give a direct measure of the feedwater concentrations within the boiler. Specific conductance is affected by the loss of carbon dioxide with steam and by the introduction of solids as internal chemical treatment. Moreover, the specific conductance of feedwater (a dilute solution) and boiler water (a concentrated solution) cannot be compared directly.

The specific conductance of a sample is caused by ionization of the various salts present. In dilute solutions, the dissolved salts are almost completely ionized, so the specific conductance increases proportionally to the dissolved salt concentration. In concentrated solutions, ionization is repressed and the ratio of specific conductance to dissolved salts decreases. The relationship between specific conductance and dissolved solids is determined most accurately through measurement of both parameters and the establishment of a correlation factor for each system. However, the factor may be estimated. The solids content of very dilute solutions such as condensate may be calculated with a factor of 0.5–0.6 ppm of dissolved solids per microsiemens (micromho) of specific conductance. For a more concentrated solution such as boiler water, the factor can vary between 0.55 and 0.90 ppm of dissolved solids per microsiemens of specific conductance. The hydroxide ion present in many boiler waters is highly conductive compared to the other ions. Therefore, it is common practice to neutralize the caustic with an organic acid prior to measuring conductivity. Although gallic acid is conventionally used to neutralize the phenolphthalein alkalinity in samples with high specific conductance, boric acid may be used in samples of low and high specific conductance with minimal impact on the correlation factor between dissolved solids and specific conductance.

Silica, Alkalinity, Sodium, Lithium, and Molybdate

Under certain circumstances, measurement of the silica and alkalinity content of boiler water can be used to control blowdown. Sodium, lithium, and molybdate have been used for accurate calculation of blowdown rates in high-pressure units where demineralized water is used as feedwater.

Chloride

If the chloride concentration in the feedwater is high enough to measure accurately, it can be used to control blowdown and to calculate the rate of blowdown. Because chlorides do not precipitate in the boiler water, the relative chloride concentrations in feedwater and boiler water provide an accurate basis for calculating the rate of blowdown.

The chloride test is unsuitable for this calculation when the feedwater chloride is too low for accurate determination. A slight analytical error in determining feedwater chloride content will cause an appreciable error in calculating blowdown rate.

Specific Gravity

The specific gravity of a boiler water is proportional to the dissolved solids. However, determination of dissolved solids by hydrometer measurement of specific gravity is so inaccurate that it cannot be recommended for proper blowdown control.

BOILER SYSTEM FAILURES

Successful, reliable operation of steam generation equipment requires the application of the best available methods to prevent scale and corrosion. When equipment failures do occur, it is important that the cause of the problem be correctly identified so that proper corrective steps can be taken to prevent a recurrence. An incorrect diagnosis of a failure can lead to improper corrective measures; thus, problems continue.

There are times when the reasons for failures are obscure. In these instances, considerable investigation may be required to uncover the causes. However, in most cases the problem area displays certain specific, telltale signs. When these characteristics are properly interpreted, the cause of a problem and the remedy become quite evident.

DEAERATOR CRACKING

In numerous deaerators, cracks have developed at welds and heat-affected zones near the welds. The cracking most commonly occurs at the head-to-shell weld below the water level in the storage compartment. However, it may also occur above the water level and at longitudinal welds. Because cracks can develop to the point of equipment failure, they represent a potential safety hazard requiring periodic equipment inspection and, when warranted, repair or replacement. Wet fluorescent magnetic particle testing is recommended for. identification of cracks.

The mechanism of most deaerator cracking has been identified as environmentally assisted fatigue cracking. Although the exact causes are not known, steps can be taken to minimize the potential for cracking (e.g., stress-relieving of welds and minimization of thermal and mechanical stress during operation). In addition, water chemistry should be designed to minimize corrosion.

FEEDWATER LINE EROSION

High-velocity water and especially water/steam mixtures cause erosion in feedwater systems. The most commonly encountered erosion problems occur at the hairpin bends in steaming economizers. Here, the mixture of steam and water thins the elbows, leaving a characteristic reverse horseshoe imprint.

Similar problems can be encountered in feedwater lines where high velocities create the familiar thinning pattern. These problems can occur even at moderate average flow velocities when a sequence of bends causes a significant increase in local velocity.

In order to mitigate erosion problems, it is helpful to maintain water chemistry conditions that form the most tenacious oxide layer. However, the problems cannot be completely resolved without design or operational changes.

ECONOMIZER TUBES

Water tube economizers are often subject to the serious damage of oxygen pitting (see Figure 14-1). The most severe damage occurs at the economizer inlet and, when present, at the tube

Figure 14-1. Economizer tube severely damaged by oxygen.

weld seams. Where economizers are installed, effective deaerating heater operation is absolutely essential. The application of a fast-acting oxygen scavenger, such as catalyzed sodium sulfite, also helps protect this vital part of the boiler.

While oxygen pitting is the most common form of waterside corrosion that causes economizer tube failures, caustic soda has occasionally accumulated under deposits and caused caustic gouging. Usually, this type of attack develops in an area of an economizer where steam generation is taking place beneath a deposit and free caustic soda is present in the feedwater. The best solution to this problem is improved treatment that will eliminate the deposition.

Other common causes of economizer failure include fatigue cracking at the rolled tube ends and fireside corrosion caused by the condensation of acid from the boiler flue gas.

FAILURES DUE TO OVERHEATING

When tube failures occur due to overheating and plastic flow (conditions commonly associated with deposits), the cause is usually identified by the deposits which remain, as shown in Figure 14-2. An accurate analysis of the deposits indicates the source of the problem and the steps needed for correction. Metallographic analyses are useful, at times, in confirming whether a short- or long-term exposure to overheating conditions existed prior to failure. Such analyses are helpful also when metal quality or manufacturing defects are suspected, although these factors are significant only in isolated instances.

When tube failures occur due to overheating, a careful examination of the failed tube section reveals whether the failure is due to rapid escalation

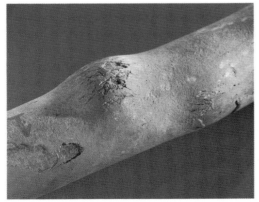

Figure 14-2. Deposit accumulation restricted heat transfer, leading to long-term overheating.

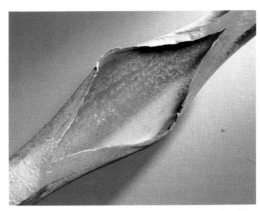

Figure 14-3. Thin-lipped burst caused by rapid overheating.

in tube wall temperature or a long-term, gradual buildup of deposit. When conditions cause a rapid elevation in metal temperature to 1600 °F or above, plastic flow conditions are reached and a violent rupture occurs. Ruptures characterized by thin, sharp edges are identified as "thin-lipped" bursts (see Figure 14-3).

Violent bursts of the thin-lipped variety occur when water circulation in the tube is interrupted by blockage or by circulation failure caused by low water levels. In some steam drum designs, water level is extremely critical because the baffling may isolate a generating section of the boiler when the steam drum water level falls below a certain point.

Thin-lipped bursts also occur in superheater tubes when steam flow is insufficient, when deposits restrict flow, or when tubes are blocked by water due to a rapid firing rate during boiler start-up.

Interruptions in flow do not always result in rapid failure. Depending on the metal temperature reached, the tube can be damaged by corrosive or thinning mechanisms over a long period of time before bulges or blisters or outright failures develop. In such instances, a metallurgical examination in addition to an examination of the contributing mechanical factors can be helpful in identifying the source of the problem.

A long-term scaling condition which will lead to a tube leak is usually indicated by a wrinkled, bulged external surface and a final thick-lipped fissure or opening. This appearance is indicative of long-term creep failure created by repetitive scale formation, causing overheating and swelling of the tube surface in the form of a bulge or blister. The scale, in such instances, tends to crack off; water contacts the metal and cools it until further scaling repeats the process. The iron oxide coating on the external surface cracks during the process, giving rise to the characteristic longitudinal creep cracks.

FAILURES DUE TO CORROSION

Stress Corrosion Cracking

Various corrosion mechanisms contribute to boiler tube failure. Stress corrosion may result in either intercrystalline or transgranular cracking of carbon steel. It is caused by a combination of metal stress and the presence of a corrosive. A metallurgical examination of the failed area is required to confirm the specific type of cracking. Once this is determined, proper corrective action can be taken.

Caustic Embrittlement

Caustic embrittlement, a specific form of stress corrosion, results in the intercrystalline cracking of steel. Intercrystalline cracking results only when all of the following are present: specific conditions of stress, a mechanism for concentration such as leakage, and free NaOH in the boiler water. Therefore, boiler tubes usually fail from caustic embrittlement at points where tubes are rolled into sheets, drums, or headers.

The possibility of embrittlement may not be ignored even when the boiler is of an all-welded design. Cracked welds or tube-end leakage can provide the mechanism by which drum metal may be adversely affected. When free caustic is present, embrittlement is possible.

An embrittlement detector may be used to determine whether or not a boiler water has embrittling tendencies. The device, illustrated in Figure 14-4, was developed by the United States Bureau of Mines. If boiler water possesses embrittling characteristics, steps must be taken to protect the boiler from embrittlement-related failure.

Sodium nitrate is the standard treatment for inhibiting embrittlement in boilers operating at low pressures. The ratios of sodium nitrate to sodium hydroxide in the boiler water recommended by the Bureau of Mines depend on the boiler operating pressure. These ratios are as follows:

Pressure, psi	NaNO$_3$/NaOH Ratio
Up to 250	0.20
Up to 400	0.25
Up to 1000	0.40–0.50

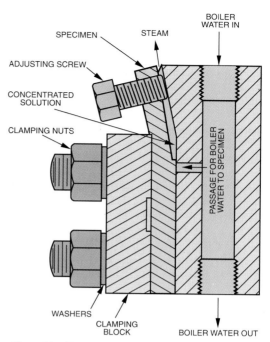

Figure 14-4. Embrittlement detector.

The formula for calculating the sodium nitrate/ sodium hydroxide ratio in the boiler water is:

$$\frac{NaNO_3}{NaOH} = \frac{ppm\ nitrate\ (as\ NO_3^-)}{ppm\ M\ alkalinity\ -\ ppm\ phosphate \atop (as\ CaCO_3) \qquad (as\ PO_4^{3-})}$$

At pressures above 900 psig, coordinated phosphate/pH control is the usual internal treatment. When properly administered, this treatment method precludes the development of high concentrations of caustic, eliminating the potential for caustic embrittlement.

Fatigue and Corrosion Fatigue

Transgranular cracking primarily due to cyclic stress is the most common form of cracking encountered in industrial boilers. In order to determine the cause of a transgranular failure, it is necessary to study both the design and the operating conditions of the boiler. Straight tube, shell-and-tube waste heat boilers frequently develop tube and tube sheet failures due to the imposition of unequal stresses. A primary cause of this is the uneven distribution of hot gases across the face of the tube sheet. The tubes involved tend to come loose, creating leakage problems. Even when the tubes are securely welded, imposed stresses can cause transverse cracking of the tubes.

Any design feature that allows steam pockets to form within a unit can cause cyclic overheating and quenching. This can lead to transverse cracking of tubes and, occasionally, shells. Such cracking always appears in the area of greatest stress and results in cracks that are primarily transgranular.

Some intercrystalline cracking may develop in this type of failure whether or not free NaOH is present. However, the predominant type of cracking is across the grain structure of the metal. Because it is mechanically induced, the cracking occurs irrespective of boiler water chemical concentrations. The cracks are often accompanied by a number of pits adjacent to or in line with the cracking—another specific indicator of the mechanical stresses imposed. Any corrosives present contribute to the formation of the pits. The normal reaction between iron and water is sufficient to cause pitting at breaks in the thin oxide film formed on freshly exposed surfaces under stress.

Stress-Induced Corrosion

Certain portions of the boiler can be very susceptible to corrosion as a result of stress from mechanical forces applied during the manufacturing and fabrication processes. Damage is commonly visible in stressed components, such as rolled tube ends, threaded bolts, and cyclone separators. However, corrosion can also occur at weld attachments throughout the boiler (see Figure 14-5) and can remain undetected until failure occurs. Regular inspection for evidence of corrosion, particularly in the windbox area of Kraft recovery boilers, is

14-5. Stress at a weld attachment caused localized corrosion.

recommended because of the potential for an explosion caused by a tube leak.

The potential for stress-induced corrosion can be reduced if the following factors are minimized: stresses developed in the boiler components, the number of thermal cycles, and the number of boiler chemical cleanings. In addition, it is necessary to maintain proper water chemistry control during operation and to provide protection from corrosion during shutdowns.

Dissolved Oxygen

Dissolved oxygen corrosion is a constant threat to feedwater heater, economizer, and boiler tube integrity. As deposit control treatment methods have improved, the need for effective control of oxygen has become increasingly important.

The first serious emphasis on oxygen control began when phosphate-based treatments were introduced to replace the soda ash treatments common before that time. The dense, hard calcium carbonate scale which developed with the soda ash treatments protected tubes and drums from serious oxygen corrosion. With the application of phosphate treatment, the tube and drum surfaces were cleaner. Therefore, more of the surface area was exposed to corrosives in the water. This spurred the use of improved open feedwater heaters to remove most of the oxygen prior to the entrance of water into the boiler. Today, most plants are equipped with efficiently operated deaerating heaters. The use of oxygen scavengers, such as catalyzed sodium sulfite, hydrazine, and organic scavengers, is also standard practice.

The use of chelant treatments and demineralized water has improved the cleanliness of boiler heat transfer surfaces to such an extent that essentially bare-metal conditions are common. Only a thin, protective, magnetic oxide film remains in such instances. As a result, oxygen control has become even more essential today. The use of catalyzed sulfite, where applicable, is a standard recommendation in chelant applications.

The control of downtime corrosion has become increasingly important in recent years to prevent or inhibit pitting failures. Often, cold water that has not been deaerated is used for rapid cooling or start-up of a boiler. This is a risky operating practice, usually chosen for economical reasons. Severe pitting can occur in such instances, especially in boilers that have been maintained in a deposit-free condition. Therefore, it is usually more economical to maintain clean heat transfer surfaces and eliminate the use of cold water containing dissolved oxygen during cool-down and start-up periods. This practice can result in fuel savings and improved boiler reliability.

Chelant Corrosion

During the early years of chelant use, nearly all internal boiler corrosion problems were labeled "chelant corrosion." However, other corrosives such as oxygen, carbon dioxide, caustic, acid, copper plating, and water are still common causes of boiler corrosion. In addition, mechanical conditions leading to caustic embrittlement, film boiling, and steam blanketing are even more prevalent today than ever, as a result of increasing heat transfer rates and the more compact design of steam generators. Chelant corrosion, or chelant attack, has some specific characteristics, and develops only under certain conditions.

Chelant corrosion of boiler metal occurs only when excess concentration of the sodium salt is maintained over a period of time. The attack is of a dissolving or thinning type—not pitting—and is concentrated in areas of stress within the boiler. It causes thinning of rolled tube ends, threaded members, baffle edges, and similar parts of stressed, unrelieved areas. Normally, annealed tubes and drum surfaces are not attacked. When tube thinning occurs in a chelant-treated boiler, evidence of steam blanketing and/or film boiling is sometimes present. In such instances, failure occurs regardless of the type of internal treatment used.

Pitting is often thought to be a result of chelant attack. However, pitting of carbon steel boiler tubes is almost always due to the presence of uncontrolled oxygen or acid. Infrequently, copper plating (usually the result of an improper acid cleaning operation) may lead to pitting problems.

Caustic Attack

Caustic attack (or caustic corrosion), as differentiated from caustic embrittlement, is encountered in boilers with demineralized water and most often occurs in phosphate-treated boilers where tube deposits form, particularly at high heat input or poor circulation areas. Deposits of a porous nature allow boiler water to permeate the deposits, causing a continuous buildup of boiler water solids between the metal and the deposits.

Because caustic soda does not crystallize under such circumstances, caustic concentration in the trapped liquid can reach 10,000 ppm or more. Complex caustic–ferritic compounds are formed when the caustic dissolves the protective film of magnetic oxide. Water in contact with iron attempts to restore the protective film of magnetite (Fe_3O_4). As long as the high caustic concentrations remain, this destructive process causes a continuous loss of metal.

The thinning caused by caustic attack assumes irregular patterns and is often referred to as caustic gouging (see Figure 14-6). When deposits are removed from the tube surface during examination, the characteristic gouges are very evident, along with the white salts deposit which usually outlines the edges of the original deposition area. The whitish deposit is sodium carbonate, the residue of caustic soda reacting with carbon dioxide in the air.

Inspections of boilers with caustic attack often show excessive accumulations of magnetic oxide in low flow areas of drums and headers. This is caused by the flaking off, during operation, of deposits under which the complex caustic–ferritic material has formed. When contacted and diluted by boiler water, this unstable complex immediately reverts to free caustic and magnetic oxide. The suspended and released magnetic oxide moves to and accumulates in low flow or high heat flux areas of the boiler.

While caustic attack is sometimes referred to as caustic pitting, the attack physically appears as irregular gouging or thinning and should not be confused with the concentrated, localized pit penetration representative of oxygen or acid attack.

Figure 14-6. Typical gouging caused by caustic attack developed under an original adherent deposit. Note irregular depressions and white (Na_2CO_3) deposits remaining around edges of original deposit area.

Steam Blanketing

A number of conditions permit stratified flow of steam and water in a given tube, which usually occurs in a low heat input zone of the boiler. This problem is influenced by the angle of the affected tubes, along with the actual load maintained on the boiler. Stratification occurs when, for any reason, velocity is not sufficient to maintain turbulence or thorough mixing of water and steam during passage through the tubes. Stratification most commonly occurs in sloped tubes (Figure 14-7) located away from the radiant heat zone of the boiler, where heat input is low and positive circulation in the tubes may be lacking.

Examination of the affected tubes usually reveals a prominent water line with general thinning in the top area of the tube or crown. In rare instances, the bottom of the tube is thinned. When the boiler water contains caustic, high concentrations accumulate and lead to caustic corrosion and gouging under the deposits that accumulate at the water line.

In certain instances, stratification may occur together with input of heat to the top or crown of the tube. This creates a high degree of superheat in the steam blanket. Direct reaction of steam with the hot steel develops if the metal temperature reaches 750 °F or higher. Corrosion of the steel will proceed under such circumstances whether or not caustic is present. When there is doubt about the exact cause, a metallographic analysis will show if abnormal temperature excursions contributed to the problem. Deposits usually found under such circumstances are composed primarily of magnetic iron oxide (Fe_3O_4). Hydrogen is also formed as a result of the reaction and is released with the steam.

A somewhat unusual problem related to circulation and heat input problems has been encountered in roof tubes. These tubes are usually designed to pick up heat on the bottom side only. Problems generally develop when the tubes sag or break away from the roof, causing exposure of the entire surface of the tube to the hot gases. The overheating that usually develops, along with the internal pressure, causes a gradual enlargement of the tube, sometimes quite uniformly. Failure occurs when the expanded tube can no longer withstand the combined effects of the thermal stress and internal pressure.

Superheater tubes often show the same swelling or enlargement effect. In such instances, steam flow has been restricted for some reason, leading to overheating and eventually to failure.

Acidic Attack

Acid attack of boiler tubes and drums is usually in the form of general thinning of all surfaces. This results in a visually irregular surface appearance, as shown in Figure 14-8. Smooth surfaces appear at areas of flow where the attack has been intensified. In severe occurrences, other components, such as baffling, nuts and bolts, and other stressed areas, may be badly damaged or destroyed, leaving no doubt as to the source of the problem.

Severe instances of acid attack can usually be traced to either an unsatisfactory acid cleaning operation or process contamination. Some industrial plants encounter periodic returned condensate contamination, which eliminates boiler water

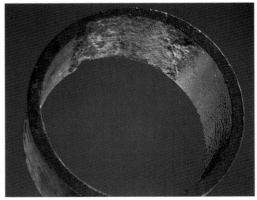

Figure 14-7. Steam blanketing caused metal wastage on top of sloped tube.

Figure 14-8. Boiler tube shows effect of acid attack.

alkalinity. Occasionally, regeneration acid from an ion exchange process is discharged accidentally into the boiler feedwater system. Cooling water contamination of condensate can depress boiler water pH and cause severe deposition and pitting in areas of high heat flux. Damage can be quite severe if immediate steps are not taken to neutralize the acid.

In the case of industrial process contamination, it is possible for organic contaminants to decompose under boiler temperature and pressure to form organic acids. Sugar is an excellent example of an organic which, when returned in a large quantity, can cause rapid loss of boiler water alkalinity and reduce pH of the boiler water to 4.3 and lower. Most sugar refining plants maintain standby pumping systems, to add caustic soda to neutralize these acids as quickly as possible.

Corrosion Due to Copper

Pitting of boiler drums and tube banks has been encountered due to metallic copper deposits, formed during acid cleaning procedures which do not completely compensate for the amount of copper oxides in the original deposits. Dissolved copper may be plated out on freshly cleaned steel surfaces, eventually establishing anodic corrosion areas and forming pits very similar in form and appearance to those caused by oxygen.

In such instances, metallic copper plating is quite evident. In most cases, it is localized in certain tube banks, giving rise to random pitting

Figure 14-9. Liquid metal embrittlement of boiler tube caused by copper deposits and high temperature (greater than 1600 °F).

in those particular areas. Whenever deposits are found containing large quantities of copper or its oxide, special precautions are required to prevent the plating out of copper during cleaning operations.

Copper deposits and temperatures over 1600 °F can cause liquid metal embrittlement. Weld repair of a tube containing copper deposits leads to the failure shown in Figure 14-9.

Hydrogen Attack or Embrittlement

Since around 1960, hydrogen attack, or embrittlement, has been encountered with increasing frequency in high-pressure, high-purity systems. It is not encountered in the average industrial plant because the problem usually occurs only in units operating at pressures of 1500 psig or higher.

In systems of this type, the alkalinity of the boiler water is maintained at values that are quite low compared to usual standards for lower-pressure operation. At the operating pressures indicated and water conditions imposed, either coordinated pH/phosphate or total volatile programs are used. Because the boiler water is relatively unbuffered, total volatile programs are more affected by contaminants that may lower boiler water alkalinity or pH.

When contaminants lower the boiler water pH sufficiently, the acid attack of the steel generates hydrogen. If this occurs under hard, adherent, nonporous tube deposits, the hydrogen pressure within the deposit can build up to the point at which the hydrogen penetrates the steel tubing.

When atomic hydrogen permeates the metal structure, it reacts with the carbon content to form methane. Because the methane molecule is too large to diffuse through the steel, excessive pressure develops within the metal structure, causing the metal to rupture along the crystalline boundaries where methane has formed. The cracking that develops is primarily intercrystalline or intergranular in nature, the metallic area affected becoming decarburized in the process. Failure occurs when the ruptured section can no longer withstand the internal pressure. Ruptures are violent and sudden, and can be disastrous (see Figure 14-10). Failed sections of tubing are cracked in the intergranular mode and decarburized, but usually retain the original dimensions or thickness of the tubing material.

Although there are many causes of low boiler water pH, it most frequently occurs when brackish

Figure 14-10. Violent rupture caused by hydrogen embrittlement.

water is used for condenser cooling. Small quantities of magnesium chloride, in particular, have caused extremely low pH excursions, requiring very close supervision and detection of very low levels of contamination in the condensate.

To summarize, hydrogen embrittlement occurs only when a hard, dense scale is present on the tube surfaces, permitting hydrogen to concentrate under the deposit and permeate the metal. Acidic contamination or low pH excursions commonly produce the conditions for generation of hydrogen. This type of attack may develop very quickly; therefore, constant surveillance of condensate purity is required.

As indicated, hydrogen embrittlement usually occurs in high-purity systems that operate at 1500 psig or higher. Although it is sometimes confused with intergranular creep cracking, this type of failure can be positively identified by the distinctive intergranular nature of the cracking and decarburized condition of the metal.

Surveys of units operating at these pressures and under these conditions have generally indicated that the application of a coordinated/pH phosphate control will lessen the possibility of hydrogen embrittlement. This is due primarily to the improved buffering of the boiler water with phosphate present.

Superheater Tubes

Superheater tube failures are caused by a number of conditions, both mechanical and chemical. In any instance of superheater tube failure, analysis

of the deposits found is an important factor in solving the problem. Magnetic oxide deposits at the point of failure are a direct indication of oxidation of the tube metal (see Figure 14-11). This oxidation occurs during overheating where metal temperatures exceed the design temperature and the steel enters into a direct reaction with the steam to form magnetic iron oxide with hydrogen release. When the deposits found in the area of failure are primarily iron oxide, it may be necessary to explore a number of operating conditions in order to determine the initial cause.

Oxidation may occur if the flow of steam through the tubes is restricted or if the heat input is excessive, permitting overheating. In the case of insufficient steam flow, the restriction may be due to conditions prevalent during the transition periods of boiler start-up or shutdown. This occurs if adequate precautions have not been taken to protect the superheater during the transition periods. At no time should gas temperatures exceed 900 °F in the area of the superheater until the boiler is up to operating pressure and all superheater tubes are clear of any water which may have accumulated during the downtime. Overheating conditions may develop during times of low-load operation when adequate distribution of saturated steam across the tube bank at the inlet header has not been achieved.

Soluble-salt deposits may form at a superheater tube inlet as a result of excessive entrainment of boiler water solids with the steam. This can result in restricted flow. However, overheating and direct oxidation failures may occur in areas distinctly removed from the blockage, such as the bottom loops or the hottest areas of the superheater tubes.

Figure 14-11. Oxygen intrusion during shutdowns caused this pitting of superheater.

In some cases, there is a very clear delineation between oxidation products in the hot area and soluble-salts deposits at the inlet. However, in most occurrences, a high percentage of sodium salt deposits is found in the hot areas along with the oxidation products. There is little doubt in such instances that boiler water carryover has contributed to the problem.

Periodic overheating of superheaters, caused by insufficient control of firebox temperatures during start-up and shutdown periods, usually results in thick-lipped fissures and blistering with all the evidence of creep failure. As in the case of water tubes, a superheater tube will fail rapidly (often violently) when flow is blocked for a short period of time and tube temperature escalates rapidly to plastic flow temperatures. Determination of whether a failure is due to a long- or short-term situation depends essentially on the same general characteristics that apply to water tube examination.

Oxygen pitting of superheater tubes, particularly in the pendant loop area, is rather common and occurs during downtime. It is caused by the exposure of water in these areas to oxygen in the air.

It is essential that the manufacturer's instructions be followed rigidly to prevent overheating problems during start-up or shutdown and to prevent oxygen corrosion during downtime.

When soluble-salts deposits are found in superheater tubes, steam purity is of paramount concern. It has been the experience of Betz Laboratories, after conducting thousands of steam purity studies over many years, that soluble-salts deposits in superheaters can be expected, with attendant problems, whenever steam solids exceed 300 ppb. Therefore, when soluble-salts deposits are found, a thorough investigation of steam purity (and reasons for poor purity) is necessary.

Boiler Design Problems

Certain basic design flaws can contribute to tube failures. Problems which occur as a result of a design flaw may be intensified by the boiler water chemistry. The boiler water often contains elements that become corrosive when concentrated far beyond normal values as a result of design problems.

Many industrial boilers, for example, are treated in such a manner that low concentrations of caustic soda are present in the boiler water. The caustic can become corrosive to steel when the

boiler water is allowed to concentrate to abnormally high values as a result of poor design. Even in the absence of caustic, conditions which permit stratification or steam blanketing and localized elevation of metal temperatures in excess of 750 °F allow direct oxidation or corrosion of the steel in contact with water or steam. This causes loss of metal and eventual rupture of the tube.

Roof tubes, nose arch tubes, and convection pass tubes with slopes of less than 30 degrees from the horizontal are more subject to deposition and stratification problems and tube failures than vertical tubes. Whenever chelant is present in boiler water, the sodium salts of ethylenediaminetetraacetic acid (EDTA), in particular, are destroyed at high temperature, leaving a residue of caustic soda. The caustic soda residue from the chelant is usually an insignificant additive to whatever caustic may be present normally.

A frequent contributor to waste heat boiler problems is the uneven distribution of gases across the inlet tubes at the hot end. This causes unequal stresses and distortion and leads to mechanical stress and fatigue problems.

The use of horizontal hairpin tube configurations with inadequate forced circulation of water through the tubes often permits stratification of steam and water. This often leads to steam blanketing or caustic corrosion problems.

Procedures for Boiler Tube Failure Analysis

At times, the cause of a failure cannot be readily determined, making it difficult to determine the appropriate corrective action. A detailed examination of the failure and associated operating data is usually helpful in identifying the mechanism of failure so that corrective action may be taken.

Proper investigative procedures are needed for accurate metallurgical analyses of boiler tubes. Depending on the specific case, macroscopic examination combined with chemical analysis and microscopic analysis of the metal may be needed to assess the primary failure mechanism(s). When a failed tube section is removed from a boiler, care must be taken to prevent contamination of deposits and damage to the failed zones. Also, the tube should be properly labeled with its location and orientation.

The first step in the lab investigation is a thorough visual examination. Both the fireside and the waterside surfaces should be inspected for failure or indications of imminent failure. Photographic documentation of the as-received condition of tubing can be used in the correlation and interpretation of data obtained during the investigation. Particular attention should be paid to color and texture of deposits, fracture surface location and morphology, and metal surface contour. A stereo microscope allows detailed examination under low-power magnification.

Dimensional analysis of a failed tube is important. Calipers and point micrometers are valuable tools that allow quantitative assessment of failure characteristics such as bulging, wall thinning at a rupture lip, and corrosion damage. The extent of ductile expansion and/or oxide formation can provide clues toward determining the primary failure mechanism. External wall thinning from fireside erosion or corrosion mechanisms can result in tube ruptures which often mimic the appearance of overheating damage. In those cases, dimensional analysis of adjacent areas can help to determine whether or not significant external wall thinning occurred prior to failure. A photograph of a tube cross section taken immediately adjacent to a failure site can assist in dimensional analysis and provide clear-cut documentation.

The extent, orientation, and frequency of tube surface cracking can be helpful in pinpointing a failure mechanism. While overheating damage typically causes longitudinal cracks, fatigue damage commonly results in cracks that run transverse to the tube axis. In particular, zones adjacent to welded supports should be examined closely for cracks. Nondestructive testing (e.g., magnetic particle or dye penetrant inspection) may be necessary to identify and assess the extent of cracking.

When proper water chemistry guidelines are maintained, the waterside surfaces of boiler tubes are coated with a thin protective layer of black magnetite. Excessive waterside deposition can lead to higher-than-design metal temperatures and eventual tube failure. Quantitative analysis of the internal tube surface commonly involves determination of the deposit-weight density (DWD) value and deposit thickness. Interpretation of these values can define the role of internal deposits in a failure mechanism. DWD values are also used to determine whether or not chemical cleaning of boiler tubing is required. In addition, the tube surface may be thoroughly cleaned

by means of glass bead blasting during DWD testing. This facilitates accurate assessment of waterside or fireside corrosion damage (e.g., pitting, gouging) that may be hidden by deposits.

The presence of unusual deposition patterns on a waterside surface can be an indication that nonoptimal circulation patterns exist in a boiler tube. For example, longitudinal tracking of deposits in a horizontal roof tube may indicate steam blanketing conditions. Steam blanketing, which results when conditions permit stratified flow of steam and water in a given tube, can lead to accelerated corrosion damage (e.g., wall thinning and/or gouging) and tube failure.

When excessive internal deposits are present in a tube, accurate chemical analyses can be used to determine the source of the problem and the steps necessary for correction. Whenever possible, it is advisable to collect a "bulk" composition, by scraping and crimping the tube and collecting a cross section of the deposit for chemical analysis. Typically, a loss-on-ignition (LOI) value is also determined for the waterside deposit. The LOI value, which represents the weight loss obtained after the deposit is heated in a furnace, can be used to diagnose contamination of the waterside deposit by organic material.

In many cases, chemical analysis of a deposit from a specific area is desired. Scanning electron microscope-energy dispersive spectroscopy (SEM-EDS) is a versatile technique that allows inorganic chemical analysis on a microscopic scale. SEM-EDS analyses are shown in Figures 14-12 and

Figure 14-12. Scanning electron microscope (SEM) reveals crystalline structure of highly reflective precipitated magnetite on boiler tube surface, 500X.

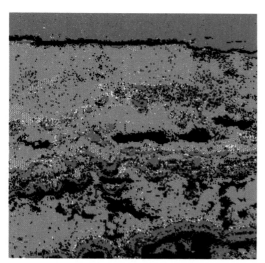

Figure 14-13. SEM-EDS analysis of a boiler tube cross section reveals layers of iron in a boiler tube deposit consisting primarily of calcium (red = Fe, green = Ca, yellow = Fe + Ca, light blue = P, purple = Fe + P).

14-13. For example, SEM-EDS can be useful in the following determinations:

■ differences in deposit composition between corroded and noncorroded areas on a tube surface

■ the extent to which under-deposit concentration of boiler salts on heat transfer surfaces is promoting corrosion damage

■ elemental differences between visually different tube surface deposits

Inorganic analyses through SEM-EDS can also be performed on ground and polished cross sections of a tube covered with thick layers of waterside deposit. This testing is called elemental mapping and is particularly valuable when the deposits are multilayered. Similar to the examination of rings on a tree, cross-sectional analysis of boiler deposits can identify periods when there have been upsets in water chemistry, and thereby provides data to help determine exactly how and when deposits formed. With elemental mapping, the spatial distribution of elements in a deposit cross section is represented by color-coded dot maps. Separate elements of interest can be represented by individual maps, or selected combinations of elements can be represented on composite maps.

A scanning electron microscope (SEM) can also be utilized to analyze the topography of surface deposits and/or morphology of fracture surfaces. Fractography is particularly helpful in classifying

a failure mode. For example, microscopic features of a fracture surface can reveal whether the steel failed in a brittle or ductile manner, whether cracks propagated through grains or along grain boundaries, and whether or not fatigue (cyclic stress) was the primary cause of failure. In addition, SEM-EDS testing can be used to identify the involvement of a specific ion or compound in a failure mechanism, through a combination of fracture surface analysis and chemical analysis.

Most water-bearing tubes used in boiler construction are fabricated from low-carbon steel. However, steam-bearing (superheater and reheater) tubes are commonly fabricated from low-alloy steel containing differing levels of chromium and molybdenum. Chromium and molybdenum increase the oxidation and creep resistance of the steel. For accurate assessment of metal overheating, it is important to have a portion of the tube analyzed for alloy chemistry. Alloy analysis can also confirm that the tubing is within specifications. In isolated instances, initial installation of the wrong alloy type or tube repairs using the wrong grade of steel can occur. In these cases, chemical analysis of the steel can be used to determine the cause of premature failure.

At times, it is necessary to estimate the mechanical properties of boiler components. Most often, this involves hardness measurement, which can be used to estimate the tensile strength of the steel. This is particularly useful in documenting the deterioration of mechanical properties that occurs during metal overheating. Usually, a Rockwell hardness tester is used; however, it is sometimes advantageous to use a microhardness tester. For example, microhardness measurements can be used to obtain a hardness profile across a welded zone to assess the potential for brittle cracking in the heat-affected zone of a weld.

Microstructural analysis of a metal component is probably the most important tool in conducting a failure analysis investigation. This testing, called metallography, is useful in determining the following:

■ whether a tube failed from short-term or long-term overheating damage

■ whether cracks initiated on a waterside or fireside surface

■ whether cracks were caused by creep damage, corrosion fatigue, or stress-corrosion cracking (SCC)

■ whether tube failure resulted from hydrogen damage or internal corrosion gouging

Proper sample orientation and preparation are critical aspects of microstructural analysis. The orientation of the sectioning is determined by the specific failure characteristics of the case. After careful selection, metal specimens are cut with a power hacksaw or an abrasive cut-off wheel and mounted in a mold with resin or plastic. After mounting, the samples are subjected to a series of grinding and polishing steps. The goal is to obtain a flat, scratch-free surface of metal in the zone of interest. After processing, a suitable etchant is applied to the polished metal surface to reveal microstructural constituents (grain boundaries, distribution and morphology of iron carbides, etc.)

Metallographic analysis of the mounted, polished, and etched sections of metal is performed with a reflective optical microscope (Figure 14-14). This is followed by a comparison of microstructures observed in various areas of a tube section—for example, the heated side versus the unheated side of a waterwall tube. Because the microstructure on the unheated side often reflects the as-

manufactured condition of the steel, comparison with the microstructure in a failed region can provide valuable insight into the degree and extent of localized deterioration (Figure 14-15).

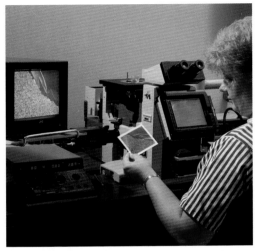

Figure 14-14. A reflective optical microscope is used to compare the microstructures of metal specimens.

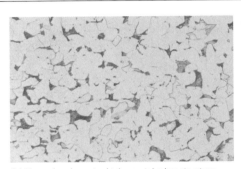

(a) Normal carbon steel tube metal microstructure. ASTM A178-73 Grade "A" tubing.

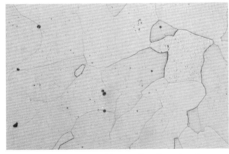

(b) Grain growth caused by overheating of carbon steel. Temperature in the range of 1575 °F and higher.

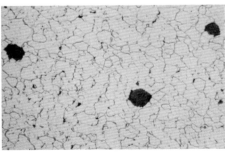

(c) Carbide spheroidization and graphitization indicating very long mild overheating in the 900–1025 °F range.

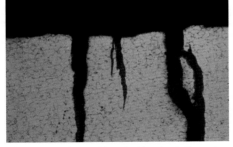

(d) Transgranular cracking of tubing due to thermal cyclic stress.

Figure 14-15. Typical photomicrographs for carbon steel boiler tubing.

CHEMICAL CLEANING OF STEAM GENERATOR SYSTEMS

Internal surfaces of steam generator systems are cleaned to remove contaminants that impair heat transfer and may ultimately cause tube failure. In the decision to clean chemically and the selection of the cleaning method, the following factors should be considered:

- the type of design, amount of flow, operating temperature, and pressure of the steam generator
- the characteristics and quantity of deposit
- the compatibility of the cleaning solvent with system metallurgy
- the method of spent solvent disposal
- the availability of demineralized water or condensate
- safety
- cost

PREOPERATIONAL CLEANING

Traditionally, preoperational cleaning has not been a major consideration in plant design and construction. This area has recently received more attention because of costs, environmental restrictions on plant discharges, smaller boiler design margins, higher boiler operating pressures and temperatures, and noise abatement requirements.

The purpose of preoperational cleaning is to remove construction contaminants which could cause operating problems or even failure during initial operation. Preoperational contaminants include mill scale, weld slag, corrosion products, oil, grease, debris and dirt, temporary protective coatings, and other contaminants remaining after fabrication and erection of the unit.

Mill scale, a dense magnetite layer produced on steel surfaces during fabrication, is subject to fracture and erosion during system operation. Because the exposed metal is anodic to the mill scale, these surface cracks are potential corrosion sites. In addition, mill scale and weld scale can become deposit accumulation sites and may cause damage if they break loose. Removal of metallic scale and corrosion products generally requires chemical cleaning.

Frequently, sand and dirt (which add scale-forming contaminants such as silica and calcium to the boiler water) can be removed by flushing.

Oils and grease are also used in boiler fabrication. The insulating effect of even thin films of these substances can lead to overheating failure. Oil and grease, including temporary protective coatings, are generally removed by hot alkaline chemical cleaning.

Two areas merit consideration in the planning of preoperational cleaning. First, plant design should include provisions that allow field cleaning with a minimum of temporary piping, time, and cost. Second, fabrication and erection cleanliness should be maintained at the highest practical level so that only a minimum of field cleaning is necessary. Increased attention is being given to cleanliness during fabrication due to several factors:

- the trend toward more shop fabrication and less field fabrication allows better cleanliness control
- system start-up is quicker when complex cleaning methods are not required
- environmental regulations complicate the disposal of chemical cleaning solutions
- noise abatement limits the use of steam blows in some locations

The preoperational cleaning method chosen depends largely on the cleanliness maintained during fabrication and erection. If a high degree of construction cleanliness is maintained, a water flush of condensate and feedwater systems may be used, along with an alkaline boil-out followed by a steam blow of the steam generator. This can provide sufficient cleaning in most boilers that operate at less than 900 psig.

A hot alkaline flush of the condensate and feedwater systems and an alkaline boil-out of the boiler and economizer are required if certain contaminants are present. In addition to removing oil, grease, and temporary protective coatings, the alkaline conditions and the flushing action may cause dislodging of some flaky surface oxide and mill scale. If no other cleaning is required, the alkaline cleaning should be followed by steam blowing.

Contaminated boiler and steam systems, as well as systems that operate at over 900 psig, may require a more thorough chemical cleaning. This should be preceded by an alkaline cleaning to prevent oil or grease from interfering with the cleaning process unless a single-stage alkaline/chelant cleaning solution is used.

Preoperational acid or chelant cleaning procedures vary greatly depending on individual circumstances. The equipment to be cleaned may include any combination of the following:

■ boiler/economizer

■ condensate/feedwater systems

■ superheater/reheater

■ main steam lines

The other variable, *when* to clean, also depends on the individual application. Typical sequences include the following:

1. alkaline clean before initial operation

2. alkaline clean and acid/chelant clean before initial operation

3. alkaline clean before initial operation followed by an acid/chelant cleaning after several weeks to 1 year of operation

Subcritical boilers operated at pressures greater than 900 psig are generally given a hot alkaline cleaning and an acid or chelant cleaning of the boiler and economizer. Supercritical boilers and systems are usually given a hot alkaline cleaning along with an acid or chelant cleaning of the condensate/feedwater systems, boiler, and econo-

mizer. Often the superheater and reheater are cleaned as well.

Acid or chelant cleaning of the boiler and economizer is generally performed before initial operation to prevent contaminant damage or water chemistry upsets. Sometimes, this cleaning is delayed until after initial operation (i.e., until after most condensate and feedwater system contaminants have been carried to the boiler). However, this presents the risk of contamination in the economizer and high heat zones of the boiler.

Acid or chelant cleaning can be beneficial in condensate/feedwater systems because of the equipment's large surface area. However, the necessity for additional chemical cleaning is reduced when these systems are made primarily of copper alloys and stainless steel. Acid or chelant cleaning removes contaminants in these systems, and there is no reason to delay boiler and economizer cleaning until after initial operation.

Acid or chelant cleaning of the superheater and reheater increases cleaning costs and complexity but minimizes possible turbine problems by removing contaminants which could cause solid particle erosion. Although it was not common in the past, chemical cleaning of superheaters or reheaters is becoming more frequent.

POSTOPERATIONAL CLEANING

Unless a proper inspection program is in place, a blown tube may be the first indication of a need for chemical cleaning of a boiler. The type and frequency of inspection required vary with the boiler design, its operating requirements, and the history of operation and water treatment. Visual inspection of the fireside may reveal blistered tubes, while waterside inspection of the drums, tube internals, and headers may reveal deposit accumulations that indicate a need for boiler cleaning.

Boroscopes and miniature video cameras (Figure 15-1) can be used to monitor conditions in tubes and headers not otherwise visible. Tube samples are often cut from the highest heat flux area of the boiler to permit visual examination of conditions at this critical location, along with quantitative measurement of the deposit accumulation. Chordal thermocouples (Figure 15-2) may be used to measure the effect of any boiler deposition on the resistance to heat transfer in certain critical tubes and thereby identify the need for cleaning.

Figure 15-1. *Miniature video equipment permits the inspection of otherwise inaccessible boiler tubes.*

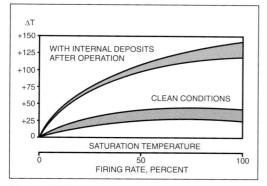

Figure 15-2. *Chordal thermocouple measurements detect the buildup of water-side deposits.*

After a boiler has been in operation for some time, its cleaning schedule may be based on number of years in operation or amount of steam generated.

Deposits may include corrosion products, mineral scale, sludge, or process contaminants, in any combination. Deposits form from low levels of accumulation for long periods of operation, due to improperly controlled water treatment or process contamination. Deposit characteristics are influenced by raw water characteristics, type of external treatment, feedwater treatment methods and control, and the nature and degree of external contaminants which have entered the feedwater during operation. Power and water plant operators play a big part in the deposit control of a unit.

Corrosion product deposits are the major contaminant in high-pressure boilers using demineralized makeup water. Condensate and feedwater system corrosion products can be carried into the boiler, where they can deposit on the metal surfaces. Boiler tube metal corrosion can add to this deposit, but with good boiler operation and proper water treatment practices, tube metal corrosion is not usually significant. The insulating effect of a corrosion product deposit can cause tube metal overheating and subsequent failure. Because they are usually porous, corrosion product deposits can also provide sites for boiler water concentration and, consequently, the potential for caustic attack.

In boilers that receive some hardness in the makeup water, typical deposits are calcium, magnesium, and phosphate compounds. These mineral scales, along with sludge and process contaminants such as oils, can also cause an insulating effect which leads to overheating. The principal cause of mineral scale formation in boilers is the fact that scale-forming salt solubility decreases with increasing temperature. Thus, as boiler water is heated, the solubility of scale-forming salts is exceeded and crystallization takes place on the boiler heating surfaces.

Removal of deposits has advantages other than minimization of failures. Cleaning reveals the true boiler metal condition, allowing more complete inspection. Previously unnoticed pits or gouges may become clearly visible. Cracks resulting from fatigue, embrittlement, or corrosion may be revealed. Also, a clean unit may be operated more efficiently.

Most internal deposits can be removed with mineral acid, organic acid, or chelant. The most widely used solvent is a 5% hydrochloric acid solution with inhibitor and complexors. Phosphoric, sulfuric, and sulfamic acids are also used in boilers that operate at less than 2000 psig. Inhibited ammonium citrate and a combination of hydroxyacetic and formic acids are commonly used organics. Salts of EDTA (ethylenediamine-tetraacetic acid) have been used for both off-line and on-line cleaning.

Environmental regulations must be considered in solvent selection. Plants with no waste treatment facilities must have spent cleaning solution trucked away, or they must use a process in which the spent solution can be evaporated.

A new boiler or an operating boiler that is subject to oil contamination should be given an alkaline boil-out before chemical cleaning. Otherwise, oil-based contaminants will interfere with cleaning. Brass or bronze parts should be replaced temporarily with steel or steel alloy parts before a chemical cleaning. Proper provisions should be made for venting of chemical vapors.

Connections to steam headers or other equipment should be valved shut with provisions for backfilling and draining.

There are two basic methods of acid or chelant cleaning: the circulation method and the static (sometimes called fill-and-soak) method.

In the circulation method, the cleaning solution is intermittently circulated through the unit until the chemical content and the iron and copper content of the cleaning solution reach a constant value, indicating the end of the reaction between the solvent and deposits. This method is suitable for units with positive liquid flow paths, and it allows monitoring of the cleaning operation at multiple sample points.

In the static method, the cleaning solution soaks in the unit for a prescribed length of time. This method is used when positive circulation of the cleaning solution is impractical.

In addition to inhibited solvent, intensifier chemicals are required in some instances to help remove particular scale constituents. The addition of certain wetting agents with the inhibitor can reduce corrosion rates during cleaning.

WATER FLUSHING AND STEAM BLOWING

Alkaline water flushes are usually performed with potable-quality tempered water. Flush velocity should be greater than system design velocity. After the flush, the condenser hot well, deaerator, and deaerator storage tank are usually cleaned by hand, because these units do not drain completely.

Steam blowing normally follows the last boiler and economizer chemical cleaning operation. When reheaters are present, this step is usually done in two stages. First, the main steam piping and superheater are blown. Then, the main steam piping, superheater, and reheater are blown. The most effective cleaning is accomplished when the blowing steam force exceeds maximum operating steam force. A 1.6:1 force ratio has been found to provide satisfactory cleaning of particles from the piping. Continuous blowing is used, but intermittent blow periods of several minutes each are more common.

In order to assess blowing effectiveness, blow steam may be directed onto a target plate in the main steam line. This practice is most common in high-pressure boilers generating superheated steam for a turbine drive. Successive blows are made until solid particle erosion is no longer visible on the target plate. The required blow time varies, but in general is approximately 24 hr unless

chemical cleaning has preceded the blow. Chemical cleaning can significantly reduce required blow times.

ALKALINE CLEANING

Alkaline cleaning (flush/boil-out) solutions are basically trisodium phosphate and surfactant. Trisodium phosphate concentrations typically range from 0.1 to 1%. Disodium phosphate and/or sodium hydroxide may be added. Sodium nitrate is often added when caustic is used, as a precaution against caustic embrittlement. Sodium sulfite is occasionally used to prevent oxygen corrosion.

Although mill scale removal is not one of the purposes of alkaline cleaning, the addition of chelating agents may remove enough mill scale to eliminate the need for further chemical cleaning. However, where mill scale removal is specifically desired, acid or chelant cleaning should follow alkaline cleaning. Commercial alkaline cleaning products may include chelants and other compounds specifically formulated to provide effective cleaning.

Alkaline cleaning should follow procedures that boiler manufacturers furnish for their specific boilers. During cleaning, boilers are fired at a low rate, but not enough to establish positive circulation. Boil-out pressure is usually about half the operating pressure. Several times during the alkaline boil-out, half of the boiler water should be blown down as shown on the gauge glass through the bottom blowdown valves. Alternate valves are used where there is more than one blowdown connection. The boiler is refilled close to the top of the glass after each blow. Superheaters and reheaters are protected by backfilling with a properly treated condensate. Each chemical cleaning stage lasts for approximately 6 to 24 hr.

When alkaline cleaning is completed, the unit is slowly cooled, the alkaline solution is drained to the disposal area, and the unit is flushed with good-quality condensate. Flushing, with intermediate blowdown, should be continued until flush water conductivity is less than 50 mhos and phosphate is less than 1 ppm. Alkaline cleaning should be repeated if organic or residual oil-based contaminants are found upon inspection of the unit.

SOLVENT CLEANING

The choice of solvent should be based on laboratory studies of the deposit sample found inside the tube. This will help ensure that the expected results of the chemical cleaning are achieved at minimum expense and risk to the system.

Inhibited hydrochloric acid is the most commonly used solvent. It is effective for removing most calcium and iron deposits. Hydrofluoric acid or ammonium bifluoride is often added to the hydrochloric acid to aid in the removal of silicate-containing scales. This mixture is also used to accelerate the removal of certain complex scales.

When copper oxide is present in the deposit, it may dissolve and then plate out as metallic copper, causing pitting of the system metal. For this reason, a copper complexing agent such as thiourea is often added to the hydrochloric acid to keep the copper in solution.

Chelating agents, such as EDTA and citric acid, are used to dissolve iron oxide deposits. They can also be used to dissolve deposits containing copper oxides by the injection of an oxidizing agent through the boiler circuits.

In superheaters and the water–steam circuits of once-through supercritical boilers, magnetite and complex scales may be removed by inhibited hydroxyacetic and formic acid, inhibited EDTA, or inhibited ammoniated citrate. Because these materials are highly volatile, they are useful in units that are difficult to fill, circulate, and drain.

After the cleaning, the unit must be drained under a nitrogen blanket and flushed with pure water until it is free of solvent and soluble iron. If a mineral acid solvent is used, a boil-out should follow to repassivate the metal surface. If a chelant is used, oxygen injection into the cleaning solution provides passivation.

Following cleaning, the boiler should be inspected to verify that the expected results have been achieved. Visual inspection of the drums and headers may reveal the accumulation of loose particulate matter, which signifies a need for additional flushing. Video or boroscope inspection of the tubes or cutting of tube samples may reveal the need for further cleaning.

Because of possible harm to personnel and equipment, only qualified personnel should be involved in solvent selection, planning, and supervision of the cleaning.

STEAM PURITY

Precise system control is required for the operation of modern superheated steam turbines. Solids in the steam leaving a boiler can deposit in the superheater and turbines, causing costly damage. For this reason, close control of steam purity is critical.

Steam purity refers to the amount of solid, liquid, or vaporous contamination in the steam. High-purity steam contains very little contamination. Normally, steam purity is reported as the solids content.

Steam purity should not be confused with steam quality. Steam quality is a measure of the amount of moisture in the steam. It is expressed as the weight of dry steam in a mixture of steam and water droplets. For example, steam of 99% quality contains 1% liquid water.

Carryover is any solid, liquid, or vaporous contaminant that leaves a boiler steam drum with the steam. In boilers operating at pressures of less than 2000 psig, entrained boiler water is the most common cause of steam contamination. The entrained boiler water contains dissolved solids and can also contain suspended solids.

There are many causes of boiler water entrainment in steam. A few of the more common mechanisms have been given specific names, such as "spray carryover," "priming," "foaming," and "leakage carryover."

EFFECTS OF CARRYOVER

Boiler water solids carried over with steam form deposits in nonreturn valves, superheaters, and turbine stop and control valves. Carryover can contaminate process streams and affect product quality. Deposition in superheaters can lead to failure due to overheating and corrosion, as shown in Figure 16-1.

Superheated steam turbines are particularly prone to damage by carryover. Sticking of gov-

Figure 16-1. Overheating of this superheater tube was caused by deposits that resulted from boiler water carryover into the steam.

ernor and stop valves due to deposits can cause turbine overspeed and catastrophic damage. Solid particles in steam can erode turbine parts, while deposition on turbine blades can reduce efficiency and capacity. Losses of 5% in turbine efficiency and 20% in turbine capacity have occurred due to deposition. When large slugs of boiler water carry over with steam, the resulting thermal and mechanical shock can cause severe damage.

Loss of production may result from reduced capacity or equipment failure caused by carryover. In some instances, the effect of carryover on production overshadows all other considerations.

Steam can be contaminated with solids even where carryover is not occurring. Contaminated spray attemperating water, used to control superheated steam temperature at the turbine inlet, can introduce solids into steam. A heat exchanger coil may be placed in the boiler mud drum (Figure 16-2) to provide attemperation of the superheated steam. Because the mud drum is at a higher pressure than superheated steam, contamination will occur if leaks develop in the coil. Failure to check for these possible

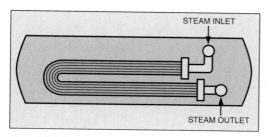

Figure 16-2. Leaks in drum attemperator coils contaminate steam with boiler water.

sources of contamination may necessitate time-consuming steam purity studies.

CAUSES OF CARRYOVER

Carryover results from incomplete separation of steam from the steam–water mixture in a boiler. Many factors, both mechanical and chemical, contribute to incomplete separation. Mechanical factors include boiler design, inadequate or leaking separating equipment, high water levels, method of firing, and load characteristics.

Among the chemical causes are high total solids concentrations (dissolved and/or suspended solids), excessive alkalinity, and the presence of oily materials and other organic contaminants. Methods of external and internal treatment can also affect steam purity. In certain instances, vaporization of solids may occur—another form of chemical carryover.

Mechanical Causes

In the modern water-tube boiler, separation of the steam–water mixture entering a relatively small steam drum is a complex process. For every pound of steam generated, as many as 15–20 lb of water may circulate through a drum. As a result, 99.97% or more of the circulating water must be removed from steam to ensure the desired steam purity.

Boiler Design. Certain types of boilers are known for their ability to produce clean steam constantly; other types are traditionally recognized as troublesome. Factors that affect carryover include design pressure, steam drum size, design generating rate, circulation rate, arrangement of downcomers and risers, and the type of mechanical separating equipment used.

In some older boiler designs, the steam-carrying or riser tubes discharge below the water level, causing severe turbulence within the steam drum. This condition is minimal for units in which the steam generating tubes discharge above the working water level or beneath a baffle that separates them from the drum water.

The use of suitable steam separation equipment also effectively prevents severe steam contamination. However, these devices impose a small pressure drop. Therefore, if any leaks occur in this equipment, leakage carryover occurs.

Operating Conditions. Operation at loads in excess of design rating can increase carryover. Sudden increases in load (e.g., when a safety valve blows or when soot blowing begins) also promote carryover.

Sudden increases in process steam demand may lower the steam header pressure and, in turn, the boiler drum pressure, causing rapid expansion of steam–water mixture in the boiler. This can significantly raise the drum water level and cause carryover. Sudden changes in boiler operation should be avoided as much as possible. In a plant with more than one boiler in operation, the boiler most susceptible to carryover should be operated at a safe, constant load while others are used to accommodate load swings.

High water levels in the steam drum reduce deentrainment space, increasing carryover.

Chemical Causes

Foaming. Foaming and selective vaporous carryover are the two basic mechanisms of chemical carryover. Foaming is the formation of stable bubbles in boiler water. Because bubbles have a density approaching that of steam, they are not readily removed by steam purifying equipment. Foaming has caused a variety of carryover problems and can cause erroneous water level readings that produce swings in feedwater flow.

Foaming tendencies of boiler water are increased with increases in alkalinity and solids content. Boiler water solids have a dual effect on carryover. First, for a given boiler and static operating conditions, high solids in the boiler water result in high solids in each drop of boiler water carried over. Second, foaming potential increases with increasing boiler water solids. If boiler water solids are allowed to double (without any foaming), carryover will be doubled. If the higher solids cause foaming, the carryover may be increased even more.

Organic and Synthetic Contamination. Oil and other organic contaminants in boiler water can cause severe carryover conditions. The alkalinity of the boiler water saponifies fatty acids, producing a crude soap that causes foaming.

Conventional mineral analysis of water does not reveal the foaming tendencies originating from organic contamination. Even determination of the organic content of the water does not necessarily provide this information, because many surface supplies from heavily wooded areas contain a relatively high concentration of harmless or beneficial lignin-type organics.

Wastes discharged into many surface water supplies contain synthetic detergents and wetting agents. Contamination of surface supplies by these agents has caused difficulty with boiler foaming.

Selective Vaporous Carryover. Selective vaporous carryover occurs as a result of variances in the solvent properties of steam for different water phase impurities. Boiler water salts, such as sodium sulfate, sodium chloride, sodium hydroxide, and sodium phosphate, are soluble in water and (to varying degrees) in steam. However, the solubility of these salts in steam is low and is usually not a problem when boiler pressure is less than 2400 psig.

Selective vaporous carryover of silica can occur at pressures as low as 400 psig. However, silica vaporization is not usually a problem below 900 psig.

PREVENTION OF CARRYOVER

Carryover can never be eliminated completely. Even the best boiler designs operating with well controlled water chemistry produce trace amounts (0.001–0.01 ppm total solids) of carryover. However, the primary consideration in the selection of a boiler and its operating conditions is the amount of carryover that can be tolerated.

Whenever superheated steam is required for process use or turbines, steam purity as low as 10–30 ppb total solids may be necessary to prevent deposits. These limits apply to most industrial applications within a pressure range of 300–1500 psig, to ensure uninterrupted service of superheaters and turbines.

Although boiler manufacturers do not ordinarily guarantee less than 0.03% carryover, purity levels well below this level are routinely achieved in many systems. To obtain the desired steam purity, both the boiler designer and the operator must carefully select system equipment and operating conditions. The methods used to achieve the required steam purity can be divided into mechanical and chemical means of carryover prevention.

Mechanical Separation

Low-capacity, low-pressure boilers (usually firetube boilers) rely principally on simple gravity separation of steam and water. At 200 psig and saturation conditions, the density of water is 115 times greater than that of steam. Because the steam is typically used for heating, steam purity requirements are not very stringent. The installation of a dry pipe near the top of the drum (Figure 16-3) to enhance steam–water separation is normally satisfactory.

To meet the needs of superheated steam turbines, steam purity requirements become more stringent at higher boiler pressure. In these applications, the density difference between water and steam decreases rapidly. At 1000 psig, water density is only 20 times that of steam. In comparison with low-pressure boilers, the separating force is reduced by 83%, making entrainment possible at relatively low steam velocities. The cost of a drum adequately sized to separate steam and water at this higher pressure by gravity alone is prohibitive.

Internal mechanical separating devices may be installed to allow the use of economical drum sizes at higher pressures. These devices are classified into two categories: primary separators and secondary separators.

Primary Separation. Primary separation of steam and boiler water is secured by changes in direction of flow. In this method, the difference in density

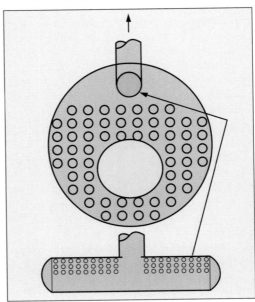

Figure 16-3. Gravity separation, often enhanced by a dry pipe, is used to produce steam of satisfactory purity in firetube boilers.

between water and steam is used as the means of separation. The major separation is effected by primary devices (such as baffle plates), which also reduce turbulence and the steam content of recirculating boiler water (Figure 16-4). Any appreciable amount of steam in a downcomer circuit (carryunder) reduces the head available for circulation, thereby lowering the boiler circulation rate. In addition, errors in water level indication caused by the presence of steam in the drum water are reduced.

Figure 16-5 shows an installation of centrifugal primary separators in a boiler drum. Steam and water from the risers enter the separator tangentially. Water moves downward in a long helical path on the inside wall of the cylinder, and steam spirals upward. Centrifugal force on the mixture whirling around the cylinder helps to separate steam from water.

Secondary Separation. Also termed "steam scrubbing" or "steam drying," secondary separation is used to separate small amounts of moisture from large amounts of steam. Steam flow is directed in a frequently reversing pattern through a large contact surface. A mist of boiler water collects on the surface and is drained from the separating unit.

Closely fitted screens or corrugated plates are normally used. Steam velocity is kept to a low level to prevent reentrainment of separated boiler water and to ensure maximum surface contact. Figure 16-6 depicts a typical arrangement of primary separators and steam scrubbers in a

boiler steam drum. Figure 16-7 shows a steam drum with typical steam purity equipment.

Although steam separation is usually performed in a steam drum, external separating devices are also available. They are usually centrifugal separators, similar to those used for primary separation in many boilers. They are particularly useful where only a portion of the steam

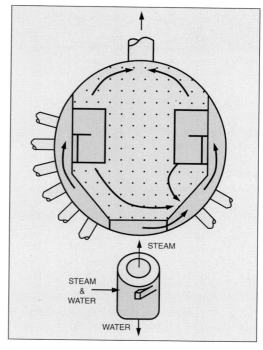

Figure 16-5. Cyclones are commonly used for primary separation.

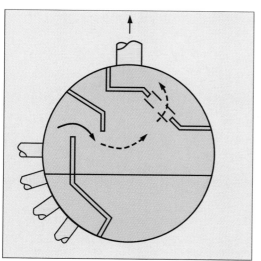

Figure 16-4. Baffle plates in the steam drum reduce turbulence, improving steam purity and reducing steam in the circulating boiler water.

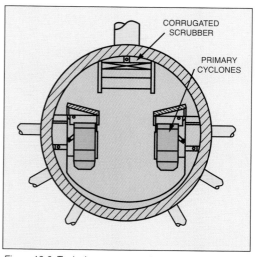

Figure 16-6. Typical arrangement of primary and secondary separators in a high-pressure boiler steam drum.

Figure 16-7. Typical steam drum with generating tube baffle plates, centrifugal separators, and secondary scrubbers.

production must be purified for a single application. In some instances, the cost of external purification may be lower than that of modifying boiler drum internals to improve steam purity.

Chemical Control

The principal chemical factors that cause carryover are the concentrations of total dissolved solids, alkalinity, silica, and organic contaminants in the boiler water.

The American Boiler Manufacturers Association (ABMA) recommendations for boiler water limits are provided in Table 16-1. These guidelines should not be considered absolute. Some systems

cannot tolerate operation at these concentrations; others operate continuously at significantly higher concentrations.

Operating conditions also have an effect. It is difficult to predict the maximum boiler water solids that can be tolerated without significant carryover under most operating conditions. The maximum specific limits for a set of operating conditions can be established only after a steam purity study has been conducted for those conditions.

Whenever carryover is being caused by excessive boiler water concentrations, an increase in boiler blowdown rate is normally the simplest and most expedient solution. When high concentrations are caused by high levels of feedwater impurities, adjustment or upgrading of the external treatment may provide the most economical solution.

Because various suspended solids and organic matter in boiler feedwater have different effects on carryover, no generalization of their permissible concentration can be made. Concentrations should be kept as close to zero as possible. No method of internal treatment can be relied on to overcome carryover problems caused by oil and other organics. In order to prevent the carryover difficulties caused by these contaminants, they must be removed from the boiler feedwater.

Table 16-1. Recommended water tube boiler water limits and associated steam purity at steady state full load operation.[a]

	Drum Pressure, psig	Total Dissolved Solids[b] in Boiler Water, ppm (max.)	Total Alkalinity[c] in Boiler Water, ppm	Suspended Solids in Boiler Water, ppm (max.)	Total Dissolved Solids[c,e] in Steam, ppm (max. expected value)
Drum-type boilers	0–300	700–3500	140–700	15	0.2–1.0
	301–450	600–3000	120–600	10	0.2–1.0
	451–600	500–2500	100–500	8	0.2–1.0
	601–750	200–2000	40–400	3	0.1–0.5
	751–900	150–1500	30–300	2	0.1–0.5
	901–1000	125–1250	25–250	1	0.1–0.5
	1001–1800	100	variable[d]	1	0.1
	1801–2350	50	variable[d]	N/A	0.1
	2351–2600	25	variable[d]	N/A	0.05
	2601–2900	15	variable[d]	N/A	0.05
Once-through boilers	1400 & above	0.05	N/A	N/A	0.05

[a] Reprinted (with permission) from "Boiler Water Limits and Steam Purity Recommendations for Watertube Boilers," American Boiler Manufacturers Association, 1982.
[b] Actual values within the range reflect the TDS in the feedwater. Higher values are for high solids; lower values are for low solids in the feedwater.
[c] Actual values within the range are directly proportional to the actual value of TDS of boiler water. Higher values are for high solids; lower values are for low solids in the boiler water.
[d] Dictated by boiler water treatment.
[e] These values are exclusive of silica.

The organic compounds and blends used as boiler water and condensate treatment chemicals are selected on the basis of two factors:

■ the ability to prevent deposition, corrosion, and carryover, as well as condensate system corrosion

■ low tendency to cause foaming in the boiler water

Steam Purity Studies. The design of modern steam turbines is such that the tolerance for steam impurities is very low. There is an ever-increasing demand not only for higher steam purity, but also for techniques to measure impurities at very low levels. The sodium tracer and cation conductivity techniques are commonly used to detect impurities in the parts per billion range.

Carryover can be a serious steam plant problem, and frequently the cause of steam contamination can be determined only through extensive studies employing sensitive sampling and testing techniques. A water treatment engineer, through proper use of these tools, is able to help plant operators obtain maximum steam purity with minimum blowdown while maintaining clean boiler waterside surfaces.

Antifoam Agents. Frequently, the cause of a carryover problem cannot be economically corrected through adjustment of boiler water balances or installation of additional external treatment facilities. In many of these instances, the use of an effective antifoam agent can reduce carryover tendencies significantly (see Figure 16-8).

The primary purpose of antifoam application is the generation of high-purity steam. However, antifoam agents also contribute to reduced blowdown requirements. Antifoam feed allows boiler water concentrations to be carried at much higher values without compromising steam purity.

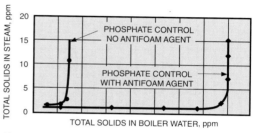

Figure 16-8. Test results showing the effect of antifoam agents on steam purity in experimental boilers.

MEASUREMENT OF STEAM PURITY

Accurate measurement of steam purity is essential to identifying the cause of potential or existing steam purity problems in modern boiler plants. One reason for this is that superheated steam turbines have an extremely low tolerance for solids contamination in the steam. Fortunately, techniques are available to determine steam contamination in the parts per billion range to satisfy the demands of most systems. The test results make it possible to determine the effect of changing boiler operation on steam purity.

IMPURITIES

Impurities present in steam can be solid, liquid, or gaseous. Solids are usually dissolved in water droplets or are present as dust. Because water treatment practices are such that most soluble chemical constituents of boiler feedwater are converted to sodium salts, most solids present in steam are sodium salts, with minor amounts of calcium, magnesium, iron, and copper also present.

Gaseous constituents commonly found in low-pressure steam (less than 2000 psig) are ammonia, carbon dioxide, nitrogen, amines, and silica. Of these, only silica contributes to the difficulties commonly associated with impure steam; the other constituents are of concern only where they interfere with the measurement of steam purity.

METHODS OF STEAM PURITY MEASUREMENT

Several methods of measuring steam purity have been available and used for many years. Each offers its own distinct advantages.

Specific Conductance

Specific conductance is one of the most commonly used methods. The specific conductance of a sample, measured in microsiemens (μS) or micromhos (μmho), is proportional to the concentration of ions in the sample. When boiler water is carried over in steam, the dissolved solids content of the boiler water contaminates the steam, and steam sample conductivity increases. Measurement of this increase provides a rapid and reasonably accurate method for determining steam purity.

One of the disadvantages of using specific conductance is that some gases common to steam (such as carbon dioxide and ammonia) ionize in water solution. Even at extremely low concentrations, they interfere with measurement of dissolved solids by increasing conductivity. This interference can be appreciable in a high-purity steam sample.

For example, in a sample containing less than 1 ppm dissolved solids, specific conductance may be in the range of 1.0–2.0 μS. The presence of any ammonia or carbon dioxide in this sample significantly increases the conductance reading:

■ ammonia by 8.0–9.0 μS per ppm of ammonia

■ carbon dioxide by 5.0 μS per ppm of carbon dioxide

Neither of these gases is a dissolved solid. In order to obtain a proper measure of dissolved solids, the influence of each gas must be determined, and conductivity readings must be corrected for their presence. When the ammonia and carbon dioxide contents of the sample are known, accurate conductivity correction curves may be obtained to allow proper corrections to be made.

Equipment is available to degas a sample prior to measurement of conductance. Hydrogen-form cation exchange resin columns are used to reduce ammonia and amines to negligible levels. Cation conductivity analyzers apply this technology to detect acid-producing anions, such as chlorides, sulfates, and acetates. They also take advantage of the high conductance of solutions containing hydrogen ions. These solutions have a conductivity several times greater than that of a solution with an equal concentration of ions formed by a neutral salt (Figure 17-1).

In a Larson-Lane analyzer (Figure 17-2), a condensed steam sample is passed through a hydrogen-form cation exchange resin column. This resin column removes ammonia, amines, and sodium hydroxide from the sample. The sample then flows through a reboiler, which removes carbon dioxide. Conductivity is mea-

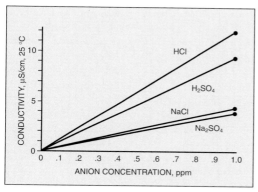

Figure 17-1. Cation conductivity increases the sensitivity of detecting contaminants.

sured after this process and may also be measured at the analyzer inlet and ion exchange column outlet. When conductivity is measured at all three points, a fairly complete picture of steam composition is provided.

Sodium Tracer Technique

Another commonly used method for measuring steam purity is the sodium tracer technique. This technique is based on the fact that the ratio of total solids to sodium in the steam is the same as the ratio of total solids to sodium in the boiler water for all but the highest-pressure (greater than 2400 psig) boiler systems. Therefore, when the boiler water total solids to sodium ratio is known, the total solids in the steam can be accurately assessed by measurement of sodium content. Because sodium constitutes approximately one-third of the total solids in most boiler waters and can be accurately measured at extremely low concentrations, this method of steam purity testing has been very useful in a large number of plants.

Sodium Ion Analyzer. The instrument most frequently used for sodium measurement is the sodium ion analyzer (Figure 17-3). A selective ion

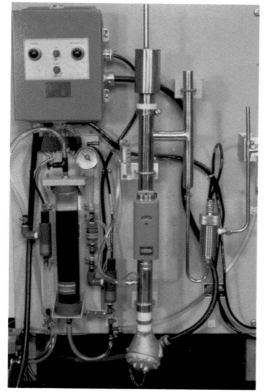

Figure 17-2. Larson-Lane analyzer monitors conductivity, cation conductivity, and degassed cation conductivity.

Figure 17-3. Sodium ion analyzer provides sensitive measurement of steam solids contamination.

electrode similar to a pH electrode is used to measure the sodium content of the steam sample.

In typical operation, a regulated amount of an agent such as ammonium hydroxide is added to a regulated amount of condensed steam sample to raise pH and eliminate the possibility of hydrogen ion interference. A reservoir stores the conditioned sample and feeds it at a constant flow rate to the tip of the sodium ion electrode and then to a reference electrode. The measured electrode signal is compared to the reference electrode potential and translated into sodium ion concentration, which is displayed on a meter and supplied to a recording device.

Good results have been reported with various sodium ion analyzers. According to the manufacturers, the instruments operate in a concentration range of 0.1 ppb to 100,000 ppm of sodium ion with a sensitivity of 0.1 ppb. Recalibrated on a weekly basis, these instruments are valuable for continuous, long-term testing and monitoring.

The acceptance of the sodium ion analyzer as an accurate, reliable steam purity evaluation instrument is evidenced by its widespread use for continuous monitoring as well as for field testing. Many steam generating plants have switched from previously accepted methods to sodium ion analysis in order to improve accuracy.

Although sodium ion analyzers measure total contamination, they do not show rapid changes in sodium concentration. This is due to a lag in electrode response and the dilution effect of the reservoir, which dampens sharp, momentary changes in conditions. Because of this, peaks that exceed boiler guarantee limits or a known carry-over range may not show up on the analyzer. This would affect interpretation of test results.

Flame emission spectroscopy and flame spectrophotometer testing. Flame spectrophotometer testing is much more sensitive to quick changes in operating conditions and detects peaks in solids concentration. Flame emission spectroscopy also provides accurate measurement in the low parts per billion range despite quick variations. Neither method is suitable for continuous, unattended monitoring.

Interpretation of Sodium Test Results. The exact ratio of sodium to dissolved solids in the boiler water and consequently in the steam can be determined for each plant but is approximately 1:3 for most plants (i.e., for each 0.1 ppm of

sodium in the steam there is approximately 0.3 ppm of dissolved solids present).

Initially, the use of the sodium tracer technique for steam purity evaluation required collecting sample bottles and transporting them to the laboratory for analysis. This technique is still a valuable tool for steam purity measurement. Samples are gathered in special laboratory-prepared, polyethylene bottles, and care is taken to protect against contamination.

In the preferred sampling procedure, three or four samples are drawn within a 15-minute period to ensure representative sampling. If there are excess solids in the steam, the bottle samples are used to define the range of the problem before implementation of an in-plant study with continuous analyzers. Bottle samples can also be used to monitor steam purity on a periodic basis.

Experience has shown that solids levels as low as 0.003 ppm can be measured with either shipped bottle samples or in-plant testing.

Anion Analysis

Occasionally, it is of interest to determine the amount of anions (chlorides, sulfates, acetates, etc.) in the steam. Degassed cation conductivity provides a measure of the total anion concentration in the sample. In addition, chloride-specific ion electrodes and ion chromatography are used to detect low levels of specific contaminants.

REPRESENTATIVE STEAM SAMPLING

In order to ensure accurate analysis, samples must be truly representative of the steam being generated. When the sampling procedures are not followed properly, the steam purity evaluation is of little or no value.

Sampling nozzles recommended by the ASTM and ASME have been in use for many years. The nozzles have ports spaced in such a way that they sample equal cross sectional areas of the steam line. Instructions for these nozzles can be found in ASTM Standard D 1066, "Standard Method of Sampling Steam" and ASME PTC 19.11. Field steam studies have shown that sampling nozzles of designs other than these often fail to provide a reliable steam sample.

Isokinetic flow is established when steam velocity entering the sampling nozzle is equal to the velocity of the steam in the header. This condition helps to ensure representative sampling for more reliable test results. The isokinetic sampling rate for many nozzles that do not conform to ASME or ASTM specifications cannot be determined.

Accurate sampling of superheated steam presents problems not encountered in saturated steam sampling. The solubility of sodium salts in steam decreases as steam temperature decreases. If a superheated steam sample is gradually cooled as it flows through the sample line, solids deposit on sample line surfaces. To eliminate this problem, the steam can be desuperheated at the sampling point.

STEAM TURBINE DEPOSITION, EROSION, AND CORROSION

The development of modern, high-efficiency steam turbines has led to an increase in deposition, erosion, and corrosion problems. Close tolerances in the turbines, the use of high-strength steels, and impure steam all contribute to these conditions.

TURBINE DEPOSITION

Although several factors influence the formation of deposits on turbine components, the general effect is the same no matter what the cause. Adherent deposits form in the steam passage, distorting the original shape of turbine nozzles and blades. These deposits, often rough or uneven at the surface, increase resistance to the flow of steam. Distortion of steam passages alters steam velocities and pressure drops, reducing the capacity and efficiency of the turbine. Where conditions are severe, deposits can cause excessive rotor thrust. Uneven deposition can unbalance the turbine rotor, causing vibration problems.

As deposits accumulate on turbine blades, stage pressures increase. Figure 18-1 shows the effect of gradual deposit buildup on stage pressure. The deposits were caused by the use of contaminated water to attemperate the steam. In a fouled condition, this 30-MW turbine lost over 5% of its generating capacity.

Turbine deposits can accumulate in a very short time when steam purity is poor. The turbine shown in Figure 18-2 was forced off-line by deposition only 3 months after it was placed in operation. Carryover of boiler water, resulting from inadequate steam–water separation equipment in the boiler, caused this turbine deposit problem.

The nature of silica deposits found on turbine blades varies greatly. Table 18-1 lists a number of silica compounds that have been identified in various studies of turbine blade deposition. Of these, amorphous silica (SiO_2) is the most prevalent.

Causes of Turbine Deposition

Entrainment. Some mechanical entrainment of minute drops of boiler water in the steam always occurs. When this boiler water carryover is excessive, steam-carried solids produce turbine blade deposits. The accumulations have a

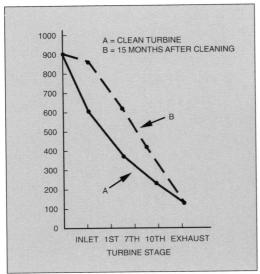

Figure 18-1. Contaminated steam attemperating water fouled this 30-MW turbine, increasing stage pressures and decreasing capacity by over 5% during a 15-month period.

Figure 18-2. Steam contamination, due to inadequate steam separating equipment in the boiler, forced this turbine off-line after only 3 months of operation.

Table 18-1. Silicate deposits found in steam turbines.

SiO_2	silica
Na_2SiO_3	sodium metasilicate
$Na_2SiO_3 \cdot 5H_2O_3$	sodium metasilicate pentahydrate
$Na_2SiO_3 \cdot 9H_2O_3$	sodium metasilicate nonahydrate
$NaAlSiO_4$	sodium aluminum silicate
$Na_4AlSi_3O_{12}(OH)$	sodium aluminum silicate hydroxide
$Na_4Al_6SO_4(SiO_4)_8$	sodium aluminum sulfate silicate
$NaFeSi_2O_6$	sodium iron silicate
$Na_3[Cl(AlSiO_4)_6]$	sodium chlorohexaaluminum silicate
$KAlSi_3O_8$	potassium aluminum silicate
$KNa_3(AlSiO_4)_6$	potassium trisodium aluminum silicate
$Mg_6[(OH)_8Si_4O_{10}]$	magnesium octahydride silicate
$Mg_3Si_4O_{10}(OH)_2$	magnesium silicate hydrate
$Ca_2Si_2O_4$	calcium silicate
$Ca_2Al_2Si_3O_{10}(OH)$	calcium aluminum silicate hydroxide
$3Al_2O_3 \cdot 4Na_2O \cdot 6SiO_2 \cdot SO_3$	noselite
$(Fe,Mg)_7Si_3O_{22}(OH)_2$	iron magnesium hydroxide silicate
$Na_8Al_6Si_6O_{24}MoO_4$	sodium aluminum molybdenum oxide silicate

composition similar to that of the dissolved solids in the boiler water. Priming and foaming are common causes of high levels of boiler water carryover. Because of the high levels of carryover often encountered, these conditions often lead to superheater tube failures as well.

Attemperating Water Impurity. Turbine deposits are also caused by the use of impure water for steam attemperation and by leakage in closed heat exchangers used for attemperation. If a boiler produces pure steam and turbine deposits still occur, the attemperating system should be investigated as a possible source of contamination. Attemperating water should be of the same purity as the steam. Any chemical treatment in the attemperating water should be volatile.

Vaporization of Boiler Water Salts. Another source of turbine deposition is the vaporization of salts present in boiler water. With the exception of silica, vaporization of boiler water salts is usually not significant at pressures below 2400 psig. Silica can vaporize into the steam at operating pressures as low as 400 psig. This has caused deposition problems in numerous turbines. The solubility of silica in steam increases with increased temperature; therefore, silica becomes more soluble as steam is superheated. As steam is cooled by expansion through the turbine, silica solubility is reduced and deposits are formed, usually where the steam temperature is below that of the boiler water. To minimize this problem,

the quantity of silica in the steam must be controlled. Silica deposits are not a problem in most turbines where the silica content in the steam is below 0.02 ppm. Therefore, it has become customary to limit silica to less than 0.02 ppm in the steam. Sometimes, because of the more stringent operating conditions of certain turbines, vendors specify that steam silica be maintained at less than 0.01 ppm.

The conditions under which vaporous silica carryover occurs have been thoroughly investigated and documented. Researchers have found that for any given set of boiler conditions using demineralized or evaporated quality makeup water, silica is distributed between the boiler water and the steam in a definite ratio. This ratio, called the distribution ratio, depends on two factors: boiler pressure and boiler water pH. The value of the ratio increases almost logarithmically with increasing pressure and decreases with increasing pH. The effect of boiler water pH on the silica distribution ratio becomes greater at higher pH values. A pH increase from 11.3 to 12.1 reduces the ratio by 50%, while a pH increase from 7.8 to 9.0 has no measurable effect. For any boiler pressure and pH, the distribution ratio for silica can be determined from Figure 18-3. The amount of silica vaporized with the steam can be determined by measurement of boiler water silica. The proper boiler water silica level necessary to maintain less than 0.02 ppm silica in the steam is shown in Figure 18-4.

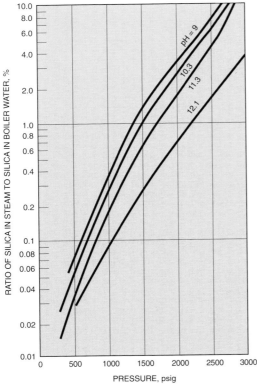

Figure 18-3. Effect of silica and boiler water pH on the volatility of silica.

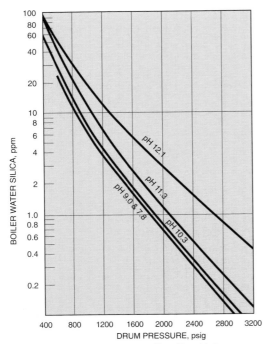

Figure 18-4. Maximum boiler water silica allowable to maintain less than 0.02 ppm silica in the steam.

When soluble, the silica present in boiler feedwater does not influence the amount of silica present in the steam. When added to boiler water in separate experiments, equivalent quantities of silicic acid and sodium silicate produced the same amount of silica in the steam. Because the amount of silica in the steam is greatly affected by pH, it is likely that silicic acids are involved in the vaporization mechanism.

Silica has a higher solubility in superheated steam than in saturated steam for any given pressure. If mechanical carryover contributes to the silica content of the saturated steam, the silica will be dissolved during superheating, provided that the total silica present does not exceed the solubility of silica in the superheated steam. Therefore, silica deposits are seldom found in superheater sections of a boiler.

After steam reaches a turbine it expands, losing pressure and temperature. As a result, the solubility of the silica decreases. Studies have shown that with a maximum of 0.02 ppm of silica in the steam, a pressure of less than 200 psig is reached in the turbine before silica starts to condense from the steam. Therefore, silica preferentially deposits in the intermediate-pressure and low-pressure sections of the turbine where the specific volume of the steam varies from approximately 1 to 10 ft³/lb. Solubility data shown in Figure 18-5 helps to explain the distribution of silica deposits in the turbine.

Localized Silica Saturation. Turbine deposits are also formed where localized silica saturation

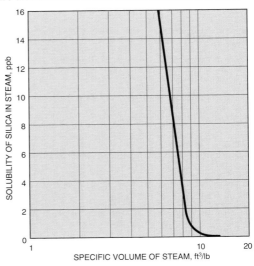

Figure 18-5. Solubility of silica in steam for conditions found in a steam turbine.

occurs and silica condenses from the steam in those areas of the turbine. Partial evaporation of the precipitated silica can then occur with only a portion of the silica being dissolved by the continuous steam flow. Deposits remain as a result.

Turbine Velocity. Another factor affecting the location of turbine deposits is the velocity in the turbine. Steam flows from the inlet to the outlet of the turbine in only a fraction of a second. Consequently, deposition is shifted downstream from the saturation point by the high steam velocities.

Prevention of Silica Deposits

The most significant factor in minimizing turbine silica deposits is the maintenance of low silica concentrations in the boiler water. External treatment equipment must be operated carefully to limit the amount of silica entering with the makeup water, and the condensate must be monitored to minimize contamination. After silica enters the boiler water, the usual corrective action is to increase boiler blowdown (to decrease the boiler water silica to acceptable levels) and then to correct the condition that caused the silica contamination.

Removal of Deposits

When a turbine becomes fouled with water-soluble salts as a result of boiler water carryover or attemperating water contamination, turbine capacity can often be restored by water washing. Because it can cause severe turbine damage, water washing should be supervised carefully and the recommendations of the turbine vendor should be followed.

When the turbine becomes fouled with compounds that are not water-soluble (including silica), water washing rarely restores capacity. Out-of-service cleaning by blasting with aluminum oxide or other soft grit material is required to remove these deposits.

EROSION

Erosion of turbine blades results in rough, uneven surfaces that alter steam flow paths. This reduces turbine efficiency and can also limit capacity. Erosion at the high-pressure end of a turbine is usually caused by solid particles (usually iron oxide) present in the steam. Iron oxide particles are present if they were not removed by steam blows during system start-up. They can also result from exfoliation of superheater or main

steam header oxides or can be introduced into the steam by contaminated attemperating water.

Erosion of intermediate and low-pressure blades is usually caused by water in the steam. Operation below design inlet steam temperature or at low load can cause condensation in these stages, leading to erosion problems.

Carbon dioxide or other acidic species present in the condensate can accelerate the damage. Some protection against erosion–corrosion can be provided by low distribution ratio amines, which neutralize the acidity and elevate the pH of the condensate.

CORROSION

Pitting, corrosion fatigue, and stress corrosion cracking problems all occur in steam turbines. The major corrodents are sodium hydroxide, chloride, sulfate, and sulfides. Usually, the level of contaminants present in steam is not high enough to corrode the system components. As steam expands through a turbine, the solubility of contaminants in the steam decreases. They condense onto surfaces at solution concentrations much higher than the original contaminant concentration in the steam. These concentrated solutions promote system corrosion.

Pitting is commonly associated with chloride deposits and occurs on rotors, disks, and buckets. Pitting attack often occurs when a moist, oxygen-laden atmosphere develops in out-of-service turbines. Damage is most severe when chloride deposits are also present. An oxygen-free or condensate-free atmosphere should be maintained to protect out-of-service turbines from corrosion.

Corrosion fatigue and stress corrosion cracking of blades and disks are commonly associated with sulfides (see Figure 18-6), chlorides, and

Figure 18-6. Sulfide contributed to stress corrosion cracking of this turbine disk.

caustic. The problems are most common in low-pressure sections of large power plant turbines, which are characterized by high stresses, crevices, and operating temperatures conducive to the condensation of concentrated solutions of steam contaminants. Problems also occur in high-pressure sections and smaller industrial-sized turbines, usually when substantial levels of steam contamination occur. These problems can be mitigated by designs that prevent crevices, lower stresses, and/or employ lower-strength materials. It is also important to avoid unnecessary stresses and to maintain high-purity steam during operation.

CONDENSATE SYSTEM CORROSION

Problems caused by iron and copper corrosion in condensate systems are not restricted to piping and equipment damage or to the loss of high-quality water and heat energy when condensate is lost. If returned to the boiler, corrosion products and process chemicals from corrosion-caused leaks contribute to the formation of damaging boiler deposits, boiler carryover, and steam-driven equipment deposits. Their presence reduces system reliability and increases operation and maintenance costs.

CORROSION OF IRON

Iron corrodes in water even in the absence of oxygen. An iron oxide surface acts like a car battery, with the surface divided into microscopic anodes (+) and cathodes (−). In condensate systems, iron acts as an anode so that it is oxidized (i.e., gives its electrons to the cathode). The cathode in pure water is a proton or hydrogen ion (H^+). When iron metal is oxidized, electrons are passed from the iron surface to hydrogen ions as shown in the reactions below.

Oxidation:

$$\underset{\text{iron}}{Fe} \rightleftharpoons \underset{\text{ferrous ion}}{Fe^{2+}} + \underset{\text{electrons}}{2e^-}$$

Reduction:

$$\underset{\substack{\text{hydrogen}\\\text{ion}}}{2H^+} + \underset{\text{electron}}{2e^-} \rightleftharpoons \underset{\substack{\text{hydrogen}\\\text{gas}}}{H_2}$$

Overall:

$$\underset{\text{iron}}{Fe} + \underset{\substack{\text{hydrogen}\\\text{ion}}}{2H^+} \rightleftharpoons \underset{\substack{\text{ferrous}\\\text{ion}}}{Fe^{2+}} + \underset{\text{hydrogen gas}}{H_2}$$

The fate of the ferrous ion (Fe^{2+}) depends on condensate temperature, pH, and flow conditions. At low temperatures, Fe^{2+} reacts with water to form insoluble ferrous hydroxide, $Fe(OH)_2$. If the condensate stream velocity is high enough, colloidal $Fe(OH)_2$ is swept into the water and carried downstream to deposit elsewhere. In low-flow areas of the condensate system, $Fe(OH)_2$ deposits near the oxidation site, forming a porous oxide layer.

At temperatures above 120 °F the deposited ferrous hydroxide reacts further to form surface-bound magnetite (Fe_3O_4) crystals.

$$\underset{\substack{\text{ferrous}\\\text{hydroxide}}}{3Fe(OH)_2} \rightleftharpoons \underset{\text{magnetite}}{Fe_3O_4} + \underset{\text{water}}{2H_2O} + \underset{\substack{\text{hydrogen}\\\text{gas}}}{H_2}$$

At even higher temperatures (above 300 °F), Fe^{2+} spontaneously forms magnetite without first forming $Fe(OH)_2$. This magnetite forms a nonporous, tightly adherent layer on the metal surface.

$$\underset{\text{ferrous ion}}{3Fe^{2+}} + \underset{\text{water}}{4H_2O} \rightleftharpoons \underset{\text{magnetite}}{Fe_3O_4} + \underset{\text{hydrogen gas}}{4H_2}$$

In most condensate systems, two or three forms of iron oxide are present. In pure water, a tightly adherent magnetite layer is formed, which is indicative of a well passivated iron surface. In the absence of contaminants, this oxide layer greatly retards any further oxidation reactions.

Oxygen Corrosion of Iron

In the presence of oxygen, the corrosion process described above is modified. Dissolved oxygen replaces hydrogen ions in the reduction reaction. The reactions are as follows:

Oxidation:

$$\underset{\text{iron}}{Fe} \rightleftharpoons \underset{\text{ferrous ion}}{Fe^{2+}} + \underset{\text{electrons}}{2e^-}$$

Reduction:

$$\underset{\text{oxygen}}{^1/_2O_2} + \underset{\text{electrons}}{2e^-} \rightleftharpoons \underset{\text{oxide ion}}{O^{2-}}$$

Overall:

$$Fe + \tfrac{1}{2}O_2 + 2H^+ \rightleftharpoons Fe^{2+} + H_2O$$
iron   oxygen   hydrogen   ferrous   water
ion   ion

This reaction occurs more readily than the direct reaction between iron and protons. Therefore, corrosion rates are accelerated in the presence of oxygen.

Two types of corrosion can occur with oxygen present. The first, generalized corrosion on the metal surface, causes a loss of metal from the entire surface. The second, oxygen pitting (Figure 19-1), causes a highly localized loss of metal that results in catastrophic failure in a short time.

Oxygen pitting begins at weak points in the iron oxide film or at sites where the oxide film is damaged. Instead of growing along the metal surface, the corrosion penetrates into the surface, effectively drilling a hole into (or through) the metal.

Pits are active only in the presence of oxygen. There is a visible difference between active and inactive pits. An active oxygen pit contains reduced black oxide along its concave surface, while the surrounding area above the pit is covered with red ferric oxide. If a pit contains red iron oxide, it is no longer active.

Sources of Oxygen. Oxygen usually enters the condensate by direct absorption of air. It can also flash over with the steam when the feedwater contains oxygen. With effective mechanical deaeration and chemical oxygen scavenging, all but a trace of oxygen is eliminated from boiler feedwater, so this source is not significant in most systems.

In a good system design, the air–condensate contact is minimized to prevent oxygen absorption. The condensate receiving tank can be designed with a cover to reduce air contact and a steam heating coil within the tank to elevate condensate temperature and thereby reduce oxygen solubility.

Under certain conditions, gross oxygen contamination of the condensate may be unavoidable. For example, condensate from warm-up steam for equipment used only intermittently should not be saved. Its dissolved oxygen attacks systems between the point of condensation and the deaerating heater. This contaminated condensate can return large amounts of corrosion products to the boiler.

In most cases, proper feedwater deaeration and elimination of air infiltration into the condensate substantially reduce oxygen corrosion.

CORROSION OF COPPER

Although copper is similar to iron in chemistry, the form of the resulting oxide layer is very different. Both copper and iron are oxidized in the presence of hydrogen ions and oxygen and can undergo oxygen pitting.

$$2Cu + 2H^+ \rightleftharpoons 2Cu^+ + H_2$$
copper   hydrogen   cuprous   hydrogen gas
ion

or in alkaline solution:

$$2Cu + H_2O \rightleftharpoons Cu_2O + H_2$$
copper   water   cuprous   hydrogen gas
oxide

Iron forms intact oxide layers. The oxide layers formed by copper and its alloys are porous and "leaky," allowing water, oxygen, and copper ions to move to and from the metal surface (Figure 19-2).

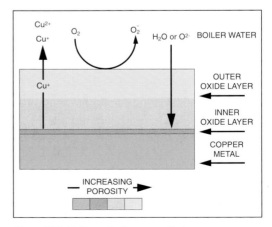

Figure 19-2. Unlike protective magnetite layers, copper oxide layers are porous and allow water, oxygen, and copper ions to move to and from the metal surface.

Figure 19-1. Typical oxygen pitting of condensate line.

The rate of movement is controlled by the copper oxide film thickness. As the oxide layer grows in thickness, the copper oxidation rate is slowed. As the oxide layer becomes thicker, the outer layers begin to slough off as particles of copper oxide. The resulting oxide layer is a much more dynamic system than that of iron. Soluble copper ions and particulate copper oxides are also formed by the normal oxidation processes.

EFFECT OF pH ON CORROSION OF IRON AND COPPER

The stability of the passivating iron or copper oxide layer is critically dependent on condensate pH. Any contaminants in the condensate system that cause the pH to decrease cause dissolution of the oxide layer and increased corrosion.

Carbon dioxide (CO_2) is the primary cause of decreased condensate pH. Carbon dioxide enters the system with air leaking into the condenser or from decomposition of feedwater alkalinity. Although part of the feedwater alkalinity is removed by a properly operated deaerating heater, most is converted to CO_2 in the boiler and released into the steam.

In boilers, carbon dioxide is liberated as shown by the following reactions:

$$2NaHCO_3 \;+\; heat \;\rightleftharpoons\; Na_2CO_3 \;+\; CO_2 \;+\; H_2O$$
sodium bicarbonate   sodium carbonate   carbon dioxide   water

$$Na_2CO_3 \;+\; H_2O \;+\; heat \;\rightleftharpoons\; 2NaOH \;+\; CO_2$$
sodium carbonate   water   sodium hydroxide   carbon dioxide

The first reaction proceeds to completion while the second is only (approximately) 80% completed. The net results are release of 0.79 ppm of carbon dioxide for each part per million of sodium bicarbonate as $CaCO_3$ and 0.35 ppm of carbon dioxide for each part per million of sodium carbonate as $CaCO_3$.

As the steam is condensed, carbon dioxide dissolves in water and depresses the pH by increasing the hydrogen ion concentration as shown in the following reaction sequence:

$$CO_2 \;+\; H_2O \;\rightleftharpoons\; H_2CO_3$$
carbon dioxide   water   carbonic acid

$$H_2CO_3 \;\rightleftharpoons\; H^+ \;+\; HCO_3^-$$
carbonic acid   hydrogen ion   bicarbonate ion

$$HCO_3^- \;\rightleftharpoons\; H^+ \;+\; CO_3^{2-}$$
bicarbonate ion   hydrogen ion   carbonate ion

Figure 19-3. Section of condensate line destroyed by carbon dioxide (low pH) corrosion. Metal destruction is spread over a relatively wide area, resulting in thinning.

Carbonic acid promotes the iron corrosion reaction by supplying a reactant, H^+. The overall reaction is:

$$2H^+ \;+\; 2HCO_3^- \;+\; Fe \;=\; Fe(HCO_3)_2 \;+\; H_2$$
hydrogen ion   bicarbonate ion   iron   ferrous bicarbonate   hydrogen

Low pH causes a generalized loss of metal rather than the localized pitting caused by oxygen corrosion. Pipe walls are thinned, particularly in the bottom of the pipe. This thinning often leads to failures, especially at threaded sections (Figure 19-3).

In order to reduce low pH-induced condensate system corrosion, it is necessary to lower the concentration of acidic contaminants in the condensate. Feedwater alkalinity can be reduced by means of various external treatment methods. Less feedwater alkalinity means less carbon dioxide in the steam and condensate. Venting at certain points of condensation can also be effective in removing carbon dioxide.

EFFECT OF OTHER CONTAMINANTS

Other contaminants in the condensate system can affect corrosion rates of iron and copper even when the pH is correctly maintained. By complexing and dissolving iron and copper oxides, contaminants such as chloride, sulfide, acetate, and ammonia (for copper) can dissolve part or all of the oxide layer.

Ammonia is the most common contaminant and is usually present in low concentrations. Ammonia contamination is usually caused by the breakdown of nitrogenous organic contaminants, hydrazine, or amine treatment chemicals. Sometimes, ammonia is fed to control condensate pH.

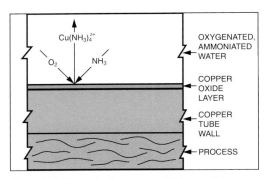

Figure 19-4. *The protective copper oxide film can be destroyed by complexing agents, such as ammonia.*

In these systems, ammonia feed rates must be carefully controlled to minimize the attack of any copper-bearing alloys (Figure 19-4).

CHEMICAL TREATMENT OF CONDENSATE SYSTEMS

Condensate systems can be chemically treated to reduce metal corrosion. Treatment chemicals include neutralizing amines, filming amines, and oxygen scavenger–metal passivators.

Neutralizing Amines

Neutralizing amines are used to neutralize the acid (H^+) generated by the dissolution of carbon dioxide or other acidic process contaminants in the condensate. These amines hydrolyze when added to water and generate the hydroxide ions required for neutralization:

$$R-NH_2 \;+\; H_2O \;\rightleftharpoons\; R-NH_3^+ \;+\; OH^-$$

neutralizing   water   amine   hydroxide
amine   ion   ion

The overall neutralization reaction can be written as shown:

$$R-NH_3^+ \;+\; OH^- \;+\; H_2CO_3 \;=$$

amine   hydroxide   carbonic
ion   ion   acid

$$R-NH_3^+ \;+\; HCO_3^- \;+\; H_2O$$

amminium   bicarbonate   water
ion   ion

By regulating the neutralizing amine feed rate, the condensate pH can be elevated to within a desired range (e.g., 8.8–9.2 for a mixed copper–iron condensate system).

Many amines are used for condensate acid neutralization and pH elevation. The ability of any amine to protect a system effectively depends on the neutralizing capacity, recycle ratio and recovery ratio, basicity, distribution ratio, and thermal stability of the particular amine.

Neutralizing Capacity. Neutralizing capacity is the concentration of acidic contaminants that is

neutralized by a given concentration of amine. The neutralizing capacity of an amine is inversely proportional to molecular weight (i.e., lower molecular weight yields higher neutralizing capacity) and directly proportional to the number of amine groups. Neutralizing capacity is important in treating systems with high-alkalinity feedwater. Table 19-1 provides a measure of the neutralizing capacity of commonly employed amines. Neutralizing capacity is not the only measure of a required product feed rate.

Recycle Ratio and Recovery Ratio. In determining product feed rates, recycle and recovery ratio are important factors. In Figure 19-5, the recycle factor is the concentration of amine at point x divided by the amine feed rate at point z. Because some amine is returned with the condensate, the total amine in the system is greater than the amount being fed. Recovery ratio is a measure of the amount of amine being returned with the condensate. It is calculated as the amine concentration at site y divided by the amine concentration at site z.

Table 19-1. Relative neutralizing capacities.

Amine	ppm Neutralizing Amine/ppm CO_2
Cyclohexylamine	2.3
Morpholine	2.0
Diethylaminoethanol	2.6
Dimethylisopropanolamine	2.3

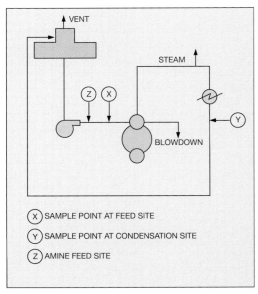

X SAMPLE POINT AT FEED SITE

Y SAMPLE POINT AT CONDENSATION SITE

Z AMINE FEED SITE

Figure 19-5. *Because a portion of the amine feed is recycled, the amine concentration in the system usually exceeds the feed rate.*

Table 19-2. Relative basicity.

Amine	Relative Basicity $K_b \times 10^6$		
	72 °F	298 °F	338 °F
Cyclohexylamine	489	61	32
Morpholine	3.4	4.9	3.8
Diethylaminoethanol	68	11.3	9.2
Ammonia	20.6	6.9	4.6

Basicity. An amine's ability to boost pH after neutralizing all of the acid species is termed "basicity." In chemical terms, it is a measure of an amine's ability to hydrolyze in pure water. The dissociation constant K_b is a common measure of basicity.

$$K_b = \frac{[R-NH_3^+][OH^-]}{[R-NH_2]}$$

As the value of K_b increases, more OH^- is formed (after all of the acid has been neutralized) and pH increases.

Examples of neutralizing amine K_b values at various temperatures are provided in Table 19-2.

Distribution of Amines between Steam and Liquid. In condensate systems, the distribution of amines between steam and liquid phases is as significant as basicity or neutralizing capacity. As the steam condenses, acidic contaminants can either remain in the steam or dissolve in the liquid phase. Some contaminants, such as carbon dioxide, stay mainly in the steam phase while others, such as hydrochloric acid, go largely into the liquid phase.

Neutralizing amines must be chosen according to their distribution characteristics to "chase" acidic contaminants. This choice must be tailored to the condensate system and the process contaminants.

For example, morpholine is an amine that primarily distributes into the liquid phase. If this amine is fed into a CO_2-laden steam system with three consecutive condensation sites, it will go into the liquid phase at the first condensation site while most of the carbon dioxide remains in the steam. With a high concentration of morpholine, the liquid phase has a high pH. At the next condensation site, the concentration of morpholine is lower, so the condensate pH is lower. At the last condensation site, where the remaining steam is condensed, little morpholine is left but most of the CO_2 is still present. The high CO_2

concentration depresses the pH, promoting acidic attack of the metal oxide layers.

An amine that is more likely to distribute into the steam, such as cyclohexylamine, is a better choice for the system described above. However, amines with a high tendency to distribute into the steam are not always the best choice.

For example, if cyclohexylamine is used in a second condensate system with two consecutive condensation sites having acetic acid as a contaminant, most of the acetic acid goes into the liquid phase at the first condensation site, while most of the cyclohexylamine remains in the steam. This results in low pH in the first condensation site liquid phase. At the second site, where total condensation takes place, the pH is high. A morpholine–cyclohexylamine blend is a better choice for this system.

In practice, the best protection is provided by blended products containing a variety of amines with differing steam/liquid distributions.

To compare the relative steam/liquid distribution of amines, the distribution ratio (DR) is traditionally used. The distribution ratio of an amine is:

$$DR = \frac{\text{amine in vapor phase}}{\text{amine in liquid phase}}$$

Amines with a DR greater than 1.0 produce a higher concentration of amine in the vapor phase than in the liquid phase. Conversely, amines with a distribution ratio less than 1.0 produce a higher concentration of amine in the liquid phase than in the vapor phase.

Distribution ratios are not true physical constants but are a function of pressure (Figure 19-6) and pH. The effect of temperature and pH of the condensation site must also be considered. In a complex condensate system, the distribution of chemicals between different condensation sites is difficult to estimate without the use of computer modeling packages specifically designed for this purpose.

Thermal Stability. All organic chemicals exposed to a high-temperature, alkaline, aqueous environment eventually degrade to some degree. Most amines eventually degrade to carbon dioxide and/or acetic acid and ammonia. The time required varies greatly with different amines. The most stable amines commonly used are morpholine and cyclohexylamine. These remain substantially intact at pressures up to 2500 psig.

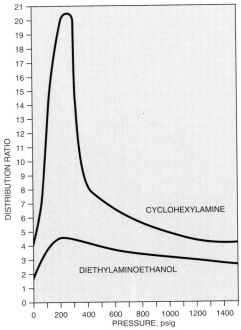

Figure 19-6. Graph shows how the distribution ratios of cyclohexylamine and diethylaminoethanol vary with pressure.

Quantity. The quantity of neutralizing amine required depends on the carbon dioxide content of the condensate at specific locations and the degree of corrosion protection desired. Complete neutralization is achieved if the condensate pH in all portions of the system is above 8.3. From a practical standpoint, it is necessary to establish a pH control range that provides the desired protection for the most critical system components.

The degree of protection can be monitored by various means. Corrosion test specimens installed in bypass racks, corrosion product analyses, corrosion rate meters, and submicron corrosion product filtration are some of the effective monitoring tools that may be employed.

The behavior of amine bicarbonate in the deaerator affects amine requirements for the system. Although they are soluble in most cases, amine bicarbonates remain associated in the condensate. In an ideal situation, the amine bicarbonate entering the deaerator breaks down with subsequent venting of carbon dioxide to the atmosphere and recirculation of the amine back to the boiler. Actual behavior involves some loss of amine additive and some recirculation of carbon dioxide. The amounts of lost amine and retained carbon dioxide are a function of the amine bicarbonate stability in the deaerator.

Filming Amines

Another approach to controlling condensate system corrosion is the use of chemicals that form a protective film on metal surfaces (Figure 19-7). This approach has come into widespread use with the development of suitable products containing long-chain nitrogenous materials.

Filming amines protect against oxygen and carbon dioxide corrosion by replacing the loose oxide scale on metal surfaces with a very thin amine film barrier. During the period of initial film formation, old, loosely adherent corrosion products are lifted off the metal surface due to the surfactant properties of the amine. The metal is cleansed of oxides, which normally cling very tightly and can build up over long periods of time. Excessive initial filming amine treatment of old, untreated or poorly treated condensate systems can cause large amounts of iron oxide to be sloughed off, plugging traps and return lines. Therefore, treatment must be increased gradually for old systems.

When contaminants are present in the condensate, filming amines have a tendency to form deposits by reacting with multivalent ions, such as sulfate, hardness, and iron. Overfeed of filming

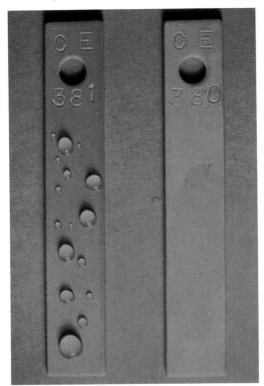

Figure 19-7. Test specimen 381 shows the nonwettable surface produced by an effective filming amine. Specimen 380 is untreated.

amines and excessive oxygen contamination can also contribute to deposit formation. For maximum efficiency, filming amines should be fed directly to the steam header.

Advances have been made in formulating filming amine treatments. Straight filming amines containing one ingredient, such as octadecylamine, are effective but often fail to cover the entire system and can produce fouling. Emulsifiers and, in some cases, small amounts of neutralizing amines can be added to improve film distribution by providing more uniform coverage. This increases system protection and reduces the fouling potential. Application experience has shown that combination amines (filming and neutralizing amines with dispersant aids) provide a superior film bond, reduce deposit problems, and provide better system coverage and thus provide more complete and economical corrosion protection (Figure 19-8).

The feed of filming amines is usually based on steam throughput. Different levels of treatment are required, depending on the particular blend in use. As in the case of neutralizing amines, various methods are used to monitor the effectiveness of the treatment, including corrosion test specimens installed in bypass coupon racks (Figure 19-9), iron analyses, corrosion rate meters, and submicron corrosion product filtration.

Oxygen Scavenging and Metal Passivation

Where oxygen invades the condensate system, corrosion of iron and copper-bearing components can be overcome through proper pH control and the injection of an oxygen scavenger.

One important factor to consider in choosing an oxygen scavenger for condensate treatment is its reactivity with oxygen at the temperature and pH of the system. A scavenger that removes

Figure 19-9. Test specimen bypass rack used to monitor amine treatment. The corrosion rate meter on the right measures relative corrosiveness instantaneously.

oxygen rapidly provides the best protection for the condensate metallurgy. Hydroquinone has been shown to be particularly effective for most systems.

Like those of neutralizing amines, the steam/liquid distribution of each scavenger has a unique temperature dependence. Some scavengers, such as ascorbic acid and hydrazine, have a very low volatility. Therefore, it is necessary to feed them close to a problem area. An example of this is the injection of hydrazine to the exhaust of a turbine to protect the condenser. Other scavengers, such as hydroquinone, are relatively volatile and can be fed well upstream of a problem area.

The use of neutralizing amines in conjunction with an oxygen scavenger/metal passivator improves corrosion control in two ways. First, because any acidic species present is neutralized and pH is increased, the condensate becomes less corrosive. Second, most oxygen scavenger/passivators react more rapidly at the mildly alkaline conditions maintained by the amine than at lower pH levels. For these reasons, this combination treatment is gaining wide acceptance, particularly for the treatment of condensate systems that are contaminated by oxygen.

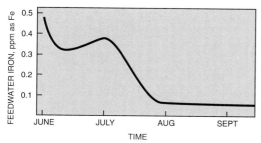

Figure 19-8. Feedwater iron determination is one method of monitoring the effectiveness of an amine treatment program. The application of a combination neutralizing–filming amine treatment reduced feedwater iron from 0.5 ppm to 0.05 ppm in just 2 months.

FIRESIDE PREBOILER SYSTEMS

The heart of any boiler is its fuel system. Fuel handling and storage problems can limit the efficiency of the entire boiler.

There are three forms of boiler fuel: liquid, solid, and gas. Methods of handling and storage vary according to the type of fuel used.

GASEOUS FUELS

Clean and relatively moisture-free gaseous fuels present little difficulty in handling. Natural gas is a prime example of clean fuel and is the one most commonly used in boiler systems.

Dirty gas, such as refinery gas, blast furnace gas, carbon monoxide gas, and other waste gases from processes, can cause significant handling problems. Special measures must be taken to prevent escape to the surrounding atmosphere, fire, fuel line deposition, moisture, and incomplete combustion. The severity of the problem depends on the specific contaminants in the gas. Methods of handling are selected according to the nature of the specific gas and local regulations.

Wet scrubbers, electrostatic precipitators, chemical dispersants, and suitable pipe and tank lagging may be used to eliminate problems. Wet scrubbers and electrostatic precipitators remove contaminants mechanically. Chemical dispersants have been used in conjunction with wet scrubbers and electrostatic precipitators to handle troublesome contaminants. Because leakage is one of the major concerns in handling gases, some method of leak detection is essential. The selected method may be as simple as detector-type sprays or as sophisticated as combustible detectors with alarms, which may be connected to automatic fire control systems.

SOLID FUELS

Solid fuels (including coal, wood, and solid waste) present some of the same handling difficulties. Problems occur unless a free-flowing,

continuous supply of fuel that is properly sized for the specific type of combustion equipment is provided. The problems include sizing, shredding or pulverizing, consistency of moisture content, freezing or lumping, dusting, fires in storage due to spontaneous combustion, and fires in the feed or ash handling systems.

Most problems can be minimized or eliminated through proper selection of fuel handling equipment. Specific types of equipment for handling, storage, and preparation depend on the characteristics of the solid fuel used.

Because the proper equipment is not always available, fuel additives or aids have been used in the attempt to minimize problems. These additives include grinding aids, moisture improvers, dusting aids, freezing inhibitors, and catalysts to minimize combustibles in ash and fly ash handling systems.

LIQUID FUELS

Liquid fuels include waste oils, light oils, heavy oils, and other combustible liquids. Because of the problems of liquid residue disposal, an increasing variety of combustible liquids is being considered and tested. Figures 20-1 and 20-2 illustrate key components found in a typical

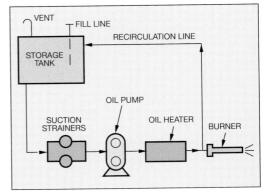

Figure 20-1. Typical preburner system flow diagram.

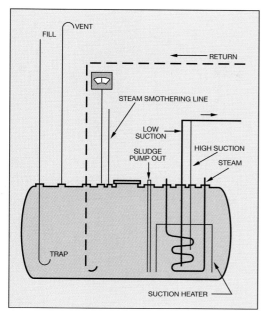

Figure 20-2. Typical arrangement of fuel oil storage tank.

liquid fuel handling system and fuel oil storage system, respectively.

Problems encountered in the handling, storage, and preparation of liquid fuels include water contamination, sludge formation, resistance to flow, biological growths, instability, and corrosive-ness. Generally, these conditions are manifested as excessive strainer plugging, poor flow, increased loading on the fuel pump, heater deposits, fuel line deposits, loss of storage space, burner tip deposits, burner fouling, leakage due to storage tank corrosion, poor atomization, and other combustion problems. Table 20-1 summarizes the nature and cause of problems associated with key liquid fuel handling system components; some of these conditions are illustrated in Figures 20-3 through 20-6.

WATER

Water can become emulsified in oil during processing and can also be introduced into oil during handling by condensation, contamination, or leakage. The presence of water can lead to many problems:

- water that separates from oil is generally acidic and can readily corrode storage tanks, particularly at the oil/water interface

- separated water occupies useful storage space

- water at the burner can cause oil flow discontinuity

- if it enters the firing system in sufficient quantity, water may cause thermal shock, leading to refractory damage

Table 20-1: Preboiler problems and their causes.

Location	Problem	Cause
Storage Tank	sludge reduces available volume of fuel oil storage	sludge may have resulted from more than 7 days' storage or from using several oil sources
	acidic corrosion at oil/water interface	water entered tank from condensation, heating coil leaks, open manholes, and/or surface drainage
Suction Strainers	frequent cleaning necessary	sludge is being carried over from storage tank
	loss of oil pump suction	sludge is being carried over from storage tank
Oil Heater	plugging	sludge is being carried over from storage tank; oil has become polymerized
	loss of oil temperature	
	varying oil temperature	
Burner	poor atomization	high oil viscosity; sludge is being carried over from storage tank; high burner temperature; water in oil
	distorted flame patterns	
	reduced maximum load	
	frequent cleaning necessary	
	difficulty cleaning burner	
	oil flow discontinuity	

Figure 20-3. Fuel strainer with deposits.

Figure 20-4. Fuel strainer after cleaning.

Most water can be eliminated by means of careful shipping and handling procedures. Proper design and maintenance of equipment can also minimize water leaks, such as those resulting from steam heater or tank leaks. To minimize condensation, proper lagging and heating of the tank are necessary.

Properly selected additives can be used to emulsify small amounts of water (up to approximately 1%) economically. Large quantities of water should be removed physically from the tank by draining or pumping. When large amounts of water are detected, the source should be identified and corrected.

Water in storage tanks can be detected by means of a Bacon Bomb. This device is lowered into a tank and opened to allow sampling at any point beneath the surface. The sample can then be evaluated by testing for bottom sediment and water (BS&W). Testing for water can be as simple as waiting for the sample to settle in a graduate.

SLUDGE

Sludge is composed of settled heavy agglomerates combined with suspended matter from an oil or liquid fuel. Sludge formation increases when fuel

oils of different crudes or liquid fuels of different sources are mixed. When fuel oils are heated in a tank to ensure good flow, the likelihood of sludge formation is increased. If the heat is high enough to break the water in oil emulsion, the heavier agglomerates may settle. Sludge formation in the tanks reduces storage space for usable fuel and removes part of the high energy containing components of the fuel. Frequent strainer cleaning is required to prevent high pressure drops and ensure good flow. Sludge formation can also cause heater burner tip fouling. Sludge can be detected in a storage tank by Bacon Bomb sampling of the tank.

In order to mix settled sludge with new fuel, it helps to fill the storage tank from the bottom. A tank with effective lagging (outer insulation) is less susceptible to sludge buildup. Long-term storage (over 7 days) should be avoided and some method of recirculation employed to keep the heavy agglomerates mixed. Where mechanical methods are not completely effective and/or some degree of help is desired, additives are effective in dispersing sludge, even at low use rates. The additives help clean fouled storage tanks and heater and burner assemblies, on-line or off-line.

Figure 20-5. Clean burner nozzle.

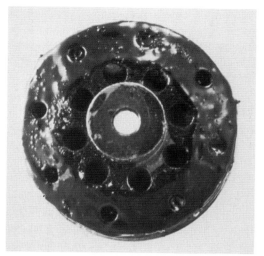

Figure 20-6. Burner nozzle affected by suspended matter.

Many advantages result when fuel sludge is minimized:

■ Increased energy content in the fuel increases efficiency.

■ Clean tanks afford maximum utilization of storage space.

■ Clean feed lines, heaters, and burners are essential to good flow and good combustion.

■ Proper fuel conditioning produces good flame patterns, reducing the chances of flame impingement.

■ Proper fuel flow allows the boiler to operate at maximum capacity and helps control the formation of burner deposits. It also allows operation at minimum excess air, which helps control slagging and high-temperature corrosion.

FUEL FLOW

The resistance to flow of a liquid fuel is a function of viscosity and pour point. A higher temperature lowers viscosity and increases the ability of the fuel to flow. It is important to maintain the correct temperature range in a storage tank to ensure good fuel flow and prevent vaporization of the light oil fraction.

One method used to control pour point involves blending of oils with differing pour points. Care must be taken because the pour point of a mixture can be higher than that of either of the two components. As a result, solidification and crystallization can occur in the fuel, which can completely plug a fuel pump. Such plugging is almost impossible to remove through normal cleaning methods. High temperature and constant motion are advisable when high pour point fuels are used.

Viscosity also affects atomization of the oil by the burner. Proper atomization requires lower fuel viscosities than those needed for good fuel flow. This lower viscosity is obtained by heating. Only preliminary heating may take place while the oil is in its storage tanks, to prevent vaporization of the lighter oil fractions. Following preheating, the oil flows to an in-line heater where it is heated to a temperature suitable for good atomization.

CORROSION OF TANKS AND BACTERIAL GROWTH

Water that separates from fuel oils is almost always acidic. Although most hydrocarbons are protective by nature, corrosion can still be found at the water/oil interface. In practice, an alkaline material or amine-type additive is added to protect metal surfaces.

Because heavier fuel oils do not provide life-sustaining nutrients, they rarely experience bacterial growths. In lighter fuel oils, such as diesel fuels, bacteriological growth has been found at the water/oil interface. In order to prevent this occurrence, it is necessary to use water-free fuel oils or to prevent separation of water from the fuel. When bacteriological growths are not prevented, they can be controlled by antimicrobials.

Corrosion test specimens can be installed at the water/oil interface to monitor corrosion, eliminating the need for periodic equipment inspections. Bacterial growth monitoring requires sampling at the water/oil interface and testing for bacteria count.

SPECIFIC EQUIPMENT PROBLEMS

Fuel strainers protect the fuel system from problems with heavy agglomerates and suspended matter. The strainers are relatively coarse, because fine straining can retard fuel flow and increase required cleaning frequency.

Fuel oil pump selection should be based on the anticipated suspended solids in the fuel, along with the type of additive to be used for fireside protection. For example, a constant differential pump operates at a constant rate, and varying amounts of unused oil are recirculated as loads vary. This recirculation, together with certain additives, may increase the amount of suspended solids in the fuel, thereby increasing the clearance tolerances needed. Naturally, the tolerances can be much closer for very light, clean oils than for heavier fuels or fuels carrying more suspended solids.

Burner nozzles are affected by suspended matter and are subject to wear. The effect of wear on the nozzles can be determined by observation of the flame pattern or by means of "go or no-go" gauges.

In-line heaters and burner tips develop clogging problems due to high temperatures, which cause some solidifying of heavier hydrocarbons. Problems with the in-line heater are revealed by pressure drop across the heaters, a decrease in fuel oil temperature, or an increase in the steam pressure required to maintain the same oil temperature. Burner tip deposits are indicated by distorted flame patterns or inability to achieve maximum load due to restricted flow.

Proper fuel handling must be maintained to ensure optimal conditions and thereby minimize these problems. In addition, periodic cleaning is often necessary. When the required frequency of cleaning is excessive, an additive may be used to help keep the heavy agglomerates dispersed and flowing easily.

SAFETY

Liquid fuels require care in handling to maximize safety. Potential problems include contamination from spills or leaks and escape of combustible vapors. Tank areas should be diked to contain any spills. To guard against fire, special tank construction is necessary and combustible vapor monitors should be used. Combustible monitors may be integrated into fire control systems.

BOILER FIRESIDE DEPOSIT AND CORROSION CONTROL

Impurities in fuels can cause deposit formation and fireside metal surface corrosion. Compounds of aluminum, barium, copper, iron, magnesium, manganese, and silica have all been used to control combustion fouling and corrosion. The most severe problems are generally found in combustion equipment firing fuels that significantly deviate in composition from the fuel on which the equipment design was based.

SOURCE AND NATURE OF FIRESIDE DEPOSITION

Fireside fouling of combustion equipment is caused by the deposition of fuel ash components. Table 21-1 shows analyses of typical high, medium, and low ash liquid fuels. Oils containing more than 0.05% ash are considered high ash oils; those containing less than 0.02% ash are considered low ash oils.

Sulfur emission regulations have severely restricted the use of high sulfur oils. Generally,

high sulfur oils (greater than 1.0% sulfur) have high ash contents. These oils are usually imported from the Caribbean area. Prior to 1972, most East Coast boilers were burning high sulfur, high ash oils.

The combustion of high ash oils produced troublesome deposits on boiler convection surfaces such as steam generating, superheat, reheat, and economizer sections. The firing of high ash oils (even in those units which were originally designed to burn coal) produced convection surface deposition that was difficult to remove by soot blowing.

The most troublesome oil ashes are those which contain a vanadium/sodium ratio of less than 10:1. In Table 21-1, the fuel with the medium ash content has a relatively low vanadium/sodium ratio and produced an extremely tenacious deposit on superheater surfaces (as shown in Figure 21-1).

Vanadium reacts with oxygen to form various oxides in the furnace. When formed, vanadium pentoxide condenses within the furnace when gas temperatures approach its solidification point.

Sodium also reacts with oxygen and sulfur trioxide to form fouling compounds. Thermo-

Table 21-1. Typical residual oil analyses.

	High Ash	Medium Ash	Low Ash
Specific Gravity, at 60 °F	0.9548	0.9944	0.9285
Viscosity SSF at 122 °F, sec	240	200	100.5
Calorific Value, Btu/gal	147,690	152,220	147,894
Bottom Sediment & Water, %	0.1	0.4	0.1
Sulfur, %	1.93	2.26	0.62
Ash, %	0.06	0.04	0.02
Vanadium, ppm	363	70	6
Sodium, ppm	16	50	9
Nickel, ppm	48	19	14
Aluminum, ppm	9	1	10
Iron, ppm	12	3	1

Figure 21-1. Pendant superheater slag with corrosion product underneath.

dynamics favor the formation of sodium sulfate rather than the relatively unstable sodium oxide.

Nickel can also contribute to deposition by forming oxides. Aluminum can be present in the oil in the form of cracking catalyst fines (aluminum that is introduced during the refining of oil). Although it is generally not a troublesome component, iron is occasionally present in fuels at relatively low levels. When waste fuels are burned, some contaminants (such as lead) can cause extremely dense and tenacious deposits.

Black liquor has been used as a boiler fuel for many years in the Kraft paper industry. It contains a significant amount of combustible material, along with sodium salts. Black liquor burning can produce relatively adherent, low bulk density sodium sulfate deposits on recovery boiler convective surfaces. In some instances, it is advisable and economically beneficial to control or limit deposition with magnesium-based additives. These materials are blended compounds which contain magnesium oxide and/or aluminum oxide.

The combustion of coal tar in boilers often produces objectionable amounts of sodium salts and/or iron compounds on boiler convective surfaces. Often, ash deposits are similar in composition to coal ash.

Combustion of solid fuels such as coal and bark (also referred to as "hog fuel") can also lead to fireside slagging. Sodium, calcium, silica, iron, and sulfur content are of primary concern in the burning of solid fuels. Other metal oxides, such as alumina, titania, and potassium oxide, can also aggravate slagging. Figure 21-2 shows a heavy accumulation of slag in a coal-fired boiler.

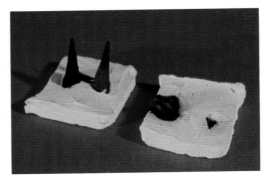

Figure 21-3. Ash fusion testing can be used to indicate slagging potential of fuel ash.

In addition to proper boiler furnace design, operational considerations should be explored to prevent fireside slagging. Ash fusion testing has been used by boiler manufacturers to assess slagging potential (see Figure 21-3). Where design and operational improvements are impractical or insufficient, chemical treatments (such as combustion catalysts and antifoulants) should be considered.

TREATMENT FOR FOULING

Additives are used to control fouling by elevating the melting point of the deposits, by physically diluting deposits, or by providing a shear plane to assist in removal by soot blowing. Additives used to control fouling contain magnesium, silica, manganese, and/or aluminum. Figure 21-4 illustrates the effect of treatment.

The melting point of untreated ash constituents can be as low as 1000 °F. The introduction of

Figure 21-2. This far wall of a coal-burning boiler is heavy with slag composed primarily of aluminum silicates, with some iron oxide.

Figure 21-4. Treatment can make deposits friable for easy removal.

appropriate metal oxides elevates the melting point of ash components by several hundred degrees. The additive components most commonly used to raise the deposit melting point are magnesium and/or aluminum. Dosages depend on ash levels in the fuel and the ratio of various ash components.

Fuel additives are intended to control fouling by forming a friable, nonadherent deposition that can be removed by soot blowing.

When melting temperatures are raised, the physical characteristics of the deposits are altered. Often, the heaviest deposition occurs in areas where the gas temperature is lower than the melting temperature of a deposit. Therefore, a treatment program designed solely to elevate the melting point of the deposit will not solve the problem, and it is necessary to introduce additive components that change the physical characteristics of the deposit, making it more friable. Additives used for this purpose contain compounds of magnesium or aluminum. Aluminum is usually the most effective material for increasing friability of deposits.

Newer technology has been developed that offers certain advantages over magnesium or magnesium/aluminum combination treatments. Silicon-based materials have been used to control fouling and form friable nonadherent oxides. The silicon component acts like a sponge to adsorb low melting point oxides, preventing their agglomeration and subsequent deposition. The silicon-based materials contribute far fewer solids than traditional treatment. This reduces fouling and minimizes loading on the solids collection system.

Additive dosage should be fixed at the minimum level necessary for proper conditioning of deposits. Overfeeding of additives, particularly magnesium oxide, can cause troublesome deposition. Magnesium oxide can react with sulfur trioxide in the flue gas to form magnesium sulfate deposits. These deposits generally form in convection areas, such as primary superheaters, steam generating sections, and economizers.

Boiler design has proven to be a very important consideration in the determination of magnesium additive levels. The addition of aluminum to magnesium-based additives reduces magnesium sulfate deposition.

The greatest need for fuel additives to control deposition is in boilers previously used to fire other fuels, such as natural gas or coal. Generally, units used to fire coal have sufficient soot blower placement for adequate removal of properly conditioned deposits.

Natural gas-burning units generally do not have the desired soot blowing equipment. It is advisable to contact the manufacturer so that the proper number of soot blowers and their location can be specified. It is virtually impossible to burn a fuel containing greater than 0.2% ash without the use of soot blowers, even with the aid of a conditioning additive.

Boilers designed for oil firing incorporate design parameters that allow for relatively trouble-free operation from a fireside standpoint. Combustion gas velocity, tube spacing, furnace gas exit temperature, and economizer configuration are adjusted to account for oil ash characteristics. The most common problems associated with oil-design boilers are air preheater fouling and corrosion. Deposition and corrosion can occur when the oil used is of lesser quality than specified for the boiler, or if multiple oils are used.

HIGH-TEMPERATURE FIRESIDE CORROSION

Fuel ash corrosion in high-temperature areas can cause extensive boiler damage. The corrosion is caused principally by complex oxide-slags with low melting points. Corrosion by slag components, such as sodium vanadyl vanadate, progresses rapidly between 1100 and 1650 °F.

Sodium sulfate is also a primary corroding medium that can be present with sodium vanadyl vanadate. It also is the primary corroding material of gas turbine blades. High-temperature corrosion can proceed only if the corroding deposit is in the liquid phase and the liquid is in direct contact with the metal. The liquid-phase deposits corrode by fluxing of the protective oxide layer on the metal surface as shown in Figure 21-1. Deposits also promote the transport of oxygen to the metal surface. Corrosion is caused by the combination of oxide layer fluxing and continuous oxidation by transported oxygen.

Tube spacers, used in some boilers, rapidly deteriorate when oil of greater than 0.02% ash content is fired. Because these spacers are not cooled and are near flue gas temperatures, they are in a liquid ash environment. Fuel oil additives can greatly prolong the life of these components.

High-temperature corrosion caused by

sulfidation is a common problem in gas turbines. The corroding medium is sodium sulfate. Sodium enters the gas turbine either with the fuel or combustion air. The sodium content of the fuel can be lowered by means of water washing. It is relatively difficult to remove all sodium-containing mists or particulates at the intake of the gas turbine. Due to salt water mist ingestion, blade sulfidation problems are common in gas turbines used for marine propulsion.

Reheating furnaces in the iron and steel industry have suffered wastage of metallic recuperators. The problem is more prominent where steel mills have converted reheat furnaces from natural gas to residual oil firing. The gas inlet temperature at the recuperator is high enough to provide an environment for high-temperature corrosion.

Firing of residual oil in refinery process heaters has caused numerous corrosion failures of convection tube support members. The tube supports are generally not cooled by water or air and, therefore, are at a temperature close to that of the gas. In some cases, tube supports fabricated from nickel and chromium alloys have been installed to alleviate this problem.

Additives containing magnesium and aluminum oxides have been successful in controlling these and other high-temperature corrosion problems. The additives function by elevating the melting point of the deposits. Enough magnesium oxide must be added to enable the metallic constituents to remain in the solid phase as they contact the metal surface.

In conjunction with an additive program to control high-temperature corrosion, it is advisable to operate the unit at minimum excess air levels. The corrosion rate is influenced by the oxygen content of the combustion gas. The second phase of the corrosion mechanism is oxidation; therefore, the partial pressure of oxygen in the combustion gas can be lowered to reduce corrosion rates.

In residual oil-fired equipment, it is often necessary to feed a 1:1 ratio of magnesium and vanadium to achieve low corrosion rates. This level of feed is generally higher than that which is necessary to control fouling.

COMBUSTION CATALYSTS

Combustion catalysts have been used for all types of fuels. The combustion catalyst functions by increasing the rate of oxidation of the fuel. Some fuels are difficult to burn within a given fixed furnace volume. Combustion catalysts are applied to these fuels to comply with particulate and opacity regulations. Combustion catalysts are also used to improve boiler efficiency by reducing carbon loss in the flue gas.

FUEL ADDITIVES

Most soluble fuel additives contain metallo-organic complexes such as sulfonates, carbonyls, and naphthenates. These additives are in a very convenient form for feeding. Most dry fuel additive preparations are used to treat fuels with high ash content, such as coal, bark, or black liquor. Metal oxides are used for this purpose.

COLD-END DEPOSITION AND CORROSION CONTROL

Cold-end corrosion can occur on surfaces that are lower in temperature than the dew point of the flue gas to which they are exposed. Air heaters and economizers are particularly susceptible to corrosive attack. Other cold-end components, such as the induced draft fan, breeching, and stack, are less frequently problem areas. The accumulation of corrosion products often results in a loss of boiler efficiency and, occasionally, reduced capacity due to flow restriction caused by excessive deposits on heat transfer equipment.

Acidic particle emission, commonly termed "acid smut" or "acid fallout," is another cold-end problem. It is caused by the production of large particulates (generally greater than 100 mesh) that issue from the stack and, due to their relatively large size, settle close to the stack. Usually, these particulates have a high concentration of condensed acid; therefore, they cause corrosion if they settle on metal surfaces.

The most common cause of cold-end problems is the condensation of sulfuric acid. This chapter addresses problems incurred in the firing of sulfur-containing fuels. Sulfur in the fuel is oxidized to sulfur dioxide:

$$\underset{\text{sulfur}}{S} + \underset{\text{oxygen}}{O_2} = \underset{\substack{\text{sulfur} \\ \text{dioxide}}}{SO_2}$$

A fraction of the sulfur dioxide, sometimes as high as 10%, is oxidized to sulfur trioxide. Sulfur trioxide combines with water to form sulfuric acid at temperatures at or below the dew point of the flue gas. In a boiler, most of the sulfur trioxide reaching the cold end is formed according to the following equation:

$$\underset{\substack{\text{sulfur} \\ \text{dioxide}}}{SO_2} + \underset{\text{oxygen}}{\tfrac{1}{2}O_2} = \underset{\substack{\text{sulfur} \\ \text{trioxide}}}{SO_3}$$

The amount of sulfur trioxide produced in any given situation is influenced by many variables, including excess air level, concentration of sulfur dioxide, temperature, gas residence time, and the presence of catalysts. Vanadium pentoxide (V_2O_5) and ferric oxide (Fe_2O_3), which are commonly found on the surfaces of oil-fired boilers, are effective catalysts for the heterogeneous oxidation of sulfur dioxide. Catalytic effects are influenced by the amount of surface area of catalyst exposed to the flue gas. Therefore, boiler cleanliness, a reflection of the amount of catalyst present, affects the amount of sulfur trioxide formed.

The quantity of sulfur trioxide in combustion gas can be determined fairly easily. The most commonly used measuring techniques involve either condensation of sulfur trioxide or adsorption in isopropyl alcohol. Figure 22-1 is a curve showing the relationship of sulfur trioxide concentration to dew point at a flue gas moisture content of 10%. Higher flue gas moisture increases the dew point temperature for a given sulfur trioxide–sulfuric acid concentration. Cold-end metal temperatures and flue gas sulfur trioxide content can be used to predict the potential for corrosion problems.

At the same sulfur content, gaseous fuels such as sour natural gas, sour refinery gas, and coke oven gas produce more severe problems than fuel oil. These gases contain more hydrogen than

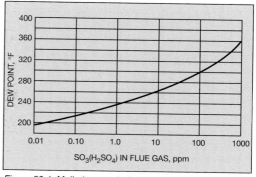

Figure 22-1. Muller's curve indicates acid dew point.

fuel oil, and their combustion results in higher flue gas moisture. Consequently, dew points are raised. With any type of fuel, corrosion and fouling potentials rapidly increase below gas temperatures of 140 °F (60 °C), which is the typical water dew point for flue gases.

Cold-end corrosion and deposition are usually much less severe in coal-fired boilers than in oil-fired units. Usually, coal ash is alkaline, so it increases the pH of the deposits formed in cold-end sections. Thus, the extent of the corrosive attack by sulfuric acid is diminished. Also, the high level of ash present when coal is fired results in a lower concentration of acid in the ash particle. At the same sulfur content, coal firing dew points are generally 20–40 °F lower than oil firing dew points.

The most common cause of deposition within air preheaters is the accumulation of corrosion products. Most air preheater deposits contain at least 60% iron sulfates formed by the corrosion of air heater tube metal. Therefore, a reduced corrosion rate frequently reduces the fouling of air preheaters.

MECHANICAL AND OPERATIONAL SOLUTIONS

Air Preheaters

A regenerative air preheater can reduce cold-end problems when installed instead of a recuperative air preheater on a new or existing boiler. In the regenerative air preheater design, heat transfer surfaces are below the acid dew point for much shorter periods of time.

Most modern regenerative air preheaters are equipped with steam or compressed air sootblowers and fixed or oscillating water washing nozzles. In boilers equipped with multiple units, individual air preheaters can be isolated and washed on-line. Suitable drain connections must be provided as well as a system for treating the wash water prior to disposal. Washing is generally continued until the pH of the wash water is above 4.5. The wash water effluent is a relatively low pH stream with a high soluble iron content. Most air preheaters are washed with untreated water. Some operators add caustic soda or soda ash to neutralize the deposits and lower the loss of air heater metal during washing.

The average cold-end temperature of an operating air preheater is the sum of combustion air inlet temperature and flue gas outlet temperature, divided by two. The average cold-end temperature

is generally used in the assessment of potential problems and the selection of air preheater size and materials of construction. The average cold-end temperature of an operating air preheater must be maintained in accordance with the manufacturer's specifications. Corrosion-resistant materials are used in some regenerative air preheater cold sections to obtain the lowest possible stack gas temperature and consequently the highest boiler efficiency.

Steam Coil Air Preheaters

In some installations, heating coils are placed between the forced draft fan outlet and the air preheater inlet to accommodate seasonal fluctuations in incoming combustion air temperature. These heat exchangers are commonly termed "steam coil air preheaters." They maintain the average cold-end temperature of the air preheater above the acid dew point. Where steam coils are used, the temperature of the combustion air entering the air heater is independent of the ambient temperature.

Steam coil air preheaters are also installed when boilers are changed from coal or gas firing to oil firing. Steam coils are installed because oil firing requires maintenance of an air preheater average cold-end temperature that is higher than that normally specified for the firing of natural gas or coal. The operation of steam coil air preheaters results in an increase in the heat rate of the steam plant. Combustion air bypasses around the air heater and hot air recirculation have also been used to control average cold-end temperatures. Both of these methods reduce boiler efficiency.

Minimizing Air Infiltration

The operation of a boiler at or below 5% excess air can result in a marked reduction in flue gas sulfur trioxide content and dew point. An experimentally determined relationship for one boiler is shown in Figure 22-2. The infiltration of air into the flame zone or into an area where the catalytic oxidation of sulfur dioxide is occurring increases the potential for cold-end problems. Therefore, maintenance and inspection procedures should be directed toward minimizing air infiltration.

Minimizing Flue Gas Moisture Content

As previously stated, the dew point is not only affected by the partial pressure of sulfuric acid in the flue gas but also by the partial pressure of water in the flue gas. The minimum obtainable flue gas moisture content is determined by the

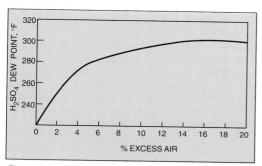

Figure 22-2. *H_2SO_4 dew point as a function of burner excess air (less than 3% sulfur fuel oil).*

moisture content of the fuel and combustion air and by the hydrogen content of the fuel.

The moisture content of coal is somewhat controllable through proper handling and storage procedures. Handling and storage specifications can be written limiting the moisture content of fuel oil. Factors that increase flue gas moisture content include:

- boiler tube leaks
- steam coil air preheater leaks
- excessive boiler or air heater soot blowing
- leaking water wash nozzles
- instrumentation leaks

When two fuels (such as coal and oil, oil and natural gas, or blast furnace gas and coke oven gas) must be fired simultaneously, certain ratios produce the highest dew points. The worst ratio on a Btu-fired basis is 1:1.

When a fuel is fired that has a higher hydrogen content than the base fuel normally used, the flue gas produced has a higher moisture content, resulting in an increased dew point. When possible, fuels of different hydrogen content should be fired separately. Figure 22-3 graphically depicts the influence on sulfuric acid dew point that results when natural gas and a sulfur-containing fuel are fired simultaneously in a single boiler.

CHEMICAL TREATMENT

Many chemical solutions have been devised to control cold-end deposition and corrosion. These solutions can be divided into two broad classifications: fuel additives and cold-end additives. Fuel additives are compounds that are added directly to the fuel or combustion process. Cold-end additives are fed into the back of the boiler after steam-generating surfaces so that they specifically treat only the lower-temperature areas.

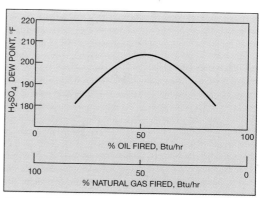

Figure 22-3. *Effect on H_2SO_4 dew point of firing natural gas and 1.0% sulfur fuel oil.*

Fuel Additives

Magnesium and magnesium/aluminum based fuel additives are used to reduce sulfur trioxide in the flue gas. These compounds function primarily by altering the effectiveness of the iron and vanadium catalysts. They are fed to liquid fuels, most commonly residual fuel oil. Alkaline fuel oil additives, such as magnesium, also increase the pH of deposits formed on cold-end surfaces, thereby reducing corrosion.

Cold-End Additives

Cold-end corrosion and deposition can be controlled more economically and effectively through the use of cold-end additives. Cold-end additives include sulfuric acid neutralizing agents and corrosion inhibitors.

Magnesium Compounds. Alkaline magnesium compounds, such as magnesium oxide and magnesium carbonate, are fed to reduce the sulfur trioxide content of flue gases. These compounds are fed in high-temperature areas, such as primary superheater sections. The reaction product formed, magnesium sulfate, often increases deposition within air preheaters.

The main benefit of magnesium compound injection is a reduction in air preheater corrosion. Often, the level of fouling is not appreciably altered, because the corrosion product fouling is replaced by magnesium sulfate fouling. Therefore, where magnesium compounds are used, suitable water wash nozzles must be present to permit periodic removal of magnesium sulfate.

Additives that remove sulfur trioxide from flue gas must be fed in stoichiometric quantities with respect to the amount of sulfur trioxide to be removed. Therefore, higher levels of sulfur in the fuel require higher feed rates for protection. Coal-fired boilers require less treatment than oil-fired boilers for a given sulfur level in the fuel.

Corrosion Inhibitors. Corrosion inhibitors can be added to the cold end of the boiler to alleviate problems associated with the condensation of sulfuric acid. These materials do not neutralize the sulfuric acid in the flue gas; they prevent surface corrosion where the sulfuric acid condenses. Fouling of the air preheater is reduced because the quantity of corrosion products is reduced. Although the dosage of inhibitor required to achieve the desired effect increases with increasing acid content in the flue gas, the relationship is not linear.

The compositions of inhibitor-type cold-end additives are usually proprietary. Products are available in powder and liquid form. Liquid solutions are injected upstream of the problem area with atomizing spray nozzles.

Justification for cold-end additives is generally based on the benefits obtained by higher unit heat rates and lower maintenance costs for the equipment in the cold-end section. The feed of cold-end additives enables the unit to operate with a lower rate of steam flow to the steam coil air preheaters, resulting in an increase in unit heat rate. If average cold-end temperature is controlled with bypasses, the bypassed air flow can be reduced so that an improvement in boiler efficiency is obtained. A smaller improvement in heat rate is gained through reduction of fan horsepower, which reduces the average pressure drop across the air preheater.

Acid Smut Control. Cold-end additives can be used to reduce acid smut problems. In some instances, it is believed that smut is created when fly ash particles agglomerate to form larger particles. These particles adsorb sulfuric acid mist and become highly acidic. Fly ash deposits often accumulate in low-temperature areas of breeching. During soot blowing or load changes, some of the deposited fly ash can be entrained in the flue gas stream and carried out the stack. The large particles then settle in the vicinity of the stack. Magnesium-based fuel additives have been beneficial in reducing acid smut problems by increasing the pH of the deposits.

Evaluation and Monitoring Techniques

Corrosion Rate Measurement. Various devices are available to assess the impact of additive application on corrosion rates. Table 22-1 shows a selection of monitoring methods. In

Table 22-1: Comparison of corrosion rate assessment methods.

Method	Variable Measured	Time Required	Instrument Cost	Advantages	Limitations
Corrosion coupon installed on surfaces	loss of weight of coupon	1–4 wk	low	simple; measures corrosion directly	not clear if location of coupon represents entire surface
Air-cooled corrosion probe for flue gas test	loss of weight of coupon	1–4 wk	low	simple; can determine corrosion as a function of temperature	temperature cannot be closely controlled without relatively high expense
Multipoint corrosion probe	short-range loss of iron	5–7 hr	low to moderate	can determine corrosion over a wide temperature range; rapid indication	difficulty in extrapolating short-range to long-range results
Sampling of flue gas	flue gas sulfur trioxide	4–5 hr	high	direct measurement of SO_3; then dew point temperature is accurate	more technical knowledge required to operate equipment; not a direct indication of corrosion
Electrical conductivity, dew point, and rate of acid buildup meter	temperature at which acid condenses on probe; rate of acid deposition below dew point	2–4 hr	high	quick; gives some idea of corrosion problem	low sulfur trioxide gives inaccurate and nonreproducible results; high dust loadings interfere with rate of acid buildup measurements; does not measure surface coating additive effect

some cases, problems in the breeching, induced draft fans, and stack can be measured by corrosion coupons placed within the flue gas stream.

For air heater corrosion and fouling problems, some provision must be made to maintain a corrosion specimen temperature that is within the ranges typically found in the operating air preheater. Corrosion probe methods (see Table 22-1) are used to simulate corrosion rates in an operating air preheater. Figure 22-4 shows a multipoint corrosion probe. The temperature to

be maintained on the specimen is determined by calculation of average cold-end temperature and measurement of dew point.

The electrical conductivity dew point meter is useful in problem assessment work and some results monitoring. In addition to corrosion monitoring, this meter provides an indication of the deposition rate by measuring the increase in the conductance of an acid-containing film with respect to time. The electrical conductivity meter is acceptable for results monitoring only where a sulfur trioxide removal additive is used. The dew point meter is shown in Figure 22-5.

Normally, direct measurement of sulfur trioxide is used only for Environmental Protection Agency tests because of its cost and complexity. All of these evaluation techniques and tools are used by suppliers of proprietary cold-end treatments. They enable the engineer to define the problem and measure results properly.

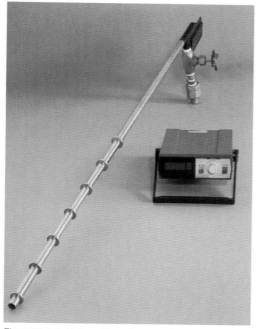

Figure 22-4. Acid deposition rate-measuring probe. Multipoint corrosion probe.

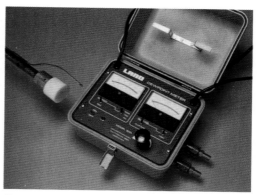

Figure 22-5. Instrument for measuring acid dew point temperature. (Courtesy of Land Combustion.)

Cooling Water Systems

Fresh water is a valuable resource that must be protected through proper management, conservation, and use.

COOLING WATER SYSTEMS—HEAT TRANSFER

The function of a cooling system is to remove heat from processes or equipment. Heat removed from one medium is transferred to another medium, or process fluid. Most often, the cooling medium is water. However, the heat transfer concepts and calculations discussed in this chapter can also be applied to other fluids.

Efficient removal of heat is an economic requirement in the design and operation of a cooling system. The driving force for the transfer of heat is the difference in temperature between the two media. In most cooling systems, this is in the range of 10–200 °F. The heat flux is generally low and in the range of 5,000 to 15,000 Btu/ft^2/hr. For exceptional cases such as the indirect cooling of molten metal, the heat flux can be as high as 3,000,000 Btu/ft^2/hr.

The transfer of heat from process fluids or equipment results in a rise in temperature, or even a change of state, in the cooling water. Many of the properties of water, along with the behavior of the contaminants it contains, are affected by temperature. The tendency of a system to corrode, scale, or support microbiological growth is also affected by water temperature. These effects, and the control of conditions that foster them, are addressed in subsequent chapters.

TYPES OF SYSTEMS

Water heated in the heat exchange process can be handled in one of two ways. The water can be discharged at the increased temperature into a receiving body (once-through cooling system), or it can be cooled and reused (recirculating cooling system).

There are two distinct types of systems for water cooling and reuse: open and closed recirculating systems. In an open recirculating system, cooling is achieved through evaporation of a fraction of the water. Evaporation results in a loss of pure water from the system and a concentration of the remaining dissolved solids. Water must be removed, or blown down, in order to control this concentration, and fresh water must then be added to replenish the system.

A closed recirculating system is actually a cooling system within a cooling system. The water containing the heat transferred from the process is cooled for reuse by means of an exchange with another fluid. Water losses from this type of system are usually small.

Each of the three types of cooling systems—once-through, open recirculating, and closed recirculating—is described in detail in later chapters. The specific approach to designing an appropriate treatment program for each system is also contained in those chapters.

HEAT TRANSFER ECONOMICS

In the design of a heat transfer system, the capital cost of building the system must be weighed against the ongoing cost of operation and maintenance. Frequently, higher capital costs (more exchange surface, exotic metallurgy, more efficient tower fill, etc.) result in lower operating and maintenance costs, while lower capital costs may result in higher operating costs (pump and fan horsepower, required maintenance, etc.). One important operating cost that must be considered is the chemical treatment required to prevent process or waterside corrosion, deposits and scale, and microbiological fouling. These problems can adversely affect heat transfer and can lead to equipment failure (see Figure 23-1).

Heat Transfer

The following is an overview of the complex considerations involved in the design of a heat exchanger. Many texts are available to provide more detail.

In a heat transfer system, heat is exchanged as two fluids of unequal temperature approach equilibrium. A higher temperature differential results in a more rapid heat transfer.

However, temperature is only one of many factors involved in exchanger design for a dynamic system. Other considerations include the area over which heat transfer occurs, the characteristics of the fluids involved, fluid velocities, and the characteristics of the exchanger metallurgy.

Process heat duty, process temperatures, and available cooling water supply temperature are

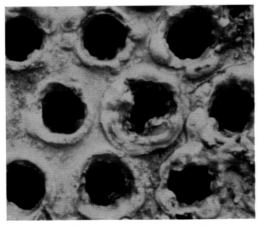

Figure 23-1. Fouling reduces exchanger's heat transfer efficiency.

usually specified in the initial stages of design. The size of the exchanger(s) is calculated according to important parameters such as process and water flow velocity, type of shell, layout of tubes, baffles, metallurgy, and fouling tendency of the fluids.

Factors in the design of a heat exchanger are related by the heat transfer equation:

$$Q = UA\,\Delta T_m$$

where

Q	=	rate of heat transfer (Btu/hr)
U	=	heat transfer coefficient (Btu/hr/ft^2/°F)
A	=	heat transfer surface area (ft^2)
ΔT_m	=	log mean temperature difference between fluids (°F)

The rate of heat transfer, Q, is determined from the equation:

$$Q = WC\Delta T + W\Delta H$$

where

W	=	flow rate of fluid (lb/hr)
C	=	specific heat of fluid (Btu/lb/°F)
ΔT	=	temperature change of the fluid (°F)
ΔH	=	latent heat of vaporization (Btu/lb)

If the fluid does not change state, the equation becomes $Q = WC\Delta T$.

The heat transfer coefficient, U, represents the thermal conductance of the heat exchanger. The higher the value of U, the more easily heat is transferred from one fluid to the other. Thermal conductance is the reciprocal of resistance, R, to heat flow.

$$U = \frac{1}{R}$$

The total resistance to heat flow is the sum of several individual resistances. This is shown in Figure 23-2 and mathematically expressed below.

$$R_t = r_1 + r_2 + r_3 + r_4 + r_5$$

where

R_t	=	total heat flow resistance
r_1	=	heat flow resistance of the process-side film
r_2	=	heat flow resistance of the process-side fouling (if any)
r_3	=	heat flow resistance of the exchanger tube wall
r_4	=	heat flow resistance of the water-side fouling (if any)
r_5	=	heat flow resistance of the water-side film

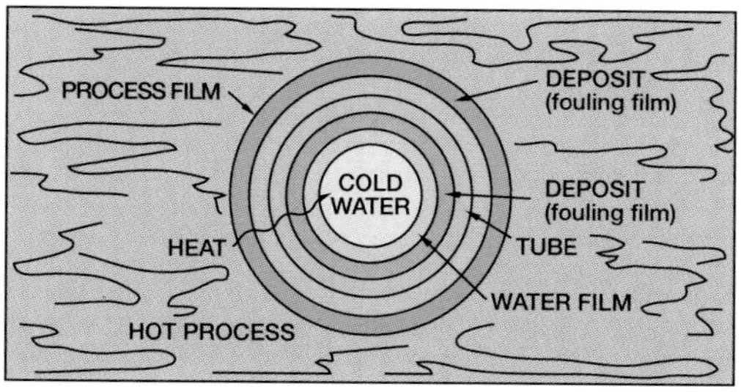

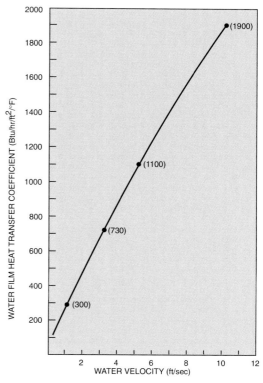

Figure 23-2. The total resistance to heat flow is the sum of several individual resistances.

The heat flow resistance of the process-side film and the cooling water film depends on equipment geometry, flow velocity, viscosity, specific heat, and thermal conductivity. The effect of velocity on heat transfer for water in a tube is shown in Figure 23-3.

Heat flow resistance due to fouling varies tremendously depending on the characteristics of the fouling layer, the fluid, and the contaminants in the fluid that created the fouling layer. A minor amount of fouling is generally accommodated in the exchanger design. However, if fouling is not kept to a minimum, the resistance to heat transfer will increase, and the U coefficient will decrease to the point at which the exchanger cannot adequately control the process temperatures. Even if this point is not reached, the transfer process is less efficient and potentially wasteful.

The resistance of the tube to heat transfer depends on the material of construction only and does not change with time. Tube walls thinned by erosion or corrosion may have less resistance, but this is not a significant change.

The log mean temperature difference (ΔT_m) is a mathematical expression addressing the temperature differential between the two fluids at each point along the heat exchanger. For true countercurrent or cocurrent flow:

$$\Delta T_m = \frac{\Delta t_2 - \Delta t_1}{\ln (\Delta t_2 / \Delta t_1)}$$

where

Δt = temperature difference at one end of the exchanger

When there is no change in state of the fluids, a countercurrent flow exchanger is more efficient for heat transfer than a cocurrent flow exchanger. Therefore, most coolers operate with a countercurrent or a variation of countercurrent flow.

Calculated ΔT_m formulas may be corrected for exchanger configurations that are not truly countercurrent.

MONITORING

Heat transfer equations are useful in monitoring the condition of heat transfer equipment or the efficacy of the treatment programs. The resistance of the tube is constant; system geometry does not change. If flow velocities are held constant on both the process side and the cooling water side, film resistance will also be held constant. Variations in measured values of the U coefficient can be used to estimate the amount of fouling taking place. If the U coefficient does not change, there is no fouling taking place on the limiting side. As the exchanger fouls, the U coefficient decreases. Therefore, a comparison of U values during operation can provide useful information about the need for cleaning and can be utilized to monitor the effectiveness of treatment programs.

The use of a cleanliness factor or a fouling factor can also be helpful in comparing the condition of the heat exchanger, during service,

Figure 23-3. Water velocity vs. heat transfer coefficient.

to design conditions. The cleanliness factor (C_f) is a percentage obtained as follows:

$$C_f = \frac{U_{final}}{U_{clean}} \times 100$$

The resistance due to fouling, or fouling factor (R_f), is a relationship between the initial overall heat transfer coefficient (U_i) and the overall heat transfer coefficient during service (U_f) expressed as follows:

$$R_f = \frac{1}{U_f} - \frac{1}{U_i}$$

Heat exchangers are commonly designed for fouling factors of 0.001 to 0.002, depending on the expected conditions of the process fluid and the cooling water.

CORROSION CONTROL—COOLING SYSTEMS

Corrosion can be defined as the destruction of a metal by chemical or electrochemical reaction with its environment. In cooling systems, corrosion causes two basic problems. The first and most obvious is the failure of equipment with the resultant cost of replacement and plant downtime. The second is decreased plant efficiency due to loss of heat transfer—the result of heat exchanger fouling caused by the accumulation of corrosion products.

Corrosion occurs at the anode, where metal dissolves. Often, this is separated by a physical distance from the cathode, where a reduction reaction takes place. An electrical potential difference exists between these sites, and current flows through the solution from the anode to the cathode. This is accompanied by the flow of electrons from the anode to the cathode through the metal. Figure 24-1 illustrates this process.

For steel, the typical anodic oxidation reaction is:

$$Fe = Fe^{2+} + 2e^-$$

This reaction is accompanied by the following:

$$Fe^{2+} + 2OH^- = Fe(OH)_2$$

The ferrous hydroxide then combines with oxygen and water to produce ferric hydroxide, $Fe(OH)_3$, which becomes common iron rust when dehydrated to Fe_2O_3.

The primary cathodic reaction in cooling systems is:

$$\tfrac{1}{2}O_2 + H_2O + 2e^- = 2OH^-$$

The production of hydroxide ions creates a localized high pH at the cathode, approximately 1–2 pH units above bulk water pH. Dissolved oxygen reaches the surface by diffusion, as indicated by the wavy lines in Figure 24-1. The oxygen reduction reaction controls the rate of

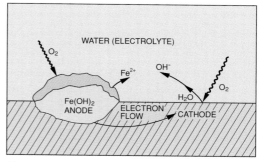

Figure 24-1. Classic corrosion cell.

corrosion in cooling systems; the rate of oxygen diffusion is usually the limiting factor.

Another important cathodic reaction is:

$$2H^+ + 2e^- = H_2$$

At neutral or higher pH, the concentration of H^+ ions is too low for this reaction to contribute significantly to the overall corrosion rate. However, as pH decreases, this reaction becomes more important until, at a pH of about 4, it becomes the predominant cathodic reaction.

TYPES OF CORROSION

The formation of anodic and cathodic sites, necessary to produce corrosion, can occur for any of a number of reasons: impurities in the metal, localized stresses, metal grain size or composition differences, discontinuities on the surface, and differences in the local environment (e.g., temperature, oxygen, or salt concentration). When these local differences are not large and the anodic and cathodic sites can shift from place to place on the metal surface, corrosion is uniform (see Figure 24-2). With uniform corrosion, fouling is usually a more serious problem than equipment failure.

Localized corrosion, which occurs when the anodic sites remain stationary, is a more serious

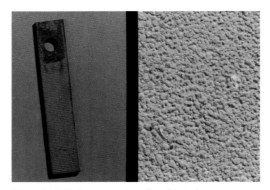

Figure 24-2. Uniform corrosion of a mild steel coupon.

industrial problem. Forms of localized corrosion include pitting, selective leaching (e.g., dezincification), galvanic corrosion, crevice or underdeposit corrosion, intergranular corrosion, stress corrosion cracking, and microbiologically influenced corrosion. Another form of corrosion, which cannot be accurately categorized as either uniform or localized, is erosion corrosion.

Pitting

Pitting (see Figure 24-3) is one of the most destructive forms of corrosion and also one of the most difficult to predict in laboratory tests. Pitting occurs when anodic and cathodic sites become stationary due to large differences in surface conditions. It is generally promoted by low-velocity or stagnant conditions (e.g., shell-side cooling) and by the presence of chloride ions. Once a pit is formed, the solution inside it is isolated from the bulk environment and becomes increasingly corrosive with time. The high corrosion rate in the pit produces an excess of positively charged metal cations, which attract chloride

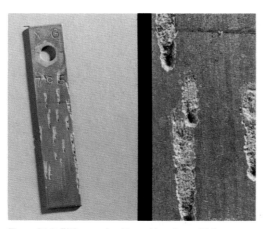

Figure 24-3. Pitting can lead to rapid equipment failure.

anions. In addition, hydrolysis produces H^+ ions. The increase in acidity and concentration within the pit promotes even higher corrosion rates, and the process becomes self-sustaining. Inhibitors can be used to control pitting, but they must be applied correctly.

Selective Leaching

Selective leaching is the corrosion of one element of an alloy. The most common example in cooling systems is dezincification, which is the selective removal of zinc from copper-zinc alloys (see Figure 24-4). The conditions that promote the pitting of steel also promote the pitting of brass, which in cooling systems usually occurs by dezincification. Low pH conditions (< 6.0) and high free chlorine residuals (> 1.0 ppm) are particularly aggressive in producing dezincification. The dezincification resistance varies with the alloy. For example, 70-30 brass is less resistant than admiralty brass (70-30 brass plus 1% tin), which is less resistant than inhibited Admiralty brass (Admiralty brass plus a small amount of arsenic, antimony, or phosphorus).

Figure 24-4. Dezincification of Admiralty brass.

Galvanic Corrosion

Galvanic corrosion occurs when two dissimilar metals are in contact in a solution. The contact must be good enough to conduct electricity, and both metals must be exposed to the solution. The driving force for galvanic corrosion is the electric potential difference that develops between two metals. This difference increases as the distance between the metals in the galvanic series increases. Table 24-1 shows a galvanic series for some commercial metals and alloys. When two metals from the series are in contact in solution, the corrosion rate of the more active (anodic) metal increases and the corrosion rate of the more noble (cathodic) metal decreases.

Table 24-1. Galvanic series of metals and alloys.[a]

CORRODED END
(anodic, or least noble)

Magnesium
Magnesium alloys

Zinc

Aluminum 2S

Cadmium

Aluminum 17ST

Steel or iron
Cast iron

Chromium-iron (active)

Ni-Resist
18-8-Cr-Ni-Fe (active)

18-8-3-Cr-Ni-Mo-Fe (active)

Hastelloy C

Lead-tin solders
Lead
Tin

Nickel (active)

Inconel (active)

Hastelloy A
Hastelloy B

Brasses
Copper
Bronzes
Copper-nickel alloys
Titanium
Monel

Silver solder

Nickel (passive)
Inconel (passive)
Chromium-iron (passive)
18-8 Cr-Ni-Fe (passive)
18-8-3 Cr-Ni-Mo-Fe (passive)

Silver

Graphite

PROTECTED END
(cathodic, or most noble)

[a]Courtesy of International Nickel Company, Inc.

Galvanic corrosion can be controlled by the use of sacrificial anodes. This is a common method of controlling corrosion in heat exchangers with Admiralty tube bundles and carbon steel tube sheets and channel heads. The anodes are bolted directly to the steel and protect a limited area around the anode. Proper placement of sacrificial anodes is a precise science.

The most serious form of galvanic corrosion occurs in cooling systems that contain both copper and steel alloys. It results when dissolved copper plates onto a steel surface and induces rapid galvanic attack of the steel. The amount of dissolved copper required to produce this effect is very small and the increased corrosion is very difficult to inhibit once it occurs. A copper corrosion inhibitor is needed to prevent copper dissolution.

Crevice Corrosion

Crevice corrosion is intense localized corrosion which occurs within a crevice or any area that is shielded from the bulk environment. Solutions within a crevice are similar to solutions within a pit in that they are highly concentrated and acidic. Because the mechanisms of corrosion in the two processes are virtually identical, conditions that promote pitting also promote crevice corrosion. Alloys that depend on oxide films for protection (e.g., stainless steel and aluminum) are highly susceptible to crevice attack because the films are destroyed by high chloride ion concentrations and low pH. This is also true of protective films induced by anodic inhibitors.

The best way to prevent crevice corrosion is to prevent crevices. From a cooling water standpoint, this requires the prevention of deposits on the metal surface. Deposits may be formed by suspended solids (e.g., silt, silica) or by precipitating species, such as calcium salts.

Intergranular Corrosion

Intergranular corrosion is localized attack that occurs at metal grain boundaries. It is most prevalent in stainless steels which have been improperly heat-treated. In these metals, the grain boundary area is depleted in chromium and therefore is less resistant to corrosion (see Figure 24-5). Intergranular corrosion also occurs in certain high-strength aluminum alloys. In general, it is not of significance in cooling systems.

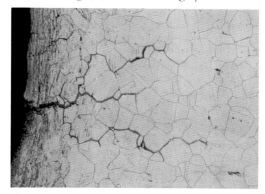

Figure 24-5. Photomicrograph illustrating intergranular corrosion of stainless steel.

Stress Corrosion Cracking

Stress corrosion cracking (SCC) is the brittle failure of a metal by cracking under tensile stress

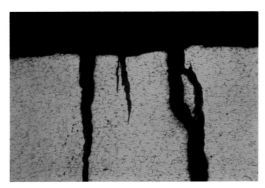

Figure 24-6. A transgranular stress corrosion cracking failure.

in a corrosive environment. Failures tend to be transgranular (see Figure 24-6), although intergranular failures have been noted. Commonly used cooling system alloys that may crack due to stress include austenitic stainless steels (300 series) and brasses. The susceptibility of stainless steels to SCC increases significantly as the temperature is increased. Most laboratory stainless steel SCC testing is done at about 300 °F, because it is very difficult to promote cracking at temperatures below 200 °F. For this reason, SCC of stainless steels has not been widely observed in cooling systems.

Chloride is the main contributor to SCC of stainless steels. High chloride concentrations, resulting from high chloride levels in the makeup water and/or high cycles of concentration, will increase susceptibility. Although low water temperatures generally preclude cracking, SCC of stainless steels can occur in cooling systems.

For brasses, the ammonium ion is the main cause of SCC. Very few service failures have been reported where ammonia is not present.

The most likely places for SCC to be initiated are crevices or areas where the flow of water is restricted. This is due to the buildup of corrodent concentrations in these areas. For example, chloride can concentrate from 100 ppm in the bulk water to as high as 10,000 ppm (1%) in a crevice. Deposits are initiating sites because of crevices formed beneath them. The low water velocities in shell-side cooling are also detrimental.

The most effective way to prevent SCC in both stainless steel and brass systems is to keep the system clean and free of deposits. An effective deposit control treatment is imperative. A good corrosion inhibitor is also beneficial. Chromate and phosphate have each been used successfully

to prevent the SCC of stainless steel in chloride solutions.

Microbiologically Influenced Corrosion (MIC)

Microorganisms in cooling water form "biofilms" on cooling system surfaces. Biofilms consist of populations of sessile organisms and their hydrated polymeric secretions. Numerous types of organisms may exist in any particular biofilm, ranging from strictly aerobic bacteria at the water interface to anaerobic bacteria such as sulfate–reducing bacteria (SRB) at the oxygen-depleted metal surface.

The presence of a biofilm can contribute to corrosion in three ways: physical deposition, production of corrosive by-products, and depolarization of the corrosion cell caused by chemical reactions.

As discussed previously, deposits can cause accelerated localized corrosion by creating differential aeration cells. This same phenomenon occurs with a biofilm. The nonuniform nature of biofilm formation creates an inherent differential, which is enhanced by the oxygen consumption of organisms in the biofilm.

Many of the by-products of microbial metabolism, including organic acids and hydrogen sulfide, are corrosive. These materials can concentrate in the biofilm, causing accelerated metal attack.

Corrosion tends to be self-limiting due to the buildup of corrosion reaction products. However, microbes can absorb some of these materials in their metabolism, thereby removing them from the anodic or cathodic site. The removal of reaction products, termed "depolarization," stimulates further corrosion.

Figure 24-7 shows a typical result of microbial corrosion. The surface exhibits scattered areas of localized corrosion, unrelated to flow pattern.

Figure 24-7. A classic case of biologically induced corrosion.

The corrosion appears to spread in a somewhat circular pattern from the site of initial colonization.

Erosion Corrosion

Erosion corrosion is the increase in the rate of metal deterioration from abrasive effects. It can be identified by grooves and rounded holes, which usually are smooth and have a directional pattern. Erosion corrosion is increased by high water velocities and suspended solids. It is often localized at areas where water changes direction. Cavitation (damage due to the formation and collapse of bubbles in high-velocity turbines, propellers, etc.) is a form of erosion corrosion. Its appearance is similar to closely spaced pits, although the surface is usually rough.

CONTROL OF CORROSION

Corrosion control requires a change in either the metal or the environment. The first approach, changing the metal, is expensive. Also, highly alloyed materials, which are very resistant to general corrosion, are more prone to failure by localized corrosion mechanisms such as stress corrosion cracking.

The second approach, changing the environment, is a widely used, practical method of preventing corrosion. In aqueous systems, there are three ways to effect a change in environment to inhibit corrosion:

- form a protective film of calcium carbonate on the metal surface using the natural calcium and alkalinity in the water

- remove the corrosive oxygen from the water, either by mechanical or chemical deaeration

- add corrosion inhibitors

Calcium Carbonate Protective Scale

The Langelier Saturation Index (LSI) is a useful tool for predicting the tendency of a water to deposit or dissolve calcium carbonate (see Chapter 25 for a thorough discussion of LSI). A uniform coating of calcium carbonate, deposited on the metal surfaces, physically segregates the metal from the corrosive environment. To develop the positive LSI required to deposit calcium carbonate, it is usually necessary to adjust the pH, alkalinity, or calcium content of the water. Soda ash, caustic soda, or lime (calcium hydroxide) may be used for this adjustment. Lime is usually the most economical alkali because it raises the calcium

content as well as the alkalinity and pH.

Theoretically, controlled deposition of calcium carbonate scale can provide a film thick enough to protect, yet thin enough to allow adequate heat transfer. However, low-temperature areas do not permit the development of sufficient scale for corrosion protection, and excessive scale forms in high-temperature areas and interferes with heat transfer. Therefore, this approach is not used for industrial cooling systems. Controlled calcium carbonate deposition has been used successfully in some waterworks distribution systems where substantial temperature increases are not encountered.

Mechanical and Chemical Deaeration

The corrosive qualities of water can be reduced by deaeration. Vacuum deaeration has been used successfully in once-through cooling systems. Where all oxygen is not removed, catalyzed sodium sulfite can be used to remove the remaining oxygen. The sulfite reaction with dissolved oxygen is:

$$\underset{\substack{\text{sodium}\\\text{sulfite}}}{Na_2SO_3} \quad + \quad \underset{\text{oxygen}}{{}^{1}/_{2}\,O_2} \quad = \quad \underset{\substack{\text{sodium}\\\text{sulfate}}}{Na_2SO_4}$$

The use of catalyzed sodium sulfite for chemical deaeration requires 8 parts of catalyzed sodium sulfite for each part of dissolved oxygen. In certain systems where vacuum deaeration is already used, the application of catalyzed sodium sulfite may be economically justified for removal of the remaining oxygen. The use of sodium sulfite may also be applicable to some closed loop cooling systems.

In open recirculating cooling systems, continual replenishment of oxygen as the water passes over the cooling tower makes deaeration impractical.

Corrosion Inhibitors

A corrosion inhibitor is any substance which effectively decreases the corrosion rate when added to an environment. An inhibitor can be identified most accurately in relation to its function: removal of the corrosive substance, passivation, precipitation, or adsorption.

- *Deaeration* (mechanical or chemical) removes the corrosive substance—oxygen.

- *Passivating (anodic) inhibitors* form a protective oxide film on the metal surface. They are the best inhibitors because they can be

Figure 24-8. Mild steel corrosion protection provided by a passivating inhibitor.

used in economical concentrations, and their protective films are tenacious and tend to be rapidly repaired if damaged (see Figure 24-8).

■ *Precipitating (cathodic) inhibitors* are simply chemicals which form insoluble precipitates that can coat and protect the surface. Precipitated films are not as tenacious as passive films and take longer to repair after a system upset.

■ *Adsorption inhibitors* have polar properties which cause them to be adsorbed on the surface of the metal. They are usually organic materials.

A discussion of each of these inhibitors follows, preceded by an overview of the role of polarization in corrosion.

Polarization. Figure 24-9 is a schematic diagram showing common potential vs. corrosion current plots. The log current is the rate of electrochemical reaction, and the plots show how the rate of

anodic and cathodic reactions change as a function of surface potential. In Figure 24-9(a), uninhibited corrosion is occurring. The corrosion potential, E_{corr}, and the corrosion current, I_{corr}, are defined by the point at which the rate of the anodic reaction equals the rate of the cathodic reaction. I_{corr} is the actual rate of metal dissolution. Figure 24-9(b) shows the condition after an anodic inhibitor has been applied. The rate of the anodic reaction has been decreased. This causes a decrease in I_{corr} accompanied by a shift in E_{corr} to a more positive (anodic) potential. Figure 24-9(c) shows the effect of a cathodic inhibitor. Here, the rate of the cathodic reaction has been decreased, accompanied again by a decrease in I_{corr}, but this time the shift in E_{corr} is in the negative (cathodic) direction.

Passivation Inhibitors. Examples of passivators (anodic inhibitors) include chromate, nitrite, molybdate, and orthophosphate. All are oxidizers and promote passivation by increasing the electrical potential of the iron. Chromate and nitrite do not require oxygen and thus can be the most effective. Chromate is an excellent aqueous corrosion inhibitor, particularly from a cost perspective. However, due to health and environmental concerns, use of chromate has decreased significantly and will probably be outlawed in the near future. Nitrite is also an effective inhibitor, but in open systems it tends to be oxidized to nitrate.

Both molybdate and orthophosphate are excellent passivators in the presence of oxygen. Molybdate can be a very effective inhibitor, especially when combined with other chemicals. Its main drawback is its high cost. Orthophosphate is not really an oxidizer per se, but becomes one in the presence of oxygen. If iron is put into a phosphate solution without oxygen present, the corrosion potential remains active and the corrosion rate is not reduced. However, if oxygen is present,

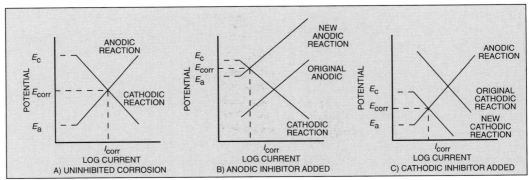

Figure 24-9. Simplified polarization curves showing potential vs. log current.

the corrosion potential increases in the noble direction and the corrosion rate decreases significantly.

A negative attribute of orthophosphate is its tendency to precipitate with calcium hardness found in natural waters. In recent years, deposit control agents that prevent this deposition have been developed. Due to its relatively low cost, orthophosphate is widely used as an industrial corrosion inhibitor.

Precipitating Inhibitors. As discussed earlier, the localized pH at the cathode of the corrosion cell is elevated due to the generation of hydroxide ions. Precipitating inhibitors form complexes which are insoluble at this high pH (1–2 pH units above bulk water), but whose deposition can be controlled at the bulk water pH (typically 7–9 pH). A good example is zinc, which can precipitate as hydroxide, carbonate, or phosphate. Calcium carbonate and calcium orthophosphate are also precipitating inhibitors. Orthophosphate thus exhibits a dual mechanism, acting as both an anodic passivator and a cathodic precipitator.

Copper Corrosion Inhibitors. The most effective corrosion inhibitors for copper and its alloys are the aromatic triazoles, such as benzotriazole (BZT) and tolyltriazole (TTA). These compounds bond directly with cuprous oxide (Cu_2O) at the metal surface, forming a "chemisorbed" film. The plane of the triazole lies parallel to the metal surface; thus, each molecule covers a relatively large surface area. The exact mechanism of inhibition is unknown. Various studies indicate anodic inhibition, cathodic inhibition, or a combination of the two. Other studies indicate the formation of an insulating layer between the water surface and the metal surface. A very recent study supports the idea of an electronic stabilization mechanism. The protective cuprous oxide layer is prevented from oxidizing to the nonprotective cupric oxide. This is an anodic mechanism. However, the triazole film exhibits some cathodic properties as well.

In addition to bonding with the metal surface, triazoles bond with copper ions in solution. Thus, dissolved copper represents a "demand" for triazole, which must be satisfied before surface filming can occur. Although the surface demand for triazole filming is generally negligible, copper corrosion products can consume a considerable amount of treatment chemical. Excessive chlorination will deactivate the triazoles and significantly increase copper corrosion rates. Due to all of these factors, treatment with triazoles is a complex process.

Adsorption Inhibitors. Adsorption inhibitors must have polar properties in order to be adsorbed and block the surface against further adsorption. Typically, they are organic compounds containing nitrogen groups, such as amines, and organic compounds containing sulfur or hydroxyl groups. The size, orientation, shape, and electrical charge distribution of the molecules are all important factors. Often, these molecules are surfactants and have dual functionality. They contain a hydrophilic group, which adsorbs onto the metal surface, and an opposing hydrophobic group, which prevents further wetting of the metal.

Glycine derivatives and aliphatic sulfonates are examples of compounds which can function in this way. The use of these inhibitors in cooling systems is usually limited by their biodegradability and their toxicity toward fish. In addition, they can form thick, oily surface films, which may severely retard heat transfer.

Silicates. For many years, silicates have been used to inhibit aqueous corrosion, particularly in potable water systems. Probably due to the complexity of silicate chemistry, their mechanism of inhibition has not yet been firmly established. They are nonoxidizing and require oxygen to inhibit corrosion, so they are not passivators in the classical sense. Yet they do not form visible precipitates on the metal surface. They appear to inhibit by an adsorption mechanism. It is thought that silica and iron corrosion products interact. However, recent work indicates that this interaction may not be necessary. Silicates are slow-acting inhibitors; in some cases, 2 or 3 weeks may be required to establish protection fully. It is believed that the polysilicate ions or colloidal silica are the active species and these are formed very slowly from monosilicic acid, which is the predominant species in water at the pH levels maintained in cooling systems.

Effect of Conductivity, pH, and Dissolved Oxygen

Figures 24-10 through 24-12 show the effects of several operating parameters on the corrosion tendency in aqueous systems. As shown in Figure 24-10, corrosion rate increases with conductivity.

Figure 24-11 shows the effect of pH on the

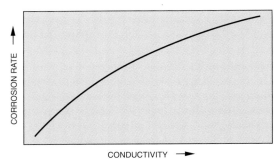

Figure 24-10. As conductivity increases, so does the corrosion rate.

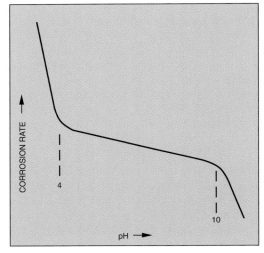

Figure 24-11. The effect of pH on mild steel corrosion rate in an open recirculating cooling system.

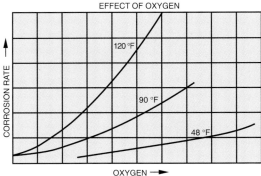

Figure 24-12. Effect of oxygen concentration on corrosion at different temperatures.

corrosion of iron. Within the acid range (pH < 4), the iron oxide film is continually dissolved. In cooling water, the potential for calcium carbonate precipitation increases with higher pH and alkalinity; thus the corrosion rate decreases slightly as pH is increased from 4 to 10. Above pH 10, iron becomes increasingly passive.

Figure 24-12 shows the effect of oxygen concentration on corrosion at different temperatures. As discussed previously, oxygen is the main driving force for corrosion of steel in cooling water. The increase in corrosion with temperature at a given oxygen concentration is due to more rapid oxygen diffusion occurring at higher temperatures.

Practical Considerations

The success of cooling water corrosion inhibitor programs is affected by the following factors:

- *Water Characteristics.* Calcium, alkalinity, and pH levels in water are important factors for reasons already cited. Further discussions are covered in the chapters on once-through and open recirculating systems.

- *Design Considerations.* Low water velocity, which occurs in shell-side cooling, increases deposition. This factor must be addressed in the design of the system.

- *Microbiological Control.* An effective microbiological control program is necessary to prevent severe fouling problems. Fouling caused by uncontrolled biological growth can contribute to corrosion by one or more mechanisms.

- *System Control.* Even the best treatment technology available will fail without a reasonable level of control. Therefore, careful consideration must be given to system control—the accuracy with which the pH, inhibitor levels, and other water characteristics are maintained.

- *Pretreatment.* Grease and/or corrosion products from previous treatment programs should be cleaned out, and the system should be treated with a high level of a good inhibitor before normal operation.

- *Contamination.* Contamination can also be a problem. Sulfide, ammonia, and hydrocarbons are among the most severe contaminants. Sulfide is corrosive to steel and copper alloys. Ammonia is corrosive to Admiralty and promotes biological growth. Hydrocarbons promote fouling and biological growth.

In the determination of treatment levels, solubility data is important. The Langelier Saturation Index, which defines the solubility of calcium carbonate, is commonly used. Solubility data for calcium orthophosphate and zinc orthophosphate may be needed if the treatment contains phosphate and zinc.

Monitoring

Every cooling water system should include a method of monitoring corrosion in the system. Tools commonly used for this purpose include metal corrosion coupons, instantaneous corrosion rate meters, and heated surfaces such as test heat exchangers and the Betz MonitAll® apparatus. Data obtained from these devices can be used to optimize an inhibitor treatment program to maintain the plant equipment in the best possible condition. When heat transfer data cannot be obtained on operating exchangers, monitoring devices can be useful for evaluating the success of a treatment program without a plant shutdown.

Corrosion Coupons. Preweighed metal coupons are still widely used as a reliable method for monitoring corrosion in cooling systems. Coupon weight loss provides a quantitative measure of the corrosion rate, and the visual appearance of the coupon provides an assessment of the type of corrosion and the amount of deposition in the system. In addition, measurement of pit depths on the coupon can indicate the severity of the pitting.

Coupons should be installed properly in a corrosion coupon bypass rack with continuous, controlled water flow past the coupons. The metallurgy should match that of the system. One disadvantage of coupons is their lack of heat transfer, resulting in a lower temperature than that of the actual heat exchanger tubes. In addition, only a time-weighted average corrosion rate is obtained.

Corrosion Rate Meters. Additional corrosion monitoring tools have been developed by various instrument manufacturers and water treatment companies. Instantaneous corrosion rate meters can measure the corrosion rate at any given point in time.

Instrument methods fall into two general categories: electrical resistance and linear polarization. With either technique, corrosion measurements are made quickly without removal of the sensing device.

The electrical resistance method is based on measuring the increase in the electrical resistance of a test electrode as it becomes thinner due to

corrosion. This method is desirable because the probes can be installed in both aqueous and nonaqueous streams. However, the electrical resistance method also has its disadvantages: conductive deposits forming on the probe can create misleading results, temperature fluctuations must be compensated for, and pitting characteristics cannot be determined accurately.

The method based on linear polarization at low applied potentials provides instantaneous corrosion rate data that can be read directly from the instrument face in actual corrosion rate units (mils per year). Systems using two or three electrodes are available. This method offers the maximum in performance, simplicity, and reliability.

Corrosion rate meters can be used to assess changes in the corrosion rate as a function of time. They are able to respond to sudden changes in system conditions, such as acid spills, chlorine levels, and inhibitor treatment levels. Coupled with recording devices, they can be powerful tools in diagnosing the causes of corrosion or optimizing inhibitor treatment programs (see Figure 24-13).

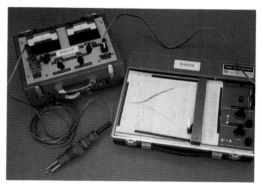

Figure 24-13. Portable corrosion rate meter with recorder.

Test Heat Exchangers. Test heat exchangers are small exchangers that can be set up to simulate operating conditions in the plant. They provide a convenient way to evaluate corrosion and fouling tendencies on heat transfer surfaces and to measure changes in heat transfer efficiency. A typical design uses cooling water on the tube side and condensing steam as a heat source on the shell side. If the test heat exchanger is insulated, a meaningful "U" (overall heat transfer coefficient) can be calculated.

BETZ MonitAll® Apparatus. The BETZ MonitAll (see Figure 24-14) is designed to measure corrosion and deposition under heat transfer conditions.

Figure 24-14. Betz MonitAll®.

Cooling water flows over a heated tube section within a glass shell. The specimen tube section is slid onto an electrical heater probe. Thermocouples measure bulk water temperature and tubeside "skin" temperature. The heat flux and flow velocity can be varied to simulate plant conditions. The tubes are available in various metallurgies and are preweighed for corrosion rate determination. The tube is visible through the glass enclosure, allowing direct observation of corrosion and scaling tendencies. Scaling/fouling can be quantified through temperature and flow measurements.

BETZ COSMOS™ **Cooling System Monitoring Station**. The Cosmos is a portable data acquisition station that monitors key parameters of a cooling system. The piping and instrumentation cabinet includes flow, pH, and conductivity sensors as well as a corrosion coupon rack, a corrosion rate probe, and a MonitAll unit. Data from all of these devices is fed into the data acquisition system (see Figure 24-15). The accumulated data can be printed directly by the built-in printer or can be downloaded to a personal computer for spreadsheet analysis.

Figure 24-15. Betz COSMOS™ data acquisition cabinet.

DEPOSIT AND SCALE CONTROL— COOLING SYSTEMS

Deposit accumulations in cooling water systems reduce the efficiency of heat transfer and the carrying capacity of the water distribution system. In addition, the deposits cause oxygen differential cells to form. These cells accelerate corrosion and lead to process equipment failure. Deposits range from thin, tightly adherent films to thick, gelatinous masses, depending on the depositing species and the mechanism responsible for deposition.

Deposit formation is influenced strongly by system parameters, such as water and skin temperatures, water velocity, residence time, and system metallurgy. The most severe deposition is encountered in process equipment operating with high surface temperatures and/or low water velocities. With the introduction of high-efficiency film fill, deposit accumulation in the cooling tower packing has become an area of concern (see Figure 25-1). Deposits are broadly categorized as scale or foulants.

SCALE

Scale deposits are formed by precipitation and crystal growth at a surface in contact with water.

Precipitation occurs when solubilities are exceeded either in the bulk water or at the surface. The most common scale-forming salts that deposit on heat transfer surfaces are those that exhibit retrograde solubility with temperature.

Although they may be completely soluble in the lower-temperature bulk water, these compounds (e.g., calcium carbonate, calcium phosphate, and magnesium silicate) supersaturate in the higher-temperature water adjacent to the heat transfer surface and precipitate on the surface.

Scaling is not always related to temperature. Calcium carbonate and calcium sulfate scaling occur on unheated surfaces when their solubilities are exceeded in the bulk water (see Figure 25-2). Metallic surfaces are ideal sites for crystal nucleation because of their rough surfaces and the low velocities adjacent to the surface. Corrosion cells on the metal surface produce areas of high pH, which promote the precipitation of many cooling water salts. Once formed, scale deposits initiate additional nucleation, and crystal growth proceeds at an accelerated rate.

Figure 25-1. Scaling of cooling tower fill by a combination of calcium carbonate and calcium phosphate.

Figure 25-2. Scaling of water distribution piping by calcium sulfate.

Scale control can be achieved through operation of the cooling system at subsaturated conditions or through the use of chemical additives.

Operational Control

The most direct method of inhibiting formation of scale deposits is operation at subsaturation conditions, where scale-forming salts are soluble. For some salts, it is sufficient to operate at low cycles of concentration and/or control pH. However, in most cases, high blowdown rates and low pH are required so that solubilities are not exceeded at the heat transfer surface. In addition, it is necessary to maintain precise control of pH and concentration cycles. Minor variations in water chemistry or heat load can result in scaling (see Figure 25-3).

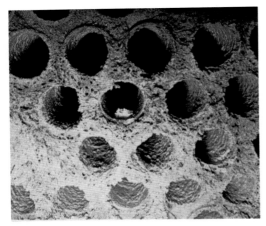

Figure 25-3. Calcium carbonate scaling of an electric utility surface condenser due to poor pH control.

Chemical Additives

Scaling can be controlled effectively by the use of sequestering agents and chelates, which are capable of forming soluble complexes with metal ions. The precipitation properties of these complexes are not the same as those of the metal ions. Classic examples of these materials are ethylenediaminetetraacetic acid (EDTA) for chelating calcium hardness, and polyphosphates for iron (Figure 25-4). This approach requires stoichiometric chemical quantities. Therefore, its use is limited to waters containing low concentrations of the metal.

Threshold Inhibitors. Deposit control agents that inhibit precipitation at dosages far below the stoichiometric level required for sequestration or chelation are called "threshold inhibitors." These materials affect the kinetics of the nucleation

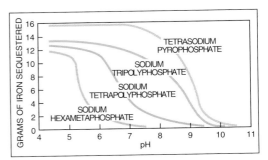

Figure 25-4. Sequestration of iron by sodium phosphates.

and crystal growth of scale-forming salts, and permit supersaturation without scale formation.

Threshold inhibitors function by an adsorption mechanism. As ion clusters in solution become oriented, metastable microcrystallites (highly oriented ion clusters) are formed. At the initial stage of precipitation, the microcrystallite can either continue to grow (forming a larger crystal with a well defined lattice) or dissolve. Threshold inhibitors prevent precipitation by adsorbing on the newly emerging crystal, blocking active growth sites. This inhibits further growth and favors the dissolution reaction. The precipitate dissolves and releases the inhibitor, which is then free to repeat the process.

Threshold inhibitors delay or retard the rate of precipitation. Crystals eventually form, depending on the degree of supersaturation and system retention time. After stable crystals appear, their continued growth is retarded by adsorption of inhibitor. The inhibitor blocks much of the crystal surface, causing distortions in the crystal lattice as growth continues. The distortions (defects in the crystal lattice) create internal stresses, making the crystal fragile. Tightly adherent scale deposits do not form, because crystals that form on surfaces in contact with flowing water cannot withstand the mechanical force exerted by the water. The adsorbed inhibitor also disperses particles, by virtue of its electrostatic charge, and prevents the formation of strongly bound agglomerates.

The most commonly used scale inhibitors are low molecular weight acrylate polymers and organophosphorus compounds (phosphonates). Both classes of materials function as threshold inhibitors; however, the polymeric materials are more effective dispersants. Selection of a scale control agent depends on the precipitating species and its degree of supersaturation. The most effective scale control programs use both a precipitation inhibitor and a dispersant. In some cases this can be achieved with a single component (e.g., polymers used to inhibit calcium phosphate at near neutral pH).

Langelier Saturation Index

Work by Professor W.F. Langelier, published in 1936, deals with the conditions at which a water is in equilibrium with calcium carbonate. An equation developed by Langelier makes it possible to predict the tendency of calcium carbonate either to precipitate or to dissolve under varying conditions. The equation expresses the relationship of pH, calcium, total alkalinity, dissolved solids, and temperature as they relate to the solubility of calcium carbonate in waters with a pH of 6.5–9.5:

$$pH_s = (pK_2 - pK_s) + pCa^{2+} + pAlk$$

where:

pH_s = the pH at which water with a given calcium content and alkalinity is in equilibrium with calcium carbonate

K_2 = the second dissociation constant for carbonic acid

K_s = the solubility product constant for calcium carbonate

These terms are functions of temperature and total mineral content. Their values for any given condition can be computed from known thermodynamic constants. Both the calcium ion and the alkalinity terms are the negative logarithms of their respective concentrations. The calcium content is molar, while the alkalinity is an equivalent concentration (i.e., the titratable equivalent of base per liter). The calculation of the pH_s has been simplified by the preparation of various nomographs. A typical one is shown in Figure 25-5.

The difference between the actual pH (pH_a) of a sample of water and the pH_s, or $pH_a - pH_s$, is called the Langelier Saturation Index (LSI). This index is a qualitative indication of the tendency of calcium carbonate to deposit or dissolve. If the LSI is positive, calcium carbonate tends to deposit. If it is negative, calcium carbonate tends to dissolve. If it is zero, the water is at equilibrium.

The LSI measures only the directional tendency or driving force for calcium carbonate to precipitate or dissolve. It cannot be used as a quantitative measure. Two different waters, one of low hardness (corrosive) and the other of high hardness (scale-forming), can have the same Saturation Index.

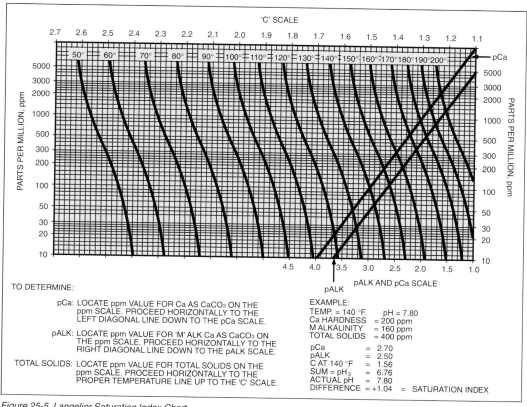

TO DETERMINE:

pCa: LOCATE ppm VALUE FOR Ca AS CaCO₃ ON THE ppm SCALE. PROCEED HORIZONTALLY TO THE LEFT DIAGONAL LINE DOWN TO THE pCa SCALE.

pALK: LOCATE ppm VALUE FOR 'M' ALK Ca AS CaCO₃ ON THE ppm SCALE. PROCEED HORIZONTALLY TO THE RIGHT DIAGONAL LINE DOWN TO THE pALK SCALE.

TOTAL SOLIDS: LOCATE ppm VALUE FOR TOTAL SOLIDS ON THE ppm SCALE. PROCEED HORIZONTALLY TO THE PROPER TEMPERATURE LINE UP TO THE 'C' SCALE.

EXAMPLE:
TEMP. = 140 °F.   pH = 7.80
Ca HARDNESS = 200 ppm
M ALKALINITY = 160 ppm
TOTAL SOLIDS = 400 ppm

pCa	= 2.70
pALK	= 2.50
C AT 140 °F	= 1.56
SUM = pH₃	= 6.76
ACTUAL pH	= 7.80
DIFFERENCE = +1.04	= SATURATION INDEX

Figure 25-5. Langelier Saturation Index Chart.

The Stability Index developed by Ryznar makes it possible to distinguish between two such waters. This index is based on a study of actual operating results with waters having various Saturation Indexes.

$$\text{Stability Index} = 2(pH_s) - pH_a$$

Where waters have a Stability Index of 6.0 or less, scaling increases and the tendency to corrode decreases. Where the Stability Index exceeds 7.0, scaling may not occur at all. As the Stability Index rises above 7.5 or 8.0, the probability of corrosion increases. Use of the LSI together with the Stability Index contributes to more accurate prediction of the scaling or corrosive tendencies of a water.

FOULING

Fouling occurs when insoluble particulates suspended in recirculating water form deposits on a surface. Fouling mechanisms are dominated by particle–particle interactions that lead to the formation of agglomerates.

At low water velocities, particle settling occurs under the influence of gravity (see Figure 25-6). Parameters that affect the rate of settling are particle size, relative liquid and particle densities, and liquid viscosity. The relationships of these variables are expressed by Stokes' Law. The most important factor affecting the settling rate is the size of the particle. Because of this, the control of fouling by preventing agglomeration is one of the most fundamental aspects of deposition control.

Foulants enter a cooling system with makeup water, airborne contamination, process leaks, and corrosion. Most potential foulants enter with makeup water as particulate matter, such as clay, silt, and iron oxides (see Figure 25-7). Insoluble aluminum and iron hydroxides enter a system from makeup water pretreatment operations.

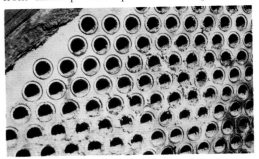

Figure 25-6. Calcium and iron phosphate fouling due to low water velocity.

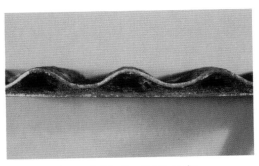

Figure 25-7. Iron and silt fouling of plate cooler.

Some well waters contain high levels of soluble ferrous iron that is later oxidized to ferric iron by dissolved oxygen in the recirculating cooling water. Because it is very insoluble, the ferric iron precipitates. The steel corrosion process is also a source of ferrous iron and, consequently, contributes to fouling.

Both iron and aluminum are particularly troublesome because of their ability to act as coagulants. Also, their soluble and insoluble hydroxide forms can each cause precipitation of some water treatment chemicals, such as orthophosphate.

Airborne contaminants usually consist of clay and dirt particles but can include gases such as hydrogen sulfide, which forms insoluble precipitates with many metal ions. Process leaks introduce a variety of contaminants that accelerate deposition and corrosion.

Foulants, such as river water silt, enter the system as finely dispersed particles, which can be as small as 1–100 nm. The particles carry an electrostatic charge, which causes similarly charged particles to repel each other, favoring their dispersion. The net charge a particle carries depends on the composition of the water. Cycling of cooling water increases the concentration of counter-charged ions capable of being electrostatically attracted to and adsorbed onto a charged particle. As counterions adsorb, the net charge of the particle decreases. Particles begin to agglomerate and grow in size as their repulsive forces are diminished.

Settling occurs when the energy imparted by fluid velocity can no longer suspend the particle, due to agglomeration and growth. After particles have settled, the nature of the deposit depends on the strength of the attractive forces between the particles themselves (agglomerate strength) and between the particles and the surface they contact. If attractive forces between particles are strong and the particles are not highly hydrated, deposits are dense and well structured; if the forces are weak, the deposits are soft and pliable. Deposition continues as long as the shear strength of the deposit exceeds the shear stress of the flowing water.

Methods of fouling control are discussed in the following sections.

Removal of Particulate Matter

The amount of particulate entering a cooling system with the makeup water can be reduced by filtration and/or sedimentation processes. Particulate removal can also be accomplished by filtration of recirculating cooling water. These methods do not remove all of the suspended matter from the cooling water. The level of fouling experienced is influenced by the effectiveness of the particular removal scheme employed, the water velocities in the process equipment, and the cycles of concentration maintained in the cooling tower.

High Water Velocities

The ability of high water velocities to minimize fouling depends on the nature of the foulant. Clay and silt deposits are more effectively removed by high water velocities than aluminum and iron deposits, which are more tacky and form interlocking networks with other precipitates. Operation at high water velocities is not always a viable solution to clay and silt deposition because of design limitations, economic considerations, and the potential for erosion corrosion.

Dispersants

Dispersants are materials that suspend particulate matter by adsorbing onto the surface of particles and imparting a high charge. Electrostatic repulsion between like-charged particles prevents agglomeration, which reduces particle growth. The presence of a dispersant at the surface of a particle also inhibits the bridging of particles by precipitates that form in the bulk water. The adsorption of the dispersant makes particles more hydrophilic and less likely to adhere to surfaces. Thus, dispersants affect both particle-to-particle and particle-to-surface interactions.

The most effective and widely used dispersants are low molecular weight anionic polymers. Dispersion technology has advanced to the point at which polymers are designed for specific classes of foulants or for a broad spectrum of materials. Acrylate-based polymers are widely used as dispersants. They have advanced from simple homopolymers of acrylic acid to more advanced copolymers and terpolymers. The performance characteristics of the acrylate polymers are a function of their molecular weight and structure, along with the types of monomeric units incorporated into the polymer backbone.

Surfactants

Surface-active or wetting agents are used to prevent fouling by insoluble hydrocarbons. They function by emulsifying the hydrocarbon through the formation of microdroplets containing the surfactant. The hydrophobic (water-hating) portion of the surfactant is dissolved within the oil drop, while the hydrophilic (water-loving) portion is at the surface of the droplet. The electrostatic charge imparted by hydrophilic groups causes the droplets to repel each other, preventing coalescence.

Through a similar process, surfactants also assist in the removal of hydrocarbon-containing deposits.

MICROBIOLOGICAL CONTROL—COOLING SYSTEMS

Cooling water systems, particularly open recirculating systems, provide a favorable environment for the growth of microorganisms. Microbial growth on wetted surfaces leads to the formation of biofilms. If uncontrolled, such films cause fouling, which can adversely affect equipment performance, promote metal corrosion, and accelerate wood deterioration. These problems can be controlled through proper biomonitoring and application of appropriate cooling water antimicrobials.

DIFFICULTIES DUE TO BIOFOULING

Microbiological fouling in cooling systems is the result of abundant growth of algae, fungi, and bacteria on surfaces. Once-through and open or closed recirculating water systems may support microbial growth, but fouling problems usually develop more quickly and are more extensive in open recirculating systems.

Once-through cooling water streams generally contain relatively low levels of the nutrients essential for microbial growth, so growth is relatively slow. Open recirculating systems scrub microbes from the air and, through evaporation, concentrate nutrients present in makeup water. As a result, microbe growth is more rapid. Process leaks may contribute further to the nutrient load of the cooling water. Reuse of wastewater for cooling adds nutrients and also contributes large amounts of microbes to the cooling system.

In addition to the availability of organic and inorganic nutrients, factors such as temperature, normal pH control range, and continuous aeration of the cooling water contribute to an environment that is ideal for microbial growth. Sunlight necessary for growth of algae may also be present. As a result, large, varied microbial populations may develop.

The outcome of uncontrolled microbial growth on surfaces is "slime" formation. Slimes typically are aggregates of biological and nonbiological materials. The biological component, known as the biofilm, consists of microbial cells and their by-products. The predominant by-product, extracellular polymeric substance (EPS), is a mixture of hydrated polymers. These polymers form a gel-like network around the cells and appear to aid attachment to surfaces. The nonbiological components can be organic or inorganic debris from many sources which have become adsorbed to or embedded in the biofilm polymer.

Slimes can form throughout once-through and recirculating systems and may be seen or felt where accessible. In nonexposed areas, slimes can be manifested by decreased heat transfer efficiency or reduced water flow. Wood-destroying organisms may penetrate the timbers of the cooling tower, digesting the wood and causing collapse of the structure. Microbial activity under deposits or within slimes can accelerate corrosion rates and even perforate heat exchanger surfaces.

MICROBIOLOGY OF COOLING WATER

Microorganisms

The microorganisms that form slime deposits in cooling water systems are common soil, aquatic, and airborne microbes (see Figure 26-1). These microbes may enter the system with makeup water, either in low numbers from fresh water sources or in high numbers when the makeup is wastewater. Significant amounts may also be scrubbed from the air as it is drawn through the cooling tower. Process leaks may contribute microorganisms as well.

Bacteria. A wide variety of bacteria can colonize cooling systems. Spherical, rod-shaped, spiral, and filamentous forms are common. Some produce spores to survive adverse environmental conditions such as dry periods or high temperatures. Both

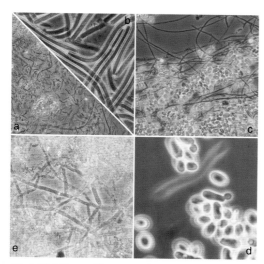

Figure 26-1. Some of the many microorganisms that colonize industrial cooling systems: (a) bacterial rods; (b) filamentous blue-green algae (now cyanobacteria); (c) unicellular green algae; (d) unicellular fungi (yeast); (e) filamentous fungi (molds)

aerobic bacteria (which thrive in oxygenated waters) and anaerobic bacteria (which are inhibited or killed by oxygen) can be found in cooling systems.

Fungi. Two forms of fungi commonly encountered are molds (filamentous forms) and yeasts (unicellular forms). Molds can be quite troublesome, causing white rot or brown rot of the cooling tower wood, depending on whether they are cellulolytic (attack cellulose) or lignin degrading. Yeasts are also cellulolytic. They can produce slime in abundant amounts and preferentially colonize wood surfaces.

Algae. Algae are photosynthetic organisms. Green and blue-green algae are very common in cooling systems (blue-green algae are now classified with the bacteria and are called cyanobacteria). Various types of algae can be responsible for green growths which block screens and distribution decks. Severe algae fouling can ultimately lead to unbalanced water flow and reduced cooling tower efficiency. Diatoms (algae enclosed by a silicaceous cell wall) may also be present but generally do not play a significant role in cooling system problems.

Differences and Similarities

Although algae, fungi, and bacteria differ in many respects, they also share many characteristics. These similarities and differences are important in understanding biofouling and its control.

- *Cell size differs according to the complexity of the cell structure.* The simpler bacteria and cyanobacteria are much smaller than molds, yeasts, and other algae. Because of their faster metabolisms and rates of growth, these smaller cells are able to reproduce much more rapidly.

- *All microorganisms require water for growth.* Although they vary in terms of absolute water requirements and ability to survive dry periods, an active, viable microbial population cannot exist without water.

- *Most microorganisms growing in cooling systems are bound by a rigid cell wall.* The cell wall gives the organism its characteristic shape and provides mechanical strength. Immediately inside the cell wall is a cell membrane which functions as a permeability barrier for the cell. The barrier allows the cell to concentrate desirable chemicals, such as nutrients, and exclude or excrete toxic or unwanted chemicals, such as waste materials. Concentration gradients of several orders of magnitude can be established across the membrane. All cells must expend considerable amounts of metabolic energy to maintain an optimal interior condition. One essential characteristic of all microbes is the ability to preserve the necessary organization and integrity of the cell in a hostile and changing environment.

- *All cells must obtain energy and chemical "building blocks" from their environment in order to survive and grow.* The ability of each type of cell to fulfill this function in different environments is discussed in the following section.

Microbiological Growth

Among the essential building blocks used by microbial cells, and those needed in largest quantity, are carbon, nitrogen, and phosphorus. Microbes differ in the method they use to obtain carbon. Green algae, cyanobacteria, and certain bacteria can utilize carbon dioxide as a sole carbon source and convert ("fix") it to cellular carbon compounds. Most bacteria, yeast, and molds require preformed carbon compounds and use organic molecules that range from very simple to very complex. In order to meet nitrogen requirements, microbes "fix" atmospheric nitrogen or utilize amines, nitrites, and nitrates present in the environment. Naturally occurring and synthetic inorganic and organic phosphates can be used to meet microbial phosphate requirements.

Microbes have developed many ways to extract energy from their surroundings. Algae and other photosynthetic organisms trap light energy from the sun. Inorganic chemicals, such as ammonia, sulfur, and hydrogen, can be oxidized by certain bacteria to release energy. More commonly, bacteria, yeasts, and molds liberate chemical energy stored in organic compounds, such as sugars, proteins, fats, oils, organic acids, and alcohols.

Aerobic organisms use oxygen to drive the oxidations that release chemical energy. Anaerobes do not use oxygen but may substitute molecules such as sulfate or nitrate in place of oxygen. In the anaerobic energy-yielding process, these oxidizing molecules are reduced, forming sulfides or nitrogen gas. When no acceptable oxidizer is available, some anaerobes can still generate energy, although less efficiently, by coupling oxidation of one half of a substrate molecule to reduction of the other half. Typically, the by-products of this "fermentative" reaction are various organic acids. All microbes extract and collect energy in small, usable packets. Once the energy is made available, there are only minor differences in how it is used.

In the presence of sufficient nutrients, growth and reproduction can occur. Bacteria and cyanobacteria multiply by binary fission, a process in which a cell divides to form two identical daughter cells. Yeasts divide by budding, with a mother cell repeatedly forming single, identical but much smaller daughter cells. Filamentous molds grow by forming new cells at the growing tip of the filament. Green algae can have several patterns of growth, depending on the species, ranging from tip extension to production of several cells from a single cell during one division cycle. As with other cell features, the complexity of growth processes also increases with increasing cell size. Under optimal conditions, some bacteria can double their numbers every 20 to 30 minutes, while molds can take many hours to double in mass.

Microorganisms are also extremely adaptable to changes in their environment. This characteristic is related to cell size and complexity. The simple forms with minimal growth needs and fast growth rates can form many cell generations within a few days. Slight random changes in cellular characteristics during those generations can produce a new cell that is more capable of surviving in a shifted environment. This new cell can soon dominate the environment. Many microbes carry information in unexpressed form for functions to be performed when needed for survival. Changes in the environment can activate this information causing all members of a microbial population to achieve new capabilities as a group, within a single generation.

Usually, cooling waters are not nutrient-rich, so microbes must expend a great deal of energy transporting and concentrating nutrients inside the cell. This process may spend energy resources already in short supply, but it is necessary to allow the biochemical machinery to run at top speed. Because there is strong competition for the available nutrients, those species most efficient at concentrating their essential nutrients will have the opportunity to grow most rapidly. The rate of growth will ultimately be limited by the nutrient which first falls below an optimal concentration, but this will not necessarily be the nutrient in the lowest concentration.

Chemicals applied to cooling systems may, at times, provide added sources of the limiting nutrient and thus contribute to microbial growth in the systems. Alterations of pH may shift a stable population balance to an unbalanced, troublesome state. Although bacteria may be under control at neutral pH, a shift to an acid pH may result in domination by molds or yeast. Because many algae grow most abundantly at an alkaline pH, an attempt to reduce corrosion by raising the pH can lead to an algal bloom.

Seasonal changes also affect growth patterns in cooling water systems. Natural algal communities in a fresh water supply are quite dynamic, and the dominant species can change rapidly with changing temperatures, nutrients, and amounts of sunlight. Cyanobacteria can often be primary colonizers in a cooling system. Seasonal changes which increase their numbers in the makeup water can lead to an algal bloom in the system. In autumn, as falling leaves increase the nutrient level and depress the pH, the bacterial population can increase at the expense of the algal population.

BIOFILMS

Microbiologists recognize two different populations of microorganisms. Free-floating (planktonic) populations are found in the bulk water. Attached (sessile) populations colonize surfaces. The same kinds of microorganisms can be found in either population, but the sessile population is responsible for biofouling.

Much is known about the formation of biofilms on wetted surfaces such as heat exchanger tubes. Microorganisms on submerged surfaces secrete polymers (predominantly polysaccharides but also proteins), which adhere firmly even to clean surfaces and prevent cells from being swept away by the normal flow of cooling water. These extracellular polymeric substances are hydrated in the natural state, forming a gel-like network around sessile microorganisms. This polymer network contributes to the integrity of the biofilm and acts as a physical barrier hindering toxic materials and predatory organisms from reaching the living cell (see Figure 26-2). Biofilm polymers can also consume oxidizers before they reach and destroy microorganisms. As a result, control of sessile microorganisms requires dosages many times greater than required to control planktonic organisms.

Biofilms develop slowly at first, because only a few organisms can attach, survive, grow, and multiply. As populations increase exponentially, the depth of the biofilm increases rapidly. Biofilm polymers are sticky and aid in the attachment of new cells to the colonized surface as well as the accumulation of nonliving debris from the bulk water. Such debris may consist of various inorganic chemical precipitates, organic flocs, and dead cell masses. Fouling results from these accumulative processes, along with the growth and replication of cells already on the surface and the generation of additional polymeric material by these cells.

When fouling occurs, even mechanical cleaning does not remove all traces of the biofilm. Previously fouled and cleaned surfaces are more rapidly colonized than new surfaces. Residual biofilm materials promote colonization and reduce the lag time before significant fouling reappears.

Biofilms on heat exchange surfaces act as insulating barriers. Heat exchanger performance begins to deteriorate as soon as biofilm thickness exceeds that of the laminar flow region. Microbes and hydrated biopolymers contain large amounts of water, and biofilms can be over 90% water by weight. As a result, biofilms have thermal conductivities very close to that of water and, in terms of heat transfer efficiency, a biofilm is the equivalent of a layer of stagnant water along the heat exchange surface.

In shell-and-tube heat exchangers, the resistance to heat transfer is least in the turbulent flow of the bulk phase, slightly greater across the metal tube walls, and greatest across laminar flow regions. As biofilm thickness increases, so does the apparent thickness of the laminar flow region. Like water, biofilms are 25 to 600 times more resistant to conductive heat transfer than many metals. A small increase in the apparent thickness of the laminar region due to biofilm growth has a significant impact on heat transfer. A thin biofilm reduces heat transfer by an amount equal to a large increase in exchanger tube wall thickness. For example, the resistance to heat transfer of a 1 mm thick accumulation of biofilm on a low carbon steel exchanger wall is equivalent to an 80 mm increase in tube wall thickness.

Biofilms can promote corrosion of fouled metal surfaces in a variety of ways (see Figure 26-3). This is referred to as microbially influenced corrosion (MIC) and is discussed further in Chapter 25. Microbes act as biological catalysts promoting conventional corrosion mechanisms:

■ the simple, passive presence of the biological deposit prevents corrosion inhibitors from reaching and passivating the fouled surface

■ microbial reactions can accelerate ongoing corrosion reactions

■ microbial by-products can be directly aggressive to the metal

The physical presence of a biofilm and biochemical activity within the film change the environment at the fouled surface. Differences between colonized and uncolonized sites may promote a galvanic-like attack. Microbes consume

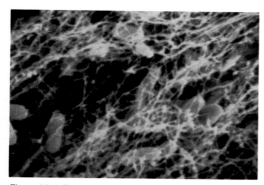

Figure 26-2. Extracellular polymers (visible in this scanning electron photomicrograph as a dehydrated fibrous network) may protect attached microorganisms.

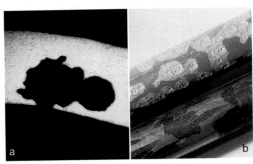

Figure 26-3. Microorganisms can contribute to corrosion of metals: (a) pitting attack visible in a cross section of stainless steel tube wall; (b) generalized surface attack associated with corrosive by-products of growing microbial colonies.

oxygen more rapidly than it can be transferred from the bulk solution, and areas beneath the biofilm become anaerobic and anodic. Repassivation of colonized surfaces is also hindered. Some microbes deprived of oxygen switch to fermentative metabolisms and produce large amounts of organic acids. This can result in local areas of low pH. Growth of anaerobes, such as sulfate-reducing bacteria, is favored in low-oxygen environments. These bacteria can oxidize hydrogen forming at the cathode and depolarize the corrosion cell. Their sulfide by-products may be directly corrosive or may contribute further to the electrochemical differential between fouled and unfouled sites.

In summary, microbes originating in the natural environment colonize cooling systems by capitalizing on favorable environmental conditions. Cooling systems are favorable environments for microorganisms because they contain water, operate in acceptable temperature and pH ranges, and provide nutrients for growth. Microbial attachment to surfaces in untreated systems produces deposits which reduce equipment efficiency and can be highly destructive to cooling equipment.

CHOICE OF SLIME CONTROL PROGRAM

Because of the speed with which microbes can grow in cooling water systems, frequent monitoring of these systems is essential for the identification of developing problems. Vigilant monitoring of operating data can identify trends, and periodic system inspections show whether or not fouling is occurring. Test coupons and test heat exchangers may be used in operating systems to facilitate monitoring without interrupting system operation.

Deposits collected from the cooling system can

be analyzed in the laboratory to determine their chemical composition and biological content. If a deposit has a significant microbiological content, its causative agents should be identified for treatment. The laboratory can identify the agents as predominantly algal, bacterial, or fungal, either microscopically (see Figure 26-4) or by routine cultural isolation and identification.

Microbial counts can also be performed to determine whether populations within the system are stable, increasing, or decreasing. Usually, planktonic populations are monitored by means of the Standard Methods plate count technique. However, not all organisms in the fouling process can be detected by this method. Anaerobic bacteria, such as sulfate-reducers which can cause under-deposit corrosion, are not revealed by aerobic cultural procedures. Special techniques must be used to ensure detection of these organisms (see Figures 26-5 and 26-6).

Sole reliance on bulk water counts will not provide sufficient information on the extent of surface fouling. Results must be interpreted in light of operating conditions at the time of sample collection. For example, in an untreated system a healthy, stable biofilm population may be present while bulk water counts are low, because few sessile organisms are being released from the fouled surface. If an antimicrobial is applied, bulk water counts may actually increase dramatically. This is due to disruption of the biofilm and sloughing of sessile organisms into the bulk water.

Figure 26-4. Microscopic examination of deposits can rapidly determine the principal microbial constituents and provide a photographic record.

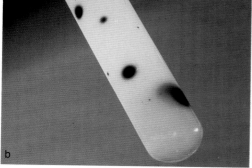

Figure 26-5. Sulfate-reducing bacteria are detected by formation of black FeS precipitates in specially prepared liquid (a) or solid (b) media.

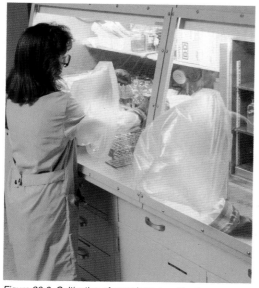

Figure 26-6. Cultivation of organisms sensitive to oxygen can be accomplished in this anaerobic glove box.

For a better diagnosis, it is necessary to use microbial monitoring techniques that allow more direct assessment of surface conditions. It is possible to clean a known surface area and suspend removed organisms in a known volume of sterile water. After this water is plated, back-calculation provides an approximation of the number of organisms on the original surface.

Another technique involves monitoring biochemical activity on a surface of a known area. A biofouled specimen is incubated with a suitable substrate. The concentration of reaction product found after a specific contact time relates to the numbers and health of organisms on the surface and consequently can be used as a measure of biofouling.

Regardless of which target population or monitoring technique is used, a single, isolated data point has little meaning. Various data must be compiled to generate a profile of microbiological trends in the system. This record should also include observations on equipment performance and operating conditions at the time of sample collection, thereby providing a meaningful context for interpretation of new data.

After it is determined that treatment is necessary to solve a fouling problem, an effective product must be chosen. Preliminary choices may be made only if the causative microbial agent is known, because the spectrum of activity of all antimicrobials is not the same. Some effectively control algae but not bacteria. For others, the reverse is true. For some, the activity spectrum, determined by inhibition of radiolabeled nutrient uptake, is quite broad, covering all common cooling water microbes (see Table 26-1 and Figure 26-7).

Knowledge of how different antimicrobials affect microorganisms is also useful in choosing the appropriate treatment. Some kill the organisms they contact. Others inhibit growth of organisms but do not necessarily kill them. These biostats can be effective if a suitable concentration is maintained in a system for a sufficient time (a continuous concentration is ideal).

A laboratory evaluation of the relative effectiveness of antimicrobials should be performed. This helps to identify those likely to work against the fouling organisms in the system and to eliminate those with little chance of success. Because the goal of antimicrobial treatment is control or elimination of biofilm organisms, it is helpful to conduct the evaluation with sessile organisms found in deposits, as well as planktonic organisms in the flowing water.

The objective of any treatment program should be to expose the attached microbial population to an antimicrobial dosage sufficient to penetrate and disrupt the biofilm. Generally, cleanup of a

Table 26-1. Antimicrobial efficacy.

	I_{50} (ppm)[a]							
	Bacteria		Fungi		Algae		Cyanobacteria	
Antimicrobial	*Klebsiella pneumoniae*	*Bacillus megaterium*	*Candida krusei*	*Trichoderma viride*	*Chlorella Pyrenoidosa*	*Scenedesmus obliquus*	*Anacystis nidulans*	*Anabaena flos-aquae*
Methylenebis-(thiocyanate)	2.7	1.5	3.1	0.7	1.4	1.2	1.0	2.5
Dibromonitrilo-propionamide	5.1	1.2	16	10	11	20	1.1	1.5
ß-bromo-ß-nitrostyrene	3.6	2.5	2.0	0.6	0.85	5.0	1.5	0.7
Bromonitro-propanediol	15	8.0	--	27	80	120	--	--

[a] I_{50} values are the concentrations that inhibit growth of the organism by 50% and are determined by use of a ^{14}C-labeled nutrient.

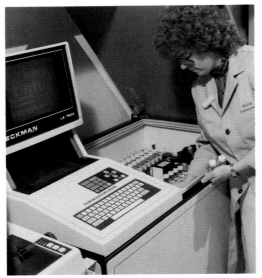

Figure 26-7. Microbial incorporation of radiolabeled nutrient, an indicator of microbial activity, is measured with this liquid scintillation counter.

fouled system requires higher concentrations of intermittently fed treatment, while maintenance of a clean system can be achieved with low-level continuous or semicontinuous feed. Given a certain level of fouling, the shorter the exposure time allowed by system operating conditions, the higher the required antimicrobial concentration. Conversely, if exposure times are long, control of the same level of fouling may be achieved with lower antimicrobial dosages.

Antimicrobial test results are most relevant when based on contact times derived from the system which is to be treated. Because once-through contact times are typically short, it can be very difficult to simulate these systems in lab testing. The longer contact times associated with recirculating cooling systems are easily duplicated in the lab.

In once-through systems, antimicrobials should be fed continuously to achieve the necessary contact time. Often, only low levels of antimicrobial are affordable on a continuous-feed basis. Semi-continuous treatment may be more economical or may be required because of effluent restrictions. Such an intermittent program for once-through systems must still be designed to achieve an effective antimicrobial concentration throughout the system, using treatment periods which range from minutes to hours per day.

Recirculating systems can also be treated continuously or intermittently, although intermittent treatment programs are more common. The purpose of intermittent treatment in these systems is to generate a high concentration of antimicrobial which will penetrate and disrupt the biofilm and eventually dissipate. When the treatment level drops below the toxic threshold, microbial growth begins again. After a period of multiplication, new growth is removed with another shock dose. As stated earlier, previously fouled surfaces can be recolonized at an accelerated rate. Therefore, the period for growth and removal may vary within a system, and certainly will vary among systems, even those using the same water source.

The time at which an antimicrobial concentration drops below a threshold concentration in a recirculating system can be determined mathematically. Such information can be very useful in planning a schedule for effective and economical slime control. The threshold concentration desired should be estimated from a toxicant evaluation. The theoretical depletion of the antimicrobial from a system can be determined from the following formula:

$$\log C_f = \log C_i - \frac{BD \times T}{2.303\,V}$$

where:

C_f = final concentration, ppm
C_i = initial concentration, ppm
BD = blowdown and windage loss, gpm
V = system capacity, gal
T = time, min

It is standard practice to repeat the shock treatment when C_f is 25% of C_i. On this basis, the time interval for antimicrobial addition can be calculated as follows:

$$T = 1.385 \times \frac{V}{BD}$$

Solving this equation for T will indicate how frequently a slug should be added to the system, but this determination is only valid for 75% depletion or two half-lives.

The equations provided are not valid for the following compounds:

■ compounds which are volatile and may be lost during passage over the tower

■ compounds which react with substances in the water (i.e., a demand)

■ compounds which degrade in water

In the planning of a slime control program, any chemical demand of process waters for the antimicrobial being used must also be considered. Failure to allow for the chemical demand may prevent attainment of the necessary threshold concentration and may lead to the failure of the treatment program. The compatibility of the antimicrobial with other treatments added to the water should also be considered.

Many system variables influence the behavior of microbes in the system, and the effects of antimicrobials can also be influenced by these variables. Therefore, careful consideration must be given to the determination of whether, when, and where to treat a cooling water system.

Cost is a primary criterion for selecting a slime control program. It cannot be determined without knowledge or estimation of the individual costs of chemicals, feed equipment, and labor required to apply and monitor the program, along with effluent treatment requirements. In addition,

possible adverse effects of implementing the program must be weighed against those which would result if no treatment is made. Knowledge of component costs can help guide implementation of the program. For example, if labor costs are high, it may be more economical to feed an antimicrobial more frequently and reduce the amount of monitoring required. Each system must be considered individually, and seasonal adjustments may also be required.

CHARACTERISTICS OF NONOXIDIZING CHEMICALS USED IN COOLING SYSTEMS

Antimicrobials used for microbiological control can be broadly divided into two groups: oxidizing and nonoxidizing. Oxidizers, such as chlorine and bromine, are addressed in Chapter 27; nonoxidizers are discussed in the balance of this chapter.

Only a relative distinction can be made between oxidizing and nonoxidizing antimicrobials, because certain nonoxidizers have weak to mild oxidizing properties. The more significant difference between the two groups relates to mode of action. Nonoxidizing antimicrobials exert their effects on microorganisms by reaction with specific cell components or reaction pathways in the cell. Oxidizing antimicrobials are believed to kill by a more indiscriminate oxidation of chemical species on the surface or within the cell.

An understanding of the chemistries and modes of action of antimicrobials is needed to ensure their proper use and an appreciation of their limitations.

Two characteristic mechanisms typify many of the nonoxidizing chemicals applied to cooling systems for biofouling control. In one, microbes are inhibited or killed as a result of damage to the cell membrane. In the other, microbial death results from damage to the biochemical machinery involved in energy production or energy utilization.

Quaternary ammonium compounds (quats) are cationic surface-active molecules. They damage the cell membranes of bacteria, fungi, and algae. As a result, compounds that are normally prevented from entering the cell are able to penetrate this permeability barrier. Conversely, nutrients and essential intracellular components concentrated within the cell leak out. Growth is hindered, and the cell dies. At low concentrations, quats are biostatic because many organisms can survive in a damaged state for some time. However, at medium to high concentrations, quats can control the organisms.

Many antimicrobials interfere with energy metabolism. Because all microbial activity ultimately depends on the orderly transfer of energy, it can be anticipated that interference with the many energy-yielding or energy-trapping reactions will have serious consequences for the cell. Antimicrobials known to inhibit energy metabolism include the following:

- organotins

- bis(trichloromethyl) sulfone

- methylenebis(thiocyanate) (MBT)

- ß-bromo-ß-nitrostyrene (BNS)

- dodecylguanidine salts

- bromonitropropanediol (BNPD)

All of these compounds are effective when applied in sufficient concentrations. Dodecylguanidine salts also have surfactant properties, which probably contribute to their effectiveness.

The exact sites or reactions which are affected by such metabolic inhibitors are frequently unknown, although laboratory experiments may provide clues or indirect evidence for specific mechanisms. Tin and other heavy metals in sufficient concentrations cause proteins to lose their characteristic three-dimensional structures, which are required for normal function. Some antimicrobials, such as methylenebis(thiocyanate) (MBT), are believed to bind irreversibly to biomolecules, preventing the sequential reduction and oxidation these molecules must undergo in order to function.

Bromonitropropanediol (BNPD), a newer cooling water control agent, can be shown to catalyze the formation of disulfide bonds (R-S-S-R) between sulfhydryl groups (R-SH). Proteins contain sulfhydryls, and because enzymes are largely protein, it is possible to speculate that the formation of disulfide bonds between adjacent -SH groups may block enzyme activity. Many different enzymes contain sulfhydryl groups, so this antimicrobial may affect a wide range of microbial activities in addition to energy generation.

The mode of action of one common nonoxidizer cannot be categorized as either a surface-active or metabolic inhibitor. The active, dibromonitrilopropionamide (DBNPA), seems to behave somewhat like an oxidizing antimicrobial, reacting quite rapidly with bacterial cells. Studies of the interaction of radioactively labeled [^{14}C]DBNPA

with bacteria have shown that the ^{14}C label never penetrates the cell, as would be necessary for it to become involved with energy metabolism. Instead, it binds strongly and rapidly to the cell walls of the bacteria. However, analogous studies with [^{14}C]-ß-bromo-ß-nitrostyrene (BNS) have shown that the nitrostyrene penetrates the bacterial cell and accumulates within at concentrations far above the external concentration. Thus, although the mode of action of DBNPA is unknown, it most probably is unlike the other mechanisms known for nonoxiding antimicrobials.

Some antimicrobials used in cooling systems are compounds that spontaneously break down in water, thereby alleviating some potential environmental hazards. This chemical breakdown is often accompanied by a reduction in the toxicity of the compound. The compound can be added to the cooling water system, accomplish its task of killing the microbes in the system, and then break down into less noxious chemicals. Among the antimicrobials which have this attribute are BNS, MBT, DBNPA, and BNPD.

PROPRIETARY NONOXIDIZING CHEMICALS

The dynamics of microbial populations in cooling water systems are complex. In situations where one microbial group or species dominates, fouling problems can occur. In other instances, a balanced population mix can exist while no fouling is evident. One explanation for such observations is that when balanced populations coexist, they compete with each other for the available nutrients and control each other's growth. When one group successfully displaces the others, its growth can proceed without competition.

Because of such considerations, some proprietary antimicrobials are formulated to contain more than one active. Proper blending of actives can compensate for limitations in the spectrum of kill shown by one or more of the actives. For example, if antimicrobial *A* is effective for bacteria but poor for fungi, large amounts of *A* might have to be used to control potential fungal problems. However, if antimicrobial *B* is fair for bacteria and good for fungi, a combination of *A* and *B* would broaden the spectrum of control and thus be preferable to high concentrations of *A* alone.

With no increase in the amount of antimicrobial used, the power of a blend can exceed that expected from a simple additive effect. This greatly enhanced performance or synergism is obtained from only certain combinations of actives. Synergism allows microbial control at much lower combined concentrations of A and B than could be achieved with either A or B alone. Use of products composed of synergistic active blends can result in reduced treatment concentrations in the blowdown water, as well as cost savings.

The spectrum of control can also be broadened by sequential feeding of antimicrobials to a system: alternate feeding of two actives can have the same outcome as blending of the actives for simultaneous feeding.

Another variation to be considered is the possible proliferation of resistant microbes in the system. The resistant forms may arise spontaneously by mutation within the cooling system but are much more likely to originate outside of the system. The antimicrobial simply functions to reduce competition by nonresistant forms and permits the unchecked growth of the newly introduced resistant organisms. This is more likely to occur during treatment with a single antimicrobial active ingredient, because the probability of a microbe being resistant to more than one active is extremely low when the actives are dissimilar. Sequentially added and synergistically blended antimicrobials are probably equally effective in eradicating antimicrobial-resistant microbes from a cooling system.

Development of an effective antimicrobial program is aided by an understanding of the mode of action of the products, the system to be treated, and environmental effects. All of these factors play a role in the selection of an effective and economical microbiological control program that is safe for the environment.

CHLORINE AND CHLORINE ALTERNATIVES

CHLORINE

Chlorine is one of the most versatile chemicals used in water and wastewater treatment. This powerful oxidizing agent is used for:

- disinfection
- control of microorganisms
- removal of ammonia
- control of taste and odor
- color reduction
- destruction of organic matter
- hydrogen sulfide oxidation
- iron and manganese oxidation

Although chlorine is beneficial for many uses, its use carries safety and environmental concerns.

Physical Properties and Aqueous Reactions

Chlorine in its gaseous state was discovered by Karl W. Scheele in 1774 and identified as an element by Humphrey Davy in 1810. Chlorine gas is greenish-yellow, and its density is about 2½ times that of air. When condensed, it becomes a clear, amber liquid with a density about 1½ times greater than water. One volume of liquid chlorine yields approximately 500 volumes of chlorine gas, which is neither explosive nor flammable. Like oxygen, chlorine gas can support the combustion of some substances. Chlorine reacts with organic materials to form oxidized or chlorinated derivatives. Some of these reactions, such as those with hydrocarbons, alcohols, and ethers, can be explosive. The formation of other chlorinated organics, specifically trihalomethanes (THM), poses an environmental threat to public drinking water supplies.

Chlorine gas is also a toxic respiratory irritant. Airborne concentrations greater than 3–5 ppm by volume are detectable by smell, and exposure to 4 ppm for more than 1 hr can have serious respiratory effects. Because chlorine gas is denser than air, it stays close to the ground when released. The contents of a 1-ton cylinder of chlorine can cause coughing and respiratory discomfort in an area of 3 square miles. The same amount concentrated over an area of ¹/₁₀ square mile can be fatal after only a few breaths.

Chlorine is generated commercially by the electrolysis of a brine solution, typically sodium chloride, in any of three types of cells: diaphragm, mercury, or membrane. The majority of chlorine produced in the United States is manufactured by the electrolysis of sodium chloride to form chlorine gas and sodium hydroxide in diaphragm cells. The mercury cell process produces a more concentrated caustic solution (50%) than the diaphragm cell. Chlorine gas can also be generated by the salt process (which employs the reaction between sodium chloride and nitric acid), by the hydrochloric acid oxidation process, and by the electrolysis of hydrochloric acid solutions. The gas is shipped under pressure in 150-lb cylinders, 1-ton cylinders, tank trucks, tank cars, and barges.

The four basic categories of chlorine treatment are defined not only by their function but also by their position in a water treatment sequence:

- *prechlorination*—the application of chlorine to water prior to other treatment processes
- *rechlorination*—the application of chlorine, at one or more points, to water that has already been chlorinated
- *post-chlorination*—the application of chlorine to a water effluent from other treatment processes
- *dechlorination*—the partial or complete removal of residual chlorine by chemical or physical methods

In chemically pure water, molecular chlorine reacts with water and rapidly hydrolyzes to hypochlorous acid (HOCl) and hydrochloric acid (HCl):

$$\underset{\text{chlorine}}{Cl_2} + \underset{\text{water}}{H_2O} \rightarrow \underset{\substack{\text{hypochlorous}\\\text{acid}}}{HOCl} + \underset{\substack{\text{hydrochloric}\\\text{acid}}}{HCl}$$

Both of the acids formed by hydrolysis react with alkalinity to reduce buffering capacity of water and lower pH. Every pound of chlorine gas added to water removes about 1.4 lb of alkalinity. In cooling water systems, this alkalinity reduction can have a major impact on corrosion rates.

At pH levels above 4.0 and in dilute solutions, the hydrolysis reaction is completed within a fraction of a second. For all practical purposes the reaction is irreversible. Hypochlorous acid is a weak acid and dissociates to form a hydrogen ion and a hypochlorite ion.

$$\underset{\substack{\text{hypochlorous}\\\text{acid}}}{HOCl} \rightleftharpoons \underset{\substack{\text{hydrogen}\\\text{ion}}}{H^+} + \underset{\substack{\text{hypochlorite}\\\text{ion}}}{OCl^-}$$

The concentration or distribution of each species at equilibrium depends on pH and temperature. Between pH 6.5 and 8.5, the dissociation reaction is incomplete, and both hypochlorous acid and hypochlorite ions are present. Figure 27-1 illustrates the relative concentrations of both species over a broad pH range. The equilibrium ratio at any given pH remains constant even if the hypochlorous acid concentration is decreasing. At constant pH and increasing temperature, chemical equilibrium favors the OCl⁻ ion over HOCl.

The primary oxidizing agents in water are hypochlorous acid and the hypochlorite ion, although hypochlorite has a lower oxidizing potential. Oxidizing potential is a measure of the tendency of chlorine to react with other materials.

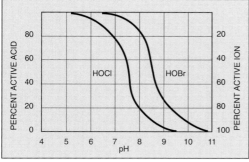

Figure 27-1. Dissociation of hypobromous and hypochlorous acid with changing pH.

The speed at which these reactions occur is determined by pH, temperature, and oxidation/reduction potential. As pH increases, the chemical reactivity of chlorine decreases; as temperature increases, reactions proceed more rapidly. The oxidation reactions of chlorine with such inorganic reducing agents as sulfides, sulfites, and nitrites are generally very rapid. Some dissolved organic materials also react rapidly with chlorine, but the completion of many organic–chlorine reactions can take hours.

Chlorine Demand. Chlorine demand is defined as the difference between the amount of chlorine added to a water system and the amount of free available chlorine or combined available chlorine remaining at the end of a specified time period. The demand is the amount of chlorine consumed by oxidation or substitution reactions with inorganic and organic materials, such as H_2S, Fe^{2+}, Mn^{2+}, NH_3, phenols, amino acids, proteins, and carbohydrates. Free available residual chlorine is the amount of chlorine which exists in the treated water system as hypochlorous acid and hypochlorite ions after the chlorine demand has been satisfied. Free residual chlorination is the application of chlorine to water to produce a free available chlorine residual.

Combined Available Residual Chlorination. Combined available residual chlorine is the chlorine residual which exists in water in combination with ammonia or organic nitrogen compounds. Combined residual chlorination is the application of chlorine to water to react with ammonia (natural or added) or other nitrogen compounds to produce a combined available chlorine residual. Total available chlorine is the total of free available chlorine, combined available chlorine, and other chlorinated compounds.

Available Chlorine. "Available chlorine" is an expression of the equivalent weights of oxidizing agents, with chlorine gas as a basis, similar to the expression of alkalinity in terms of calcium carbonate equivalents. The term originated from the need to compare other chlorine-containing compounds to gaseous chlorine. Available chlorine is based on the half-cell reaction in which chlorine gas is reduced to chloride ions with the consumption of two electrons. In this reaction, the equivalent weight of chlorine is the molecular weight of chlorine, 71 g/mole, divided by 2, or 35.5 g/mole.

$$2e^- + \underset{\text{chlorine}}{Cl_2} \rightarrow \underset{\text{chloride ions}}{2Cl^-}$$

The available chlorine of other chlorine-containing compounds is calculated from similar half-cell reactions, the formula weight of the compound, and the equivalent weight of chlorine.

Even though chlorine gas only dissociates into 50% HOCl or OCl⁻, it is considered 100% available chlorine. Because of this definition, it is possible for a compound to have more than 100% available chlorine. The active weight percent chlorine multiplied by 2 indicates available chlorine. Table 27-1 lists actual weight percent and available chlorine percent for several common compounds.

Available chlorine, like oxidation potential, is not a reliable indicator of the occurrence or extent of an oxidation reaction. It is an even poorer indicator of the antimicrobial efficacy of an oxidizing compound. For example, the antimicrobial efficacy of hypochlorous acid (HOCl) is much greater than any of the chloramines even though the chloramines have a higher available chlorine.

Chloramine Formation. One of the most important reactions in water conditioning is the reaction between dissolved chlorine in the form of hypochlorous acid and ammonia (NH_3) to form inorganic chloramines. The inorganic chloramines consist of three species: monochloramine (NH_2Cl), dichloramine ($NHCl_2$), and trichloramine, or nitrogen trichloride (NCl_3). The principal reactions of chloramine formation are:

$$NH_3 \text{ (aq)} + HOCl \rightleftharpoons NH_2Cl + H_2O$$
ammonia   hypochlorous   monochloramine   water
                     acid

$$NH_2Cl + HOCl \rightleftharpoons NHCl_2 + H_2O$$
monochloramine   hypochlorous   dichloramine   water
                          acid

$$NHCl_2 + HOCl \rightleftharpoons NCl_3 + H_2O$$
dichloramine   hypochlorous   trichloramine   water
                      acid

Table 27-1. Actual weight percent and available chlorine percent of common compounds.

Compound	Weight % Halogen	% Available as Chlorine
Cl_2 Gas	100	100
15 trade % NaOCl	6	12
$C_5H_6BrClN_2O_2$ Solid (BCDMH)	45	55
ClO_2	53	260
NH_2Cl	68	138
$NHCl$	83	165
NCl_3	89	266
70% $Ca(OCl)_2$ Solid	35	70

The relative amounts of chloramines formed are a function of the amount of chlorine fed, the chlorine/ammonia ratio, temperature, and pH. In general, monochloramine is formed above pH 7 and predominates at pH 8.3. Dichloramine predominates at pH 4.5. Between these pH values, mixtures of the two chloramines exist. Below pH 4.5, nitrogen trichloride is the predominant reaction product.

The oxidizing potential of monochloramines is substantially lower than that of chlorine, and monochloramines are slower to react with organic matter. These properties reduce the amount of trihalomethanes (THM) formed. The formation of THM is considered more detrimental in potable water than the reduction of the antimicrobial capabilities of free chlorine. Therefore, ammonia is often injected into the chlorine feed stream to form chloramines before the chlorine is fed into the potable water stream.

Combined chlorine residuals are ordinarily more chemically stable (less reactive with chlorine demand) than free chlorine residuals. This property helps maintain stable residuals in outlying pressurized water distribution systems. However, the lower antimicrobial effectiveness of chloramines compared to free chlorine requires higher combined residuals and/or longer contact times, which are often available in distribution systems.

Breakpoint Chlorination. Breakpoint chlorination is the application of sufficient chlorine to maintain a free available chlorine residual. The principal purpose of breakpoint chlorination is to ensure effective disinfection by satisfying the chlorine demand of the water. In wastewater treatment, breakpoint chlorination is a means of eliminating ammonia, which is converted to an oxidized volatile form.

A theoretical breakpoint chlorination curve is shown in Figure 27-2. The addition of chlorine to

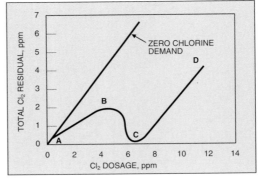

Figure 27-2. Theoretical breakpoint chlorination curve.

a water that contains ammonia or nitrogen-containing organic matter produces an increased combined chlorine residual. Mono- and dichloramines are formed between points A and B on this curve. After the maximum combined residual is reached (point B), further chlorine doses decrease the residual. Chloramine oxidation to dichloramine, occurring between points B and C, results in a decline in the combined available residuals initially formed. Point C represents the breakpoint: the point at which chlorine demand has been satisfied and additional chlorine appears as free residuals. Between points C and D, free available residual chlorine increases in direct proportion to the amount of chlorine applied.

Factors that affect breakpoint chlorination are initial ammonia nitrogen concentration, pH, temperature, and demand exerted by other inorganic and organic species. The weight ratio of chlorine applied to initial ammonia nitrogen must be 8:1 or greater for the breakpoint to be reached. If the weight ratio is less than 8:1, there is insufficient chlorine present to oxidize the chlorinated nitrogen compounds initially formed. When instantaneous chlorine residuals are required, the chlorine needed to provide free available chlorine residuals may be 20 or more times the quantity of ammonia present. Reaction rates are fastest at pH 7–8 and high temperatures.

A typical breakpoint curve is shown in Figure 27-3. The initial chlorine dosage produces no residual because of an immediate chlorine demand caused by fast-reacting ions. As more chlorine is applied, chloramines develop. These chloramines are shown in the total chlorine residual. At higher chlorine dosages, the slope to breakpoint begins. After the breakpoint, free chlorine residuals develop.

Free chlorine residuals usually destroy tastes and odors, control exposed bacteria, and oxidize

organic matter. Breakpoint chlorination can also control slime and algae growths, aid coagulation, oxidize iron and manganese, remove ammonia, and generally improve water quality in the treatment cycle or distribution system.

OXIDIZING ANTIMICROBIALS IN INDUSTRIAL COOLING SYSTEMS

The oxidizing antimicrobials commonly used in industrial cooling systems are the halogens, chlorine and bromine, in liquid and gaseous form; organic halogen donors; chlorine dioxide; and, to a limited extent, ozone.

Oxidizing antimicrobials oxidize or accept electrons from other chemical compounds. Their mode of antimicrobial activity can be direct chemical degradation of cellular material or deactivation of critical enzyme systems within the bacterial cell. An important aspect of antimicrobial efficiency is the ability of the oxidizing agent to penetrate the cell wall and disrupt metabolic pathways. For this reason, oxidation potential alone does not always correlate directly with antimicrobial efficiency.

The relative microbiological control ability of typical halogens is as follows:

$$HOCl \geq HOBr \geq NH_xBr_y >>$$

hypochlorous   hypobromous   bromamine
acid   acid

$$OCl^- > OBr^- >>> NH_xCl_y$$

hypochlorite   hypobromite   chloramine
ion   ion

Cooling water pH affects oxidizing antimicrobial efficacy. pH determines the relative proportions of hypochlorous acid and hypochlorite ion or, in systems treated with bromine donors, hypobromous acid and hypobromite ion. The acid forms of the halogens are usually more effective antimicrobials than the dissociated forms. Under some conditions, hypochlorous acid is 80 times more effective in controlling bacteria than the hypochlorite ion. Hypochlorous acid predominates below a pH of 7.6. Hypobromous acid predominates below pH 8.7, making bromine donors more effective than chlorine donors in alkaline cooling waters, especially where contact time is limited.

Antimicrobial efficacy is also affected by demand in the cooling water system, specifically demand exerted by ammonia. Chlorine reacts with ammonia to form chloramines, which are not as efficacious as hypochlorous acid or the hypochlorite ion in microbiological control.

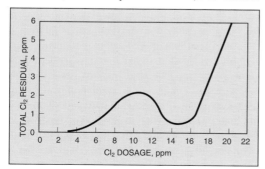

Figure 27-3. Typical breakpoint chlorination curve.

Bromine reacts with ammonia to form bromamines. Unlike chloramines, bromamines are unstable and re-form hypobromous acid.

Most microbes in cooling systems can be controlled by chlorine or bromine treatment if exposed to a sufficient residual for a long enough time. A free chlorine residual of 0.1–0.5 ppm is adequate to control bulk water organisms if the residual can be maintained for a sufficient period of time.

Continuous chlorination of a cooling water system often seems most prudent for microbial slime control. However, it is economically difficult to maintain a continuous free residual in some systems, especially those with process leaks. In some high-demand systems it is often impossible to achieve a free residual, and a combined residual must be accepted. In addition, high chlorine feed rates, with or without high residuals, can increase system metal corrosion and tower wood decay. Supplementing with nonoxidizing antimicrobials is preferable to high chlorination rates.

In once-through systems, free residuals from 0.3 to 0.8 ppm are normally maintained for $1/2$–2 hr per treatment period. The rate of recontamination determines the frequency of treatment required.

Open recirculating systems can be treated with a continuous or intermittent halogenation program. Continuous feed is the most effective and is generally affordable where chlorine gas or hypochlorite is being used and system demand is low. Free residuals of 0.1–0.5 ppm are manually maintained. Care should be taken not to feed excessive amounts of halogen that will adversely affect corrosion rates. Chlorine feed rates should not exceed 4 ppm based on recirculation rate. The use of halogen donors may be restricted to intermittent feed due to economics, although continuous feed in low-demand systems is effective. Intermittent feed requires maintaining the same free residual as in the continuous program but for only the last $1/2$ to 1 hr of the chlorine application. Up to 3 hr of chlorine addition may be required in order to achieve the free residual, depending on system demand, system cleanliness, and the frequency of chlorination.

GASEOUS CHLORINE ALTERNATIVES

Community right-to-know laws, Superfund Reauthorization, SARA Title III laws, and the release of a deadly gas in Bhopal, India, have raised serious concerns regarding the safety of gaseous chlorine. Other sources of halogens and oxidizing agents for microbiological control include:

- hypochlorites (sodium hypochlorite, sodium hypochlorite with sodium bromide, and calcium hypochlorite)

- chlorinated or brominated donor molecules, such as isocyanurates, trichloro-*s*-triazinetriones, and hydantoins

- chlorine dioxide

- ozone

Hypochlorites

Sodium hypochlorite and calcium hypochlorite are chlorine derivatives formed by the reaction of chlorine with hydroxides. The application of hypochlorite to water systems produces the hypochlorite ion and hypochlorous acid, just as the application of chlorine gas does.

$$\underset{\substack{\text{sodium} \\ \text{hypochlorite}}}{NaOCl} \rightarrow \underset{\substack{\text{hypochlorite} \\ \text{ion}}}{OCl^-} + \underset{\substack{\text{sodium} \\ \text{ion}}}{Na^+}$$

$$\underset{\substack{\text{hypochlorite} \\ \text{ion}}}{OCl^-} + \underset{\substack{\text{sodium} \\ \text{ion}}}{Na^+} + \underset{\text{water}}{H_2O} \rightleftharpoons \underset{\substack{\text{hypochlorous} \\ \text{acid}}}{HOCl} + \underset{\substack{\text{sodium} \\ \text{hydroxide}}}{NaOH}$$

$$\underset{\substack{\text{calcium} \\ \text{hypochlorite}}}{Ca(OCl)_2} \rightarrow \underset{\substack{\text{hypochlorite} \\ \text{ion}}}{2OCl^-} + \underset{\substack{\text{calcium} \\ \text{ion}}}{Ca^{2+}}$$

$$\underset{\substack{\text{hypochlorite} \\ \text{ion}}}{2OCl^-} + \underset{\substack{\text{calcium} \\ \text{ion}}}{Ca^{2+}} + \underset{\text{water}}{2H_2O} \rightleftharpoons \underset{\substack{\text{hypochlorous} \\ \text{acid}}}{2HOCl} + \underset{\substack{\text{calcium} \\ \text{hydroxide}}}{Ca(OH)_2}$$

The difference between the hydrolysis reaction of chlorine gas and hypochlorites is the reaction by-products. The reaction of chlorine gas and water increases the H^+ ion concentration and decreases pH by the formation of hydrochloric acid. The reaction of hypochlorites and water forms both hypochlorous acid and sodium hydroxide or calcium hydroxide. This causes little change in pH. Solutions of sodium hypochlorite contain minor amounts of excess caustic as a stabilizer, which increase alkalinity and raise pH at the point of injection. This can cause hardness scale to form. Addition of a dispersant (organic phosphate/polymer) to the water system is usually sufficient to control this scaling potential.

Alkalinity and pH are significantly changed when sodium or calcium hypochlorite replaces gaseous chlorine. Gaseous chlorine reduces alkalinity by 1.4 ppm per ppm of chlorine fed; hypochlorite does not reduce alkalinity. The higher alkalinity of waters treated with hypochlorite reduces the corrosion potential but can increase the deposition potential.

Sodium Hypochlorite. Sodium hypochlorite, also referred to as liquid bleach, is the most widely used of all the chlorinated bleaches. It is available in several solution concentrations, ranging from the familiar commercial variety at a concentration of about 5.3 weight percent NaOCl to industrial strengths at concentrations of 10–12%. The strength of a bleach solution is commonly expressed in terms of its "trade percent" or "percent by volume," not its weight percent: 15 trade percent hypochlorite is only 12.4 weight percent hypochlorite. Approximately 1 gal of industrial strength sodium hypochlorite is required to replace 1 lb of gaseous chlorine.

The stability of hypochlorite solutions is adversely affected by heat, light, pH, and metal contamination. The rate of decomposition of 10% and 15% solutions nearly doubles with every 10 °F rise in the storage temperature. Sunlight reduces the half-life of a 10%–15% hypochlorite solution by a factor of 3 to 5. If the pH of a stored solution drops below 11, decomposition is more rapid. As little as 0.5 ppm of iron causes rapid deterioration of 10–15% solutions. The addition of concentrated ferric chloride to a tank of sodium hypochlorite causes the rapid generation of chlorine gas.

Normal industrial grades of sodium hypochlorite may be fed neat or diluted with low-hardness water. The use of high-hardness waters for dilution can cause precipitation of calcium salts due to the high pH of the hypochlorite solution.

"High Test" Calcium Hypochlorite (HTH). The most common form of dry hypochlorite in the United States is high test calcium hypochlorite (HTH). It contains 70% available chlorine, 4–6% lime, and some calcium carbonate. Precipitates form when HTH is dissolved in hard water. For feeding calcium hypochlorite as a liquid, solutions should be prepared with soft water at 1–2% chlorine concentration. Care should be exercised in storing granular calcium hypochlorite. It should not be stored where it may be subjected to heat or contacted by easily oxidized organic material. Calcium hypochlorite decomposes exothermally, releasing oxygen and chlorine monoxide. Decomposition occurs if HTH is contaminated with water or moisture from the atmosphere. Calcium hypochlorite loses 3–5% of its chlorine content per year in normal storage.

All hypochlorites are somewhat harmful to skin and must be handled carefully. Corrosion-resistant materials should be used for storing and dispensing.

BROMINE

Bromine has been used for water treatment since the 1930's. Most bromine production in the United States occurs in the Great Lakes region and Arkansas. Bromine is generated commercially through the reaction of a bromine brine solution with gaseous chlorine, followed by stripping and concentration of the bromine liquid. Bromine is a fuming, dark red liquid at room temperature.

Bromine dissociates in water in the same manner as chlorine, by forming hypobromous acid and the hypobromite ion. Hypobromous acid is a weak acid that partially dissociates to form a hydrogen ion and a hypobromite ion. The concentration or distribution of each species at equilibrium depends on pH and temperature. Between pH 6.5 and 9, the dissociation reaction is incomplete, and both hypobromous acid and the hypobromite ion are present. Figure 27-1 (above) illustrates the relative concentrations of both species over a broad pH range. The equilibrium ratio at any given pH remains constant. At a pH above 7.5, the amount of hypobromous acid is greater than the amount of hypochlorous acid for equivalent feed rates. The higher percentage of hypobromous acid is beneficial in alkaline waters and in ammonia-containing waters.

Methods of generating hypobromous acid include:

using two liquids (or one liquid and chlorine gas)

$$NaBr + HOCl \rightarrow HOBr + NaCl$$

sodium bromide — hypochlorous acid — hypobromous acid — salt

using a compressed gas

$$BrCl + H_2O \rightarrow HOBr + HCl$$

bromine chloride — water — hypobromous acid — hydrochloric acid

using a solid

$$C_5H_6BrClN_2O_2 + 2H_2O \rightarrow HOCl +$$

bromochloro-dimethylhydantoin (BCDMH) — water — hypochlorous acid

$$HOBr + C_5H_8N_2O_2$$

hypobromous acid — dimethyl-hydantoin

Regardless of the method used to generate hypobromous acid, the goal is to take advantage of its antimicrobial activity. The liquid and solid methods do not require the storage of compressed gases—the major reason for gaseous chlorine replacement.

Bromine reacts with ammonia compounds to form bromamines, which are much more effective antimicrobials than chloramines. At pH 8.0, the hypobromous acid to bromamine ratio is 8:1 in ammonia-containing waters. Because monobromamine is unstable and because tribromamine is not formed, there is little need to proceed to breakpoint bromination.

The shorter life expectancy of bromine compounds (due to lower bond strength) lowers oxidizer residuals in plant discharges and reduces the need to dechlorinate before discharge.

HALOGEN DONORS

Halogen donors are chemicals that release active chlorine or bromine when dissolved in water. After release, the halogen reaction is similar to that of chlorine or bromine from other sources. Solid halogen donors commonly used in cooling water systems include the following:

- 1-bromo-3-chloro-5,5-dimethylhydantoin

- 1,3-dichloro-5,5-dimethylhydantoin

- sodium dichloroisocyanurate

These donor chemicals do not release the active halogen all at once, but make it slowly available; therefore, they may be considered "controlled release" oxidizing agents. Their modes of action are considered to be similar to chlorine or bromine, but they can penetrate cell membranes and carry out their oxidative reactions from within the cell. These donors are being widely used because of the simplicity, low capital cost, and low installation cost of the feed systems. In addition, because they are solids, they eliminate the handling hazards associated with gases (escapement) and liquids (spills). Evaluated on a total cost basis, halogen donors often prove to be an economical choice despite their relatively high material costs.

CHLORINE DIOXIDE

Chlorine dioxide, ClO_2, is another chlorine derivative. This unstable, potentially explosive gas must be generated at the point of application. The most common method of generating ClO_2 is through the reaction of chlorine gas with a solution of sodium chlorite.

$$2NaClO_2 \ + \ Cl_2 \ \rightarrow \ 2ClO_2 \ + \ 2NaCl$$

sodium chlorite   chlorine   chlorine dioxide   sodium chloride

Theoretically, 1 lb of chlorine gas is required for each 2.6 lb of sodium chlorite. However, an excess of chlorine is often used to lower the pH to the required minimum of 3.5 and to drive the reaction to completion. Sodium hypochlorite can be used in place of the gaseous chlorine to generate chlorine dioxide. This process requires the addition of sulfuric or hydrochloric acid for pH control.

Other methods used for chlorine dioxide generation include:

$$5NaClO_2 \ + \ 5HCl \ \rightarrow \ 4ClO_2 \ +$$

sodium chlorite   hydrochloric acid   chlorine dioxide

$$5NaCl \ + \ HCl \ + \ 2H_2O$$

sodium chloride   hydrochloric acid   water

$$10NaClO_2 \ + \ 5H_2SO_4 \ \rightarrow \ 8ClO_2 \ +$$

sodium chlorite   sulfuric acid   chlorine dioxide

$$5Na_2SO_4 \ + \ 2HCl \ + \ 4H_2O$$

sodium sulfate   hydrochloric acid   water

$$2NaClO_2 \ + \ HCl \ + \ NaOCl \ \rightarrow$$

sodium chlorite   hydrochloric acid   sodium hypochlorite

$$2ClO_2 \ + \ 2NaCl \ + \ NaOH$$

chlorine dioxide   sodium chloride   sodium hydroxide

Rather than hydrolyzing in water as chlorine does, chlorine dioxide forms a true solution in water under typical cooling tower conditions. For this reason, chlorine dioxide is volatile (700 times more volatile than HOCl) and may be easily lost from treated water systems, especially over cooling towers.

Chlorine dioxide is a powerful oxidant. It reacts rapidly with oxidizable materials but, unlike chlorine, does not readily combine with ammonia. Chlorine dioxide does not form trihalomethanes (THM) but can significantly lower THM precursors. In sufficient quantity, chlorine dioxide destroys phenols without creating the taste problems of chlorinated phenols. It is a good antimicrobial and antispore. Unlike chlorine, the antimicrobial efficiency of chlorine dioxide is relatively unaffected by changes in pH in the range of 6–9. Chlorine dioxide is also used for the oxidation of sulfides, iron, and manganese.

Complex organic molecules and ammonia are traditional chlorine-demand materials that do not react with chlorine dioxide. Because chlorine dioxide reacts differently from chlorine, a chlorine dioxide demand test must be conducted to determine chlorine dioxide feed rates. A residual must be maintained after the chlorine dioxide demand has been met, to ensure effective control of microbiological growth. The chemical behavior and oxidation characteristics of aqueous chlorine dioxide are not well understood because of the difficulty in differentiating aqueous chlorine-containing species.

Chlorine dioxide is applied to some public water supplies to control taste and odor, and as a disinfectant. It is used in some industrial treatment processes as an antimicrobial. Chlorine dioxide consumed in water treatment reactions reverts to chlorite ions (ClO_2^-), chlorate ions (ClO_3^-), and chloride ions (Cl^-). There are some concerns about the long-term health effects of the chlorite ion in potable water supplies.

As a gas, chlorine dioxide is more irritating and toxic than chlorine. Chlorine dioxide in air is detectable by odor at 14–17 ppm, irritating at 45 ppm, fatal in 44 min at 150 ppm, and rapidly fatal at 350 ppm. Concentrations greater than 14% in air can sustain a decomposition wave set off by an electric spark. The most common precursor for on-site generation of chlorine dioxide is also a hazardous material: liquid sodium chlorite. If allowed to dry, this powerful oxidizing agent forms a powdered residue that can ignite or explode if contacted by oxidizable materials. The hazardous nature of chlorine dioxide vapor and its precursor, and the volatility of aqueous solutions of chlorine dioxide, require caution in the design and operation of solution and feeding equipment.

OZONE

Ozone is an allotropic form of oxygen, O_3. Because it is an unstable gas, it must be generated at the point of use. Ozone is a very effective, clean oxidizing agent possessing powerful antibacterial and antiviral properties.

Because ozone is a strong oxidizing agent, it is a potential safety hazard. It has been reported that concentrations of 50 ppm of ozone in the air can cause oxidization of the lining of the lungs and accumulation of fluid, resulting in death by pulmonary edema. OSHA and NIOSH consider 10 ppm immediately dangerous to life or health, and the OSHA exposure limit is a time-weighted

average of 0.1 ppm. In concentrations as low as 0.02 ppm, strong ozone odors are detectable. Improper operation of ozone-generating equipment can produce 20% ozone, an explosive concentration. Ozone-generating equipment must have a destruct mechanism to prevent the release of ozone to the atmosphere where it can cause the formation of peroxyacetyl nitrate (PAN), a known air pollutant.

Ozone's short half-life may allow treated water to be discharged without harm to the environment. However, the shorter half-life reduces contact in a treated water system, so the far reaches of a water system may not receive adequate treatment.

Ozone is generated by dry air or oxygen being passed between two high-voltage electrodes. Ozone can also be generated photochemically by ultraviolet light. Ozone must be delivered to a water system by injection through a contactor. The delivery rate is dependent on the mass transfer rate of this contactor or sparger. Proper maintenance of the generator and contactor is critical.

High capital costs limit the use of ozone for microbiological growth control, particularly in systems with varying demand.

DECHLORINATION

Dechlorination is often required prior to discharge from the plant. Also, high chlorine residuals are detrimental to industrial systems, such as ion exchange resins, and some of the membranes used in electrodialysis and reverse osmosis units. Chlorine may also contribute to effluent toxicity; therefore, its concentration in certain aqueous discharges is limited.

Sometimes, dechlorination is required for public and industrial water supplies. Reducing or removing the characteristic "chlorine" taste from potable water is often desirable. Dechlorination is commonly practiced in the food and beverage processing industries. Direct contact of water containing residual chlorine with food and beverage products is avoided, because undesirable tastes can result.

Excess free residual chlorine can be lowered to an acceptable level by chemical reducing agents, carbon adsorption, or aeration.

Chemical reducing agents, such as sulfur dioxide, sodium sulfite, and ammonium bisulfite, dechlorinate water but can also promote the growth of bacteria that metabolize sulfur. Sometimes, sodium thiosulfate is used to dechlorinate water samples prior to bacteriological analysis. Common dechlorination reactions are:

$$SO_2 + Cl_2 + 2H_2O \rightleftharpoons H_2SO_4 + 2HCl$$

sulfur   chlorine   water   sulfuric   hydrochloric
dioxide   acid   acid

$$NaHSO_3 + Cl_2 + H_2O \rightleftharpoons NaHSO_4 + 2HCl$$

sodium   chlorine   water   sodium   hydrochloric
bisulfite   bisulfate   acid

$$NH_4HSO_3 + Cl_2 + H_2O \rightleftharpoons NH_4HSO_4 + 2HCl$$

ammonium chlorine water   ammonium hydrochloric
bisulfite   bisulfate   acid

Granular activated carbon (GAC) removes free chlorine by adsorption. Free chlorine in the form of HOCl reacts with activated carbon to form an oxide on the carbon surface. Chloramines and chlorinated organics are adsorbed more slowly than free chlorine.

Aeration is the least effective means of dechlorination, with effectiveness decreasing with increasing pH. The hypochlorite ion, which predominates at pH 8.3 and higher, is less volatile than hypochlorous acid.

Ultraviolet radiation dechlorinates water stored in open reservoirs for prolonged periods.

OTHER USES AND EFFECTS OF CHLORINE

In addition to serving as antimicrobials, chlorine and chlorine compounds are used to reduce objectionable tastes and odors in drinking water; improve influent clarification processes; oxidize iron, manganese, and hydrogen sulfide to facilitate their removal; reduce sludge bulking in wastewater treatment plants; and treat wastewater plant effluents.

Chlorine, along with a coagulant, is often applied to raw water in influent clarification processes. This prechlorination improves coagulation because of the effect of chlorine on the organic material in the water. It is also used to reduce taste, odor, color, and microbiological populations, and it oxidizes iron and manganese to facilitate removal by settling and filtration. One part per million of chlorine oxidizes 1.6 ppm of ferrous ion or 0.77 ppm of manganous ion. The addition of 8.87 ppm of chlorine per ppm of sulfide oxidizes sulfides to sulfates, depending on pH and temperature.

Chlorine is a successful activating agent for sodium silicate in the preparation of the coagulant aid, activated silica. The advantage of this process is that the chlorine used for activation is available for other purposes.

Low-level, intermittent chlorination of return activated sludge has been used to control severe sludge bulking problems in wastewater treatment plants.

Chlorine, injected into sewage and industrial wastes before they are discharged, destroys bacteria and such chemicals as sulfides, sulfites,

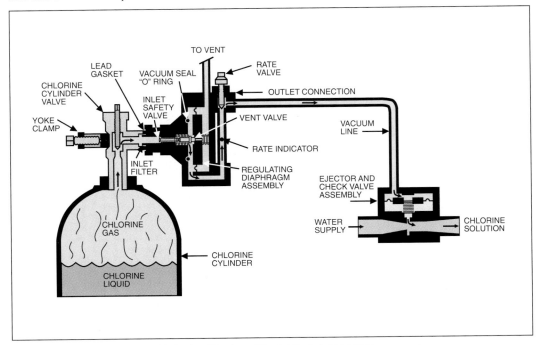

Figure 27-4. Flow diagram for gas chlorination. (Courtesy of Capital Controls Company.)

and ferrous iron. These chemicals react with and consume dissolved oxygen in the receiving body of water.

FEED EQUIPMENT

Chlorination equipment is commercially available for liquefied chlorine gas and solutions of sodium hypochlorite. Calcium hypochlorite is a solid and is usually added by shot feeding. The more recent solid halogen donors, such as 1-bromo-3-chloro-5,5-dimethylhydantoin, are fed with bypass dissolving feeders.

Liquefied chlorine gas is the least expensive form of chlorine and has generally been the antimicrobial of choice in the past. Because of the hazards of chlorine leakage, feed equipment is designed to maintain the chlorine gas below atmospheric pressure by operating under a vacuum. This causes any leaks to be directed into the feeding system rather than into the surrounding atmosphere (see Figure 27-4). Maximum solubility is about 5000 ppm at the vacuum levels currently produced by chlorine injection equipment. Chlorinator manufacturers design equipment to limit the amount of chlorine in the feed system to 3500 ppm to prevent the release of chlorine gas at the point of application. Direct injection of chlorine without the use of an appropriate eductor can be disastrous.

Sodium hypochlorite feeders include metering pumps, flow-controlled rotometers, and computerized feed systems, such as the Betz PaceSetter (see Chapter 35). The hypochlorite storage system should be protected from direct sunlight and heat to prevent degradation. Selection of appropriate storage metals is also important to prevent degradation.

Solid halogen donors, such as hydantoins, triazinetriones, and isocyanurates, are available in tablet form and, sometimes, in granular form. The solids are typically dissolved in a bypass feeder (see Figure 27-5) to regulate the dissolution rate, and the concentrated feeder effluent is applied at the appropriate point. The chemicals provided by these products are hypochlorous acid, hypobromous acid, or a combination of the two.

Figure 27-5. Solid halogen feeder.

MACROFOULING CONTROL

MACROFOULING ORGANISMS

Fouling caused by large organisms, such as oysters, mussels, clams, and barnacles, is referred to as macrofouling. Typically, organisms are a problem only in large once-through cooling systems or low cycle cooling systems that draw cooling water directly from natural water sources (rivers, lakes, coastal seas). Water that has been processed by an influent clarification and disinfection system is usually free of the larvae of macrofouling organisms.

Macrofouling has always been a concern in certain regions of the United States, especially in coastal areas. However, in the last 10 years, the incidence of problems in the United States caused by macrofouling has increased dramatically. This is due primarily to the "invasion" of two organisms that were accidentally introduced to this country—the Asiatic clam and the zebra mussel. Both organisms have flourished and represent a significant threat to system reliability. Adding to the problem is the decreased usage of chlorine and heavy metal antimicrobials. This decrease permits the infiltration and growth of macrofouling organisms in plant water systems.

Asiatic Clams

Asiatic clams are freshwater mollusks. They probably originated in China or eastern Asia and were introduced into North America and Europe in the past century. They were originally found in warm water but their territory now extends to Minnesota. They have not yet been seen in Canadian rivers or lakes.

Asiatic clams do not attach to surfaces but burrow into sediments in their natural environment. Larvae and juvenile clams easily pass through intake screens (Figure 28-1) and settle in low flow areas. Within 6 months to 1 year, the clams grow to 0.6–1.0 in. in size. When a clam

dies, the shell gapes open. Shells of living or dead organisms are carried by water flow and can wedge in condenser or heat exchanger tubes. Once a shell is wedged in a tube, other shells and debris collect and plug the tube further (Figure 28-2). The Asiatic clam reaches adulthood in about 1 year and reproduces in warm months, releasing thousands of larvae into the system.

Figure 28-1. Juvenile Asiatic clams passing through an intake screen at a large power plant.

Figure 28-2. Power plant condenser infested and numerous tubes plugged with adult Asiatic clams.

Zebra Mussels

Zebra mussels are found in cool rivers and lakes of Europe and North America. They entered the Great Lakes of North America during the 1980's in ship ballast water. This fresh-water mussel attaches to almost any hard surface, including other mussels, to form large mats or clumps (Figure 28-3). The adult zebra mussel is smaller than many fouling organisms; adults larger than 0.8 in. are rarely seen.

Zebra mussels are named for the striped coloration of their shell and make up for their small size in population densities. Colonies have grown from a few animals to several hundred thousand per square meter in 2 or 3 months. Densities of 500,000–700,000 mussels per square meter have been seen in cooling system intake bays. These organisms reduce the effective diameter of pipes and conduits, and the clumps break off and plug downstream condenser or heat exchanger tubes (Figure 28-4).

Blue Mussels

The blue mussel is a temperate marine mollusk, preferring the cool waters of the North Atlantic and North Pacific. The blue mussel attaches to hard surfaces and grows in large colonies (Figure 28-5). Population densities of more than 100,000 mussels per square meter have been seen. These organisms may reach 4 in. in size and can accumulate in mats several feet thick. They may foul systems by plugging or constricting pipes and conduits and by breaking off and wedging into condenser or heat exchanger tubes. Systems become further infested by larvae being entrained

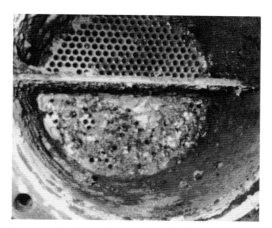

Figure 28-4. Clumps of zebra mussels foul condenser tubes.

Figure 28-5. Blue mussels grow in large colonies.

into the influent flow from the surrounding coastal waters, and from the colonies already in the system. Reproduction occurs during the warmer months. Blue mussels can reach adulthood in 1 year and each adult female can release thousands of eggs per season.

CAUSES OF MACROFOULING

The intake bays of large-volume cooling water users are protected from the entrainment of fish and other larger animals by a series of stationary and/or traveling screens. The mesh size of these screens is usually in the 0.2–0.4 in. range, preventing any but the smallest animals from entering. Any creature small enough to enter might be expected to pass, unimpeded, through condensers and other heat exchange equipment, leaving the plant through the discharge.

The larvae of the clams, mussels, and other planktonic (free-swimming) forms are far too

Figure 28-3. Zebra mussels attached to mollusk shell in clumps.

small to be removed by debris screens. They enter the system and attach to walls and other surfaces or settle in a low flow area to grow.

Cooling systems provide ideal conditions for the growth of these organisms. The pumps maintain a steady flow of water past their siphons, and the planktonic organisms and other nutrients in the water provide an ample food supply. In warm areas, the organisms may be able to grow and reproduce all year long.

It is the growth of these creatures that leads to macrofouling. There are three basic ways in which the fouling occurs.

■ The fouling organisms set and grow in pipes and conduits. As their size increases, the effective diameter of the pipe or conduit decreases and the resistance to flow increases. As fouling becomes more severe, the water flow decreases and cooling capacity drops.

■ The fouling organisms that form hard calcified shells, such as mollusks and crustaceans (oysters, blue mussels, zebra mussels, Asiatic clams, barnacles), grow in the influent side of the cooling system. When they die, their shells open, slough off the walls, and become entrained in the flow of water. As these shells are carried downstream, they often become lodged at points where there is a constriction. Condenser tubes and water spray nozzles are the primary areas where plugging commonly occurs. In extreme cases, more than 80% of the condenser tubes may be blocked by relic shells. Even if flow is not blocked, it may be so low that there is no effective cooling.

■ The third way in which fouling by aquatic or marine animals occurs is a combination of the first two. Larvae and other planktonic forms may set on the surfaces of condenser tubes or cooling jackets. As they grow they restrict a tube or passage area, forming a spot where the collection of relic shells and other debris occurs more easily. When investigating the cause of shell-plugged condenser tubes, it is often impossible to tell whether the fouling was initiated by organisms growing in the tube or by the dislodging of relic shells. In either case, the effect on the system is the same regardless of the initiating event.

CLEANING METHODS

Mechanical Cleaning

Once a cooling system becomes badly fouled, mechanical cleaning is required to restore capacity. The system is usually shut down and the walls scraped or hydroblasted to dislodge adhering organisms. In large systems, tons of macrofouling creatures may be removed. Where treatment is not used, the system will start to foul again as soon as water flow is resumed.

When the system is treated, the time between clean-outs can be extended. However, any time there is significant fouling in the system, mechanical cleaning is a prerequisite to the use of a macrofouling control program.

Thermal Treatment (Thermal Backwash)

The organisms that cause macrofouling can be killed by heated water. Some systems are designed to allow the heated water from the outlet of the condenser to be recirculated back to the intake. As the water recirculates, it is heated and improves macrofouling control. A 15–60 min exposure to water at 104 °F or higher has effectively controlled zebra mussels. Thermal treatment is not used extensively because most systems are not designed to recirculate water. Also, when heated water is recirculated the system cooling capacity is greatly diminished.

Miscellaneous Methods

Critical areas of the cooling system may be protected by additional screens and strainers. They work well to prevent the intrusion of adult organisms, but do not prevent the entrance of larvae, which grow in the system.

Other methods that have been used include ultrasonic vibration and electrical shock. They, like thermal treatment, can be effective but require relatively expensive capital equipment and are difficult to maintain and use. Therefore, these methods are not widely used.

Oxidizing Antimicrobials

The application of an oxidizing antimicrobial such as chlorine for the control of undesirable organisms is a well known and long-practiced procedure. Chlorine is toxic to all living organisms from bacteria to humans. However, in the case of hard-shelled creatures including some mollusks and crustaceans, exposure is not easily accomplished. Some mollusks (e.g., oysters, blue mussels, Asiatic clams, and zebra mussels) and crustaceans (e.g., barnacles) have sensitive chemoreceptors that detect the presence of oxidizing chemicals such as chlorine (hypochlorite), bromine (hypobromite), ozone, and hydrogen peroxide. When oxidizers are detected at life-threatening levels, the animal withdraws into its shell and closes up tightly to exclude the hostile environment. Animals like oysters and mussels can remain closed for days to weeks if necessary. There is evidence to suggest that, during extended periods of continuous chlorination, the creatures may eventually die from asphyxiation rather than chlorine toxicity.

Even when their shells are closed, the animals continue to sense their environment; and as soon as the oxidant level decreases, they reopen and resume siphoning. "Continuous chlorination" often fails to eradicate these macrofouling creatures because of interruptions in the feed, which can occur for various reasons, such as chlorine tank changeover or plugging of feed lines. If the interruption lasts long enough (1 hr or possibly less), the animals have time to reoxygenate their tissues between the extended periods of chlorination.

Any oxidant, such as chlorine, bromine, or ozone, elicits the same response from these creatures. Therefore, only continuous, uninterrupted applications are successful.

Nonoxidizing Antimicrobials

There are several categories of nonoxidizing antimicrobials that have proven to be very effective in controlling macrofouling organisms. Quaternary amine compounds and certain sur-

factants have been applied to infested systems for relatively short intervals (6–48 hr). These compounds do not trigger the chemoreceptors of the mollusks. The mollusks continue to filter feed and ingest a lethal dosage of the antimicrobial throughout the exposure period. These compounds produce a latent mortality effect—the mollusks may not die until several hours after the antimicrobial application (Figure 28-6). Cold water temperatures may extend this latent mortality effect and may also require slightly higher feed rates and longer feed durations due to slower organism metabolism.

The advantages of a nonoxidizing antimicrobial program include ease of handling, short application time, and relatively low toxicity to other aquatic organisms. In addition, some of these compounds can be readily detoxified.

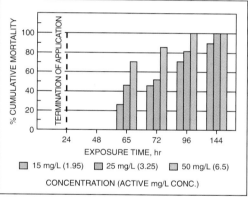

Figure 28-6. Nonoxidizing antimicrobials effectively control macrofouling. The mollusks often do not die until several hours after treatment.

COOLING TOWER WOOD MAINTENANCE

Wood continues to be widely used for the construction of cooling towers. Wood deterioration can shorten the life of a cooling tower from an anticipated 20–25 years to 10 years or less. Cooling tower operation becomes inefficient and repair and replacement costs are excessive.

In the past, redwood was selected for use in cooling towers because of its high strength-to-weight ratio, availability, ease of use, low cost, and natural resistance to decay. Pressure-treated Douglas fir and similar types of wood are replacing redwood due to cost and availability factors.

Wood is composed of three main components: cellulose, lignin, and natural extractives. Long fibers of cellulose give wood its strength. Lignin acts as the cementing agent for the cellulose. The extractives contain most of the natural compounds that enable wood to resist decay. Normally, highly colored woods are most durable.

The extractives in all woods are largely water-soluble, so circulating water leaches them from the wood. Although this leaching process does not appear to affect the strength of the wood, the loss of extractives makes the wood more susceptible to decay.

TYPES OF WOOD DETERIORATION

Cooling tower wood experiences three main types of deterioration: chemical, biological, and physical. Rarely is one present without another; usually, all three occur simultaneously. Sometimes it is difficult to determine which type of attack is the most responsible for the deterioration. Physical and chemical deterioration, which are more visible, render the wood more susceptible to biological attack.

Chemical Attack

Chemical deterioration of cooling tower wood commonly manifests itself in the form of delignification (Figure 29-1). Delignification is

Figure 29-1. Delignification, removal of some or most of the lignin from wood, may result from chemical or biological attack.

usually caused by oxidizing agents and alkaline materials. Because chemical attack removes the lignin component of wood, the residue is rich in cellulose. The deterioration is particularly severe when high chlorine residuals (more than 1 ppm free chlorine) and high alkalinity concentrations (pH more than 8) occur simultaneously.

Wood that has suffered chemical attack takes on a white or bleached appearance and its surface is fibrillated. Damage is restricted to the surface of the wood and does not impair the strength of unaffected areas. When cascading water has a chance to wash away surface fibers, the wood becomes thinned. In serious cases, the loosened fibers plug screens and tubes and serve as focal points for corrosion when fibers accumulate in heat exchange equipment.

Chemical attack occurs most frequently in the fill section and wetted portions of the tower where water contact is continuous. It also occurs where alternately wet and dry conditions develop, such as on air intake louvers and other exterior surfaces, and in the warm, moist areas of the plenum chamber of the tower. Deterioration occurs as a result of chlorine vapors and the entrainment of droplets of tower water.

Biological Attack

The organisms that attack cooling tower wood are those that can use cellulose as their source of carbon for growth and development. These organisms degrade cellulose by secreting enzymes that convert the cellulose into compounds that they absorb. This attack depletes the cellulose content of the wood and leaves a residue rich in lignin. Characteristically, the wood becomes dark in color, loses much of its strength, and may also become soft, punky, cross-checked (Figure 29-2), or fibrillated.

Principal cellulolytic organisms that have been isolated from cooling tower wood are fungi that include the classic wood destroyers (*Basidiomycetes*) and members of *Fungi imperfecti* (see

Figure 29-3). Bacterial organisms that exhibit cellulolytic properties have also been isolated, but their exact role in cooling tower wood deterioration has not been determined. Wood-destroying organisms are common air- and water-borne contaminants.

Moisture and temperature, as well as oxygen, have a marked influence on organism development. When the moisture content of the wood is between 20 and 27% and the temperature is between 88 and 105 °F, organisms usually achieve optimum growth and development.

Biological attack of cooling tower wood is of two basic types: soft or surface rot and internal decay.

Soft or Surface Rot. Soft or surface rot occurs predominantly in the flooded sections and plenum areas of the tower. Enough oxygen reaches the wood surfaces in flooded portions to support growth. Surface rot is more readily detected and less serious than internal decay.

Internal Decay. Classic internal decay, the worse of the two types of biological attack, is generally restricted to the plenum area, cell partitions, access doors, drift eliminators, decks, fan housing, and supports. Because the decay is internal, it is difficult to detect in its early stages. Even areas affected by severe decay can have a sound external appearance (see Figure 29-4).

Internal decay is rarely found in flooded portions of the tower, such as the fill section, where the wood is completely saturated with water. The water excludes oxygen, which the wood-attacking organisms need for their growth and development.

Figure 29-2. This dried wood specimen shows the darkened appearance and cross-grain cracking ("checking") typical of advanced surface attack (soft rot).

Figure 29-3. Internal decay of redwood caused by the white fungus Basidiomycetes.

Figure 29-4. The intact external surfaces of this wood beam appear sound and give no evidence of the extent of internal attack (deep rot).

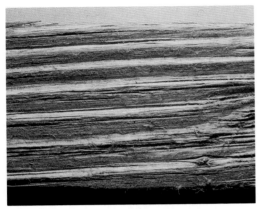

Figure 29-5. Erosion is produced by washing and/or wearing away of thin-walled cells, usually spring wood. Remaining ridges of summer wood result in a washboard effect.

Physical and Other Factors

In addition to supporting biological growth, high temperature can greatly affect the wood integrity. Continuous exposure to high water temperatures (140 °F or higher) causes significant changes in structure and accelerates loss in wood substance. This weakens the wood and predisposes it to biological attack, particularly in the plenum areas of a cooling tower.

Other factors also influence the deterioration of tower wood. Areas adjacent to iron nails and other iron hardware are susceptible to deterioration. Slime and algae growths and dust and oil deposition support the growth of soft rot organisms. Affected areas can lose much of their strength and crumble easily.

Preferential erosion of spring wood (see Figure 29-5) is relatively common in tower fill. In severe cases, significant losses occur in a very short time. Extremely high concentrations of cooling water dissolved solids should be prevented where tower wood areas are exposed to alternate wetting and drying. Although salts show little tendency to attack wood, their crystallization in dry areas can rupture wood cells.

CONTROL OF WOOD DETERIORATION

Preventive Maintenance

Preventive maintenance is the only effective method of protecting cooling towers from deterioration. Prevention is relatively easy in flooded sections of the tower, where chemical and biological attack is limited to wood surfaces. Preventive measures for the nonflooded portions of the tower, where internal decay is the primary concern, are more difficult. To ensure the success

of a program, it is important to adopt appropriate measures before infection reaches serious proportions.

Flooded Sections. Control of chemical and biological surface attack in flooded portions of a cooling tower may be accomplished through a water treatment program, which should include the use of nonoxidizing antimicrobials to control slime and prevent biological surface attack.

When chlorine is used to minimize chemical attack, it should be closely controlled. Free residuals should be restricted to less than 1 ppm, preferably to a range of 0.3–0.7 ppm.

Chlorine should be supplemented by a nonoxidizing antimicrobial to control biological surface attack. When a combination program is possible, chemical attack can be held to a minimum and biological degradation controlled effectively.

Nonflooded Sections. Although soft rot or surface attack may occasionally occur in the nonflooded portions of a tower, loss of wood structure is not as severe as it is in the flooded areas because wood is not eroded by cascading water.

Internal decay is the principal and most serious problem in nonflooded areas. Cooling towers should be inspected thoroughly at least once a year. When internal decay occurs only as white pocket rot (Figure 29-6), the affected areas can be very small and easily missed. Because internal decay usually remains undetected until extensive damage has occurred, it is important to look for signs of internal decay in structural members. Sometimes, such decay is evidenced by abnormal sagging or settling of the tower wood. At other times, it is necessary to test for soundness with a blunt probe. Unexpected softness in an apparently

Figure 29-6. Pocket internal decay is caused by organisms that are not necessarily restricted to the interior or to any structural member.

healthy wood beam is a sign of decay. When decay is not evident, samples of wood should be examined microscopically to detect the presence of internal fungi.

Infected wood must be replaced to retard the spread of infection to adjacent, structurally sound members. A weakened section shifts additional weight to sound sections, causing them to crack and become more susceptible to the spread of internal decay. Infected wood should be replaced with pretreated wood.

Several different wood preservatives are available, including:

- creosote
- ammoniacal copper arsenite
- acid copper chromate and copper naphthenate
- chromated copper arsenate
- pentachlorophenol
- fluoride chromate arsenate phenol
- chlorinated paraffin

Periodic spraying with an antifungal is an effective preventive maintenance step if performed on a regular basis. However, diffusion of antifungal typically penetrates the wood to a depth of ¼ in. or less (even when wood is incised and pressure-treated). The protection provided by spray application of antimicrobials is temporary, especially where wood surfaces are contacted by flowing water, and the protective barrier formed by the antimicrobials is easily breached by cracking of the wood.

Examination

Cooling tower inspections, conducted on a regular basis, should always include collection of tower wood samples for examination in the laboratory. Because several types of wood deterioration are likely to occur, various laboratory examinations should be employed.

Macroscopic examination of wood reveals the degree of erosion, surface structure, and depth of surface attack. Macroscopic study can also be used to assess physical aspects of the wood and to determine the presence and extent of chemical and biological decay. Breaking of specimens reveals the degree to which the wood is brash and is useful in assessing loss of structural strength.

Microscopic examination is useful for determining the extent of microbiological deterioration. A Microtome can be used to prepare thin sections

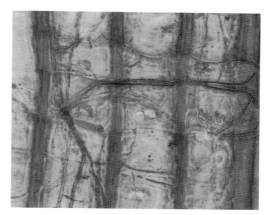
Figure 29-7. Microtome section of wood showing fungal hyphae in cell walls.

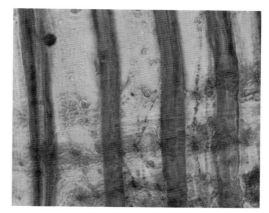
Figure 29-8. Microtome section of wood where fungal attack began. Antifungal removed traces of attack and prevented serious deterioration.

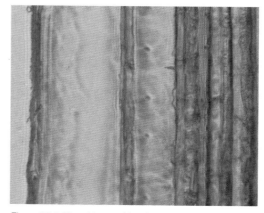
Figure 29-9. No evidence of fungi was detected in this section.

of wood (see Figures 29-7, 29-8, and 29-9). These sections, usually 25 μm thick, show the internal structure of the wood. They indicate the existence and extent of infection and show whether it was caused by bacteria or fungi. From this, it can be

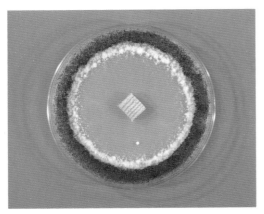

Figure 29-10. In the zone of inhibition test, residual wood preservation would be demonstrated by partial (white growth up to wood block) or complete (clear area, no growth around block) inhibition of fungal growth.

determined whether any fungi present are cellulolytic or noncellulolytic organisms.

The zone of inhibition test (Figure 29-10) can be used to determine the susceptibility of the wood to fungal growth and decay when inoculated with wood-destroying organisms.

Briefly described, a zone of inhibition test involves placing two $1/2$ in. square wood samples on nutrient agar that has been seeded with a wood-rotting organism, such as *Aspergillus niger* or *Chaetomium globosum*. Plates are then incubated for 7 days at 82 °F, after which the wood samples are evaluated for their degree of resistance or susceptibility.

A complete zone of inhibition exists when there is a clear area around the test block that is free of fungal growth. This preventive barrier is created by antifungal application or natural properties present in the wood that inhibit the growth of wood-rotting organisms.

A partial zone of inhibition exists when some growth (white zone in Figure 29-10) of the test organism occurs around the block but fungal spores (black zone in Figure 29-10) of the test organism show retarded growth. This partial zone reflects continuing presence of some residual antifungal or natural inhibitive properties in the wood.

No zone of inhibition exists when growth of the test organism or other organisms inherent in the wood occurs. This growth, on or around the block, indicates that the wood is susceptible to fungal attack, and corrective measures must be taken to prevent fungi from spreading to sound members of the tower.

A zone of inhibition test evaluates both the residual effect of antifungal treatment and the degree of resistance restored by an application of antifungal. Sometimes, pressure-treated wood can yield zone of inhibition results that suggest the need for antifungal application. This occurs when the treatment has not leached from the wood in sufficient quantities to yield a zone of inhibition in the test. If the wood possesses inhibitory properties, growth will not occur on the block.

Spraying

Direct Spraying. Manual spraying of cooling tower wood with pneumatic assist (similar to spray painting) can be an effective method of treating cooling tower wood. A concentrated antifungal with the appropriate EPA end-use registration is applied directly to the wood with spraying equipment handled by an experienced operator or team of operators. This method is most effective because small areas can be covered thoroughly with the antifungal, and close attention can be given to spraying joints, holes, and other critical areas, such as cracks or splits in wood beams.

Direct antifungal spraying is hazardous to the operator unless proper precautions are taken. This procedure should be entrusted to a competent commercial spraying company with proper equipment and experience.

Steam Spraying. Steam spraying through a permanent piping arrangement has also been used to treat cooling tower wood. The antifungal is forced into the steam and transported into the tower cell by the steam. Distribution piping must be designed properly to ensure complete coverage. Because steam dilutes the antifungal, a much more dilute solution is applied than with direct spraying. As a result, a smaller quantity of toxicant penetrates the wood and more frequent spraying is needed to keep the wood fungistatic.

Because diffusion and spray methods penetrate only the outer surfaces of wood, preventive maintenance should be started before infection begins and the interior parts of the wood lose much of their natural resistance.

Sterilization

When a tower has suffered a serious infection that the normal corrective programs are not likely to control, consideration can be given to sterilization. In this process, wood temperature is elevated to 150 °F for a period of 2 hr. Longer periods must be avoided to limit the loss of wood strength that occurs.

ONCE-THROUGH COOLING SYSTEMS

Cooling systems that use a water's cooling capacity a single time are called once-through cooling systems. These systems use large volumes of water and typically discharge the once-through water directly to waste. Large volumes of water are necessary for even the smallest once-through systems; therefore, a plentiful water supply at a suitably low temperature is needed.

Because of the large volumes involved, once-through cooling systems often use water from rivers, lakes, or (occasionally) well networks. The only external treatment generally applied to a once-through system is mechanical screening to protect downstream equipment from serious damage due to foreign material intrusion. Evaporation is negligible, so no significant change in water chemistry occurs.

Once-through cooling water systems are identified by various names. For example, in the paper industry, most mills refer to their once-through cooling water as "mill supply." The power industry often refers to a once-through cooling network as the "service water" system. The chemical and hydrocarbon processing industries typically use the descriptive "once-through" terminology for their systems. Whatever the name, once-through systems present unique challenges for water treatment.

ONCE-THROUGH COOLING PROBLEMS

Problems encountered in once-through cooling systems can be grouped into three general categories:

■ corrosion

■ scale or other deposition

■ biological fouling

Each of these problems is manifested in a particular manner in a once-through cooling system. Whatever the problem, mechanical and chemical solutions can be prescribed to minimize the consequences.

Corrosion can be defined as the wastage or loss of base metal in a system. Various types of corrosion can occur in once-through systems. However, in all cases, base metal loss is encountered and corrosion products enter the bulk water stream as troublesome suspended solids. In addition to the detrimental impact of suspended solids, serious process contamination and/or discharge problems can result from active corrosion.

Corrosion of ferrous metal piping results in the formation of iron oxide products many times the volume of the metal loss from the pipe wall. The accumulation of corrosion products, or tuberculation, on the pipe surface reduces the carrying capacity of lines and requires expensive mechanical or chemical cleaning. Also, the loss in head caused by tuberculation requires increased pump pressures and consequently higher pumping costs.

Deposits consist of two general types:

■ crystalline inorganic deposits (when solubility limits are exceeded)

■ sludge (when suspended solids settle)

The deposits insulate the metal surface from the cooling water, restricting heat transfer. Under-deposit corrosion can also occur. If the deposit formation is severe, hydraulic restrictions to flow may further impact the cooling system's ability to carry heat away from the process.

Biological concerns can be categorized as either microbiological or macrobiological. The proliferation of biological organisms in a cooling system results in many of the same problems caused by corrosion. Significant microbiological growth causes equipment fouling, heat transfer impediment, microbiologically induced corrosion (MIC), and possible flow restrictions. The infestation of U.S. fresh water with Asiatic clams and, more recently, zebra mussels has increased the emphasis on macrobiological control, because these organisms can completely plug a system in one growth season.

CORROSION CONTROL

Figure 30-1 illustrates the most prevalent mechanism for the corrosion of steel in once-through cooling systems. This corrosion is an electrochemical process much like that which takes place in a common automobile battery. Electron flow is from anode to cathode, and metal loss occurs at the anode. The control of this corrosion requires that the electron flow be greatly reduced or stopped, short-circuiting the battery. In untreated once-through cooling systems, the metal loss may reach 100 mils (0.1 in.) per year. Treatment typically reduces the corrosion to 10 mils (0.01 in.) or less per year.

In addition to metal loss, deposit problems are caused by the accumulation of corrosion products even in well treated systems. The value of a deposit control program is illustrated in Figure 30-2, which shows corrosion product accumulation in a heat exchanger tube exposed to an untreated stream of cooling water.

The weight and volume of corrosion products produced as a function of corrosion rate and pipe size are shown in Figure 30-3. The large quantities of corrosion products produced at high corrosion rates can seriously impede heat transfer and reduce flow. Consequently, the effectiveness of the corrosion control program has a large impact on the quantity of deposit that accumulates in the once-through system.

Figure 30-2. Deposition of corrosion products in a once-through system.

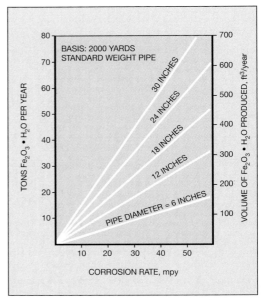

Figure 30-3. Iron corrosion products from corrosion.

Corrosion Control Chemicals

Polyphosphates and zinc, used in combination or separately, are the corrosion inhibitors most commonly employed in once-through systems. Other inhibitors, such as silicates and molybdates, may be employed for special circumstances. Usually, due to the large water volumes involved, it is not economically practical to feed enough corrosion inhibitor to stifle corrosion completely. In any treatment approach, the application rate is set to produce an acceptable level of corrosion. This level of treatment is often termed "threshold."

Polyphosphates (e.g., sodium tripolyphosphate, sodium hexametaphosphate, and potassium

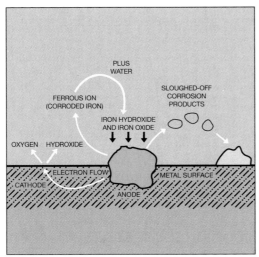

Figure 30-1. Classic steel corrosion cell.

pyrophosphate) are molecularly dehydrated forms of orthophosphate. These special phosphates possess surface-active and sequestering capabilities that make them particularly effective in controlling scale formation and minimizing tuberculation.

Polyphosphates are widely used in treating once-through industrial systems, municipal systems, and potable drinking water systems. The polyphosphates effectively reduce tuberculation in distribution lines and minimize red water caused by high iron levels in both potable and industrial systems. Effective results have been achieved with treatment levels as low as 1 ppm. Corrosion protection at this low phosphate level can be improved with the addition of a small amount of zinc (as little as 0.25 ppm). The use of zinc and polyphosphates protects the system steel from both general corrosion and pitting, while reducing the red water conditions prevalent when excessive corrosion products enter the system. Polyphosphate and zinc treatment levels are not set to stifle corrosion completely, but to provide corrosion rates of approximately 10 mils (0.01 in.) per year or less.

Silicates and molybdates are specialty inhibitors that typically do not fit the needs of once-through cooling systems. Compared to the polyphosphates, high levels of silicates and molybdates are required to control steel corrosion. A 100 ppm silicate level provides corrosion protection approximately equal to 2 ppm of polyphosphate. The cost of molybdate treatment precludes its use in most industrial once-through cooling systems.

SCALE CONTROL

The control of scale formation on heat transfer surfaces is necessary for the efficient operation of a once-through cooling system. The use of polyphosphates or organic polymers can greatly reduce the deposit potential of system streams. In addition to impeding heat transfer, deposits can alter flow characteristics, induce under-deposit corrosion, increase power consumption for pumping, and release corrosion products to contaminate process streams exposed to the once-through water.

Common scales found in once-through systems are hardness salts, such as calcium and magnesium, combined with anions, such as carbonate, sulfate, and silica. Manganese and barium are less common but equally troublesome scale formers found in certain areas of the country. Corrosion products

such as iron oxide also can be significant contributors to scale formation. Mud and silt also add to deposit formation in once-through systems.

Scale Control Chemicals

The effect of temperature on the solubility of calcium carbonate is illustrated in Figure 30-4. As temperature increases, the solubility decreases. The resulting scale formation is illustrated in Figure 30-5.

Scale control may be achieved in many ways. Specialty chemicals are usually employed to minimize calcium carbonate scale. Calcium reduction through softening is not economically

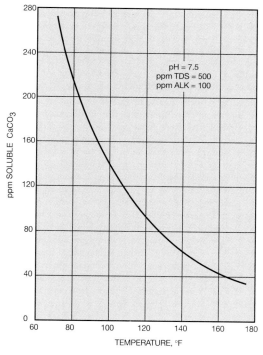

Figure 30-4. Calcium carbonate solubility vs. temperature.

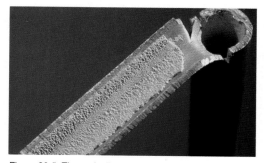

Figure 30-5. Theoretically, calcium carbonate scale can provide a protective film.

practical in once-through systems due to the large quantity of water processed.

When specialty chemicals are used, the deposit control agent must prevent crystal growth and scale formation by permitting scale-forming salts to exist in an over-saturated condition without precipitation. Commonly used agents include polyphosphates, organic phosphates, and organic polymers. Treatment levels are based on system conditions and vary from less than 1 ppm to as much as 5 ppm.

Iron deposition is another special problem in once-through systems. The iron may enter the system from wells or surface water, or as a result of system corrosion. In any case, a deposit control agent can be used to control deposition. Poly-phosphates, organic phosphates, surface-active compounds, and polyelectrolytes are used for this purpose. Treatment levels vary from less than 1 ppm to 5 ppm active compound. At even higher levels, intermittent treatment has been used to clean up old iron deposition.

Figure 30-6 illustrates the ability of hexametaphosphate to sequester calcium, thereby preventing scale formation, and the ability of

pyrophosphate to sequester iron and minimize deposition. The properties of these two common polyphosphates, as well as knowledge of the predominant depositing species, are important factors in the selection of a particular deposit control program.

The use of polyelectrolytes for deposit control in once-through systems has become more common in recent years. The power industry has pioneered the use of very low treatment levels of proprietary polymers for specific deposit control problems. Polymers in the acrylate and acrylamide families have been proven effective in large-volume once-through systems even at treatment levels measured in the parts per billion range.

BIOLOGICAL FOULING

Deposits formed as a result of microbiological activity are particularly troublesome in once-through systems. The deposits formed by living organisms restrict heat transfer and often lead to severe under-deposit corrosion. Also, the organisms serve as a matrix binder for inorganic species that increase deposit volume and tenacity. Further, severe macrobiological fouling can essentially stop water flow through heat exchange and process equipment in a short period of time.

Biological Fouling Control

Oxidizing agents such as chlorine and bromine have been used for microbiological control in once-through cooling systems. The frequency, duration, and treatment level for the oxidizing agent vary according to the nature of the system and the water source. Current and future environmental restrictions must be considered in the design of a microbiological control program for once-through systems.

Macrobiological control, particularly the control of mollusks found in fresh waters in the United States, is a serious problem in many systems. Whenever fresh water is used for once-through cooling, the threat of Asiatic clam and/or zebra mussel infestation exists. Proprietary chemicals are available for control of these organisms. Application rate and frequency are system-dependent.

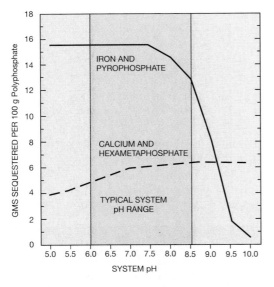

Figure 30-6. The sequestration of calcium by hexametaphosphate and iron by pyrophosphate.

OPEN RECIRCULATING COOLING SYSTEMS

An open recirculating cooling system uses the same water repeatedly to cool process equipment. Heat absorbed from the process must be dissipated to allow reuse of the water. Cooling towers, spray ponds, and evaporative condensers are used for this purpose.

Open recirculating cooling systems save a tremendous amount of fresh water compared to the alternative method, once-through cooling. The quantity of water discharged to waste is greatly reduced in the open recirculating method, and chemical treatment is more economical. However, open recirculating cooling systems are inherently subject to more treatment-related problems than once-through systems:

■ cooling by evaporation increases the dissolved solids concentration in the water, raising corrosion and deposition tendencies

■ the relatively higher temperatures significantly increase corrosion potential

■ the longer retention time and warmer water in an open recirculating system increase the tendency for biological growth

■ airborne gases such as sulfur dioxide, ammonia or hydrogen sulfide can be absorbed from the air, causing higher corrosion rates

■ microorganisms, nutrients, and potential foulants can also be absorbed into the water across the tower

COOLING TOWERS

Cooling towers are the most common method used to dissipate heat in open recirculating cooling systems. They are designed to provide intimate air/water contact. Heat rejection is primarily by evaporation of part of the cooling water. Some sensible heat loss (direct cooling of the water by the air) also occurs, but it is only a minor portion of the total heat rejection.

Figure 31-1. A natural draft ("hyperbolic") tower relies on the density difference between warm, moist air inside the tower and cooler, dryer air outside for airflow. Note the annular fill ring around the base of this crossflow model.

Types of Towers

Cooling towers are classified by the type of draft (natural or mechanical) and the direction of airflow (crossflow or counterflow). Mechanical draft towers are further subdivided into forced or induced draft towers.

Natural draft towers. Sometimes called "hyperbolic" towers due to the distinctive shape and function of their chimneys, natural draft towers do not require fans. They are designed to take advantage of the density difference between the air entering the tower and the warmer air inside the tower. The warm, moist air inside the tower has a lower density, so it rises as denser, cool air

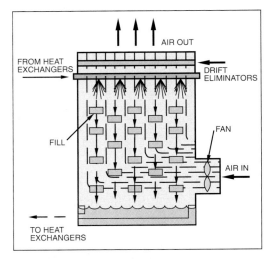

Figure 31-2. Almost all forced draft cooling towers are counterflow designs.

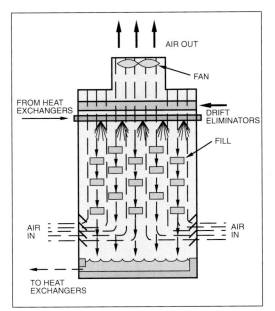

Figure 31-3. Counterflow induced draft cooling towers provide maximum heat transfer.

Figure 31-4. A six-cell, counterflow induced draft cooling tower. Air enters only at the bottom of the tower.

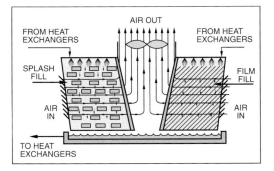

Figure 31-5. Crossflow induced draft cooling tower requires less fan horsepower than the counterflow designs.

is drawn in at the base of the tower. The tall (up to 500 ft) chimney is necessary to induce adequate airflow. Natural draft towers can be either counterflow or crossflow designs. The tower in Figure 31-1 is a crossflow model. The fill is external to the shell forming a ring around the base. In a counterflow model, the fill is inside the shell. In both models, the empty chimney accounts for most of the tower height.

Mechanical Draft Towers. Mechanical draft towers use fans to move air through the tower. In a forced draft design, fans push air into the bottom of the tower (Figure 31-2). Almost all forced draft towers are counterflow designs. Induced draft towers have a fan at the top to draw air through the tower. These towers can use either crossflow or counterflow air currents and tend to be larger than forced draft towers.

Counterflow Towers. In counterflow towers (Figures 31-3 and 31-4), air moves upward, directly opposed to the downward flow of water. This design provides good heat exchange because the coolest air contacts the coolest water. Headers and spray nozzles are usually used to distribute the water in counterflow towers.

Crossflow Towers. In crossflow towers (Figures 31-5 and 31-6), air flows horizontally across the downward flow of water. The crossflow design provides an easier path for the air, thus increasing the airflow for a given fan horsepower. Crossflow towers usually have a gravity feed system—a distribution deck with evenly spaced metering orifices to distribute the water. Often, the deck is covered to retard algae growth.

Figure 31-6. A six-cell, crossflow induced draft cooling tower.

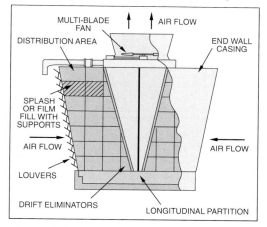

Figure 31-7. Components of a typical cooling tower. (Reprinted with permission from Power.)

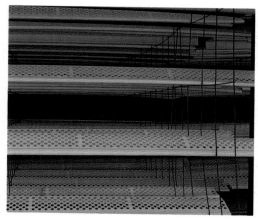

Figure 31-8. Plastic splash-type fill can replace wood slats.

Figure 31-9. Installation of corrugated film-type tower fill can increase tower capacity relative to splash fill.

Cooling Tower Components

Figure 31-7 illustrates the components of a typical cooling tower.

Fill Section. The fill section is the most important part of the tower. Packing or fill of various types is used to keep the water distributed evenly and to increase the water surface area for more efficient evaporation. Originally, fill consisted of "splash bars" made of redwood or pressure-treated fir. Splash bars are now available in plastic as well. Other types of fill include plastic splash grid (Figure 31-8), ceramic brick, and film fill.

Film fill (Figure 31-9) has became very popular in recent years. It consists of closely packed, corrugated, vertical sheets, which cause the water to flow down through the tower in a very thin film. Film fill is typically made of plastic. Polyvinyl chloride (PVC) is commonly used for systems with a maximum water temperature of 130 °F or less. Chlorinated PVC (CPVC) can withstand temperatures to approximately 165 °F.

Film fill provides more cooling capacity in a given space than splash fill. Splash fill can be partially or totally replaced with film fill to increase the capacity of an existing cooling tower. Because of the very close spacing, film fill is very susceptible to various types of deposition. Calcium carbonate scaling and fouling with suspended solids has occurred in some systems. Process contaminants, such as oil and grease, can be direct foulants and/or lead to heavy biological growth on the fill. Any type of deposition can severely reduce the cooling efficiency of the tower.

Louvers. Louvers are used to help direct airflow into the tower and minimize the amount of windage loss (water being splashed or blown out the sides of the tower).

Drift Eliminators. "Drift" is a term used to describe droplets of water entrained in the air leaving the top of the tower. Because drift has the same composition as the circulating water, it should not be confused with evaporation. Drift should be minimized because it wastes water and can cause staining on buildings and autos at some distance from the tower. Drift eliminators abruptly change the direction of airflow, imparting centrifugal force to separate water from the air. Early drift eliminators were made of redwood in a herringbone structure. Modern drift eliminators are typically made of plastic and come in many different shapes. They are more effective in removing drift than the early wood versions, yet cause less pressure drop.

Approach to Wet Bulb, Cooling Range

Cooling towers are designed to cool water to a certain temperature under a given set of conditions. The "wet bulb temperature" is the lowest temperature to which water can be cooled by evaporation. It is not practical to design a tower to cool to the wet bulb temperature. The difference between the cold sump temperature and the wet bulb temperature is called the "approach." Towers are typically designed with a 7–15 °F approach. The temperature difference between the hot return water and the cold sump water is referred to as the "cooling range" (ΔT). Cooling range is usually around 10–25 °F but can be as high as 40 °F in some systems.

CYCLES OF CONCENTRATION, WATER BALANCE

Cycles of concentration and the calculation of water flows in a cooling system are depicted in Figure 31-10.

Water circulates through the process exchangers and over the cooling tower at a rate referred to as the "recirculation rate." Water is lost from the system through evaporation and blowdown. For calculation purposes, blowdown is defined as all nonevaporative water losses (windage, drift, leaks, and intentional blowdown).

Makeup is added to the system to replace evaporation and blowdown.

Approximately 1000 Btu of heat is lost from the water for every pound of water evaporated. This

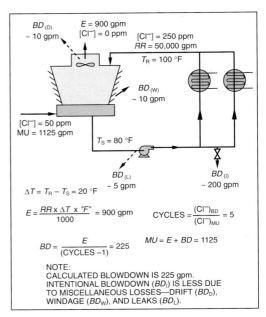

Figure 31-10. Calculation of water flows in a typical open recirculating system.

is equal to evaporation of about 1% of the cooling water for each 10 °F temperature drop across the cooling tower. The following equation describes this relationship between evaporation, recirculation rate, and temperature change:

$$E = \frac{RR\,(\Delta T)F}{1000}$$

where:

E	=	evaporation, gpm
RR	=	recirculation rate, gpm
ΔT	=	cooling range, °F
F	=	evaporation factor

The evaporation factor, F, equals 1 when all cooling comes from evaporation. For simplicity, this is often assumed to be the case. In reality, F varies with relative humidity and dry bulb temperature. The actual F value for a system is generally between 0.75 and 1.0, but can be as low as 0.6 in very cold weather.

As pure water is evaporated, minerals are left behind in the circulating water, making it more concentrated than the makeup water. Note that blowdown has the same chemical composition as circulating water. "Cycles of concentration" (or "cycles") are a comparison of the dissolved solids level of the blowdown with the makeup water. At 3 cycles of concentration, blowdown has three times the solids concentration of the makeup.

Cycles can be calculated by comparison of the concentrations of a soluble component in the makeup and blowdown streams. Because chloride and sulfate are soluble even at very high concentrations, they are good choices for measurement. However, the calculation results could be invalid if either chlorine or sulfuric acid is fed to the system as part of a water treatment program.

Cycles based on conductivity are often used as an easy way to automate blowdown. However, cycles based on conductivity can be slightly higher than cycles based on individual species, due to the addition of chlorine, sulfuric acid, and treatment chemicals.

Using any appropriate component:

$$\text{Cycles of concentration} = \frac{\text{Concentration in Blowdown}}{\text{Concentration in Makeup}}$$

Cycles of concentration can also be expressed as follows:

$$\text{Cycles of concentration} = \frac{MU}{BD}$$

where:

MU = makeup (evaporation + blowdown), gpm

BD = blowdown, gpm

Note that the relationship based on flow rate in gallons per minute is the inverse of the concentration relationship.

If $E + BD$ is substituted for MU:

$$\text{Cycles of concentration} = \frac{E + BD}{BD}$$

where:

E = evaporation

Solving for blowdown, this equation becomes:

$$BD = \frac{E}{(\text{Cycles} - 1)}$$

This is a very useful equation in cooling water treatment. After the cycles of concentration have been determined based on makeup and blowdown concentrations, the actual blowdown being lost from the system, or the blowdown required to maintain the system at the desired number of cycles, can be calculated.

Because treatment chemicals are not lost through evaporation, only treatment chemicals lost through blowdown (all nonevaporative water loss) must be replaced. Thus, calculation of blowdown is critical in determining treatment feed rates and costs.

Factors Limiting Cycles of Concentration

Physical Limitations. There is a limit to the number of cycles attainable in a cooling tower. Windage, drift, and leakage are all sources of unintentional blowdown. Drift losses of up to 0.2% of the recirculation rate in older towers can limit cycles to 5–10. Additional losses due to leaks and windage can further limit some older systems. New towers often carry drift guarantees of 0.02% of recirculation rate or less. Newly constructed systems that use towers with highly efficient drift eliminators and have no extraneous losses may be mechanically capable of achieving 50–100 cycles or more.

Chemical Limitations. As a water's dissolved solids level increases, corrosion and deposition tendencies increase. Because corrosion is an electrochemical reaction, higher conductivity due to higher dissolved solids increases the corrosion rate (see Chapter 24 for further discussion). It becomes progressively more difficult and expensive to inhibit corrosion as the specific conductance approaches and exceeds 10,000 µmho.

Some salts have inverse temperature solubility; i.e., they are less soluble at higher temperature and thus tend to form deposits on hot exchanger tubes. Many salts also are less soluble at higher pH. As cooling tower water is concentrated and pH increases, the tendency to precipitate scale-forming salts increases.

Because it is one of the least soluble salts, calcium carbonate is a common scale former in open recirculating cooling systems. Calcium and magnesium silicate, calcium sulfate, and other types of scale can also occur. Figure 31-11 shows the relative solubility of calcium carbonate and gypsum, the form of calcium sulfate normally found in cooling systems.

Calcium carbonate scaling can be predicted qualitatively by the Langelier Saturation Index (LSI) and Ryznar Stability Index (RSI). The indices are determined as follows:

$$\text{Langelier Saturation Index} = pH_a - pH_s$$

$$\text{Ryznar Stability Index} = 2(pH_s) - pH_a$$

The value pH_s (pH of saturation) is a function of total solids, temperature, calcium, and alkalinity. pH_a is the actual pH of the water.

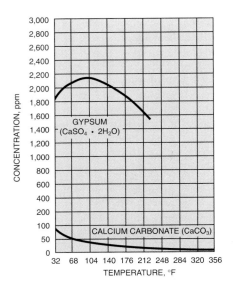

Figure 31-11. Solubility of calcium carbonate and calcium sulfate in the absence of chemical treatment.

A positive LSI indicates a tendency for calcium carbonate to deposit. The Ryznar Stability Index shows the same tendency when a value of 6.0 or less is calculated. A more complete discussion of LSI and RSI is presented in Chapter 25, Deposit and Scale Control—Cooling Systems.

With or without chemical treatment of the cooling water, cycles of concentration are eventually limited by an inability to prevent scale formation.

DEPOSITION CONTROL

As noted earlier, there are many contaminants in cooling water that contribute to deposit problems. Three major types of deposition are discussed here: scaling, general fouling, and biological fouling.

Scale Formation

Scale formation in a cooling system can be controlled by:

- minimizing cycles of concentration through blowdown control

- adding acid to prevent deposition of pH-sensitive species

- softening the water to reduce calcium

- using scale inhibitors to allow operation under supersaturated conditions

Blowdown Control. Increasing blowdown to limit cycles of concentration is an effective way to reduce the scaling potential of circulating water.

However, high rates of blowdown are not always tolerable and, depending on water quality, cannot always provide complete scale control. In many localities, supplies of fresh water are limited and costly.

Table 31-1 shows the relationship between cycles of concentration and makeup and blowdown requirements. Operating at 2 cycles of concentration, the makeup water requirement for this 50,000 gpm system is 2000 gpm. Increasing from 2 to 8 cycles decreases the makeup requirement by 43% to 1140 gpm. Makeup water savings at higher than 8 cycles are negligible.

The reduction of blowdown at higher cycles is even more significant because chemical treatment and waste disposal costs relate directly to blowdown. Increasing cycles from 2 to 8 reduces blowdown by 86%. Further increases have a much smaller effect on the blowdown rate.

Acid Feed. Sulfuric acid is relatively inexpensive and is often used in the treatment of cooling tower systems. Sufficient acid should be added to achieve the desired pH/alkalinity in the recirculating water. For acid to be used as the sole $CaCO_3$ scale control treatment, the LSI must be reduced to zero or a negative value. Acid can also be used to maintain an LSI in the range of 0 to +2.5, with a scale inhibitor to control calcium carbonate deposition. It becomes increasingly more difficult to control calcium carbonate scale as LSI exceeds +2.5. At high values, the LSI should be calculated based on theoretically cycled calcium and alkalinity values, as well as actual system concentrations. Comparison of theoretical LSI to that based on actual tower calcium and alkalinity can help to identify calcium carbonate scaling problems.

Acid treatment removes bicarbonate alkalinity from the water by the following reaction:

$$2HCO_3^- + H_2SO_4 = SO_4^{2-} + 2H_2O + 2CO_2$$

bicarbonate   sulfuric acid   sulfate   water   carbon dioxide

Table 31-1. Makeup and blowdown rates at various cycles.[a]

Cycles	Makeup, gpm	Blowdown, gpm
2	2000	1000
4	1330	333
8	1140	143
15	1070	71
20	1050	53

[a] RR = 50,000 gpm; ΔT = 20 °F.

The CO_2 formed is vented across the cooling tower, while sulfate remains as a by-product.

Lowering pH through acid feed also reduces the scaling tendencies of other pH-sensitive species such as magnesium silicate, zinc hydroxide, and calcium phosphate.

Because control of acid feed is critical, an automated feed system should be used. Overfeed of acid contributes to excessive corrosion; loss of acid feed can lead to rapid scale formation. An acid dilution system should be used for proper mixing to prevent acid attack of the concrete sump.

When makeup water sulfate is high and/or the tower is operated at high cycles, sulfuric acid feed can lead to calcium sulfate scaling. Sometimes, hydrochloric acid is used instead of sulfuric acid in such cases. However, this can result in high chloride levels, which often contribute significantly to increased corrosion rates—especially pitting and/or stress cracking of stainless steel.

Injection of carbon dioxide into the circulating water to control pH has been proposed occasionally. Such treatment reduces pH but does not reduce alkalinity. The circulating water is aerated each time it passes over the cooling tower. This reduces the carbon dioxide concentration in the water to the equilibrium value for the atmospheric conditions, causing the pH to rise. The rapid increase in pH across the tower can lead to calcium carbonate scaling on the tower fill. Because of aeration, carbon dioxide does not cycle and must be fed based on system recirculation rate. It is generally not considered a practical means of controlling pH in open recirculating systems.

Water Softening. Lime softening of the makeup or a sidestream can be used to lower the calcium and, often, alkalinity. This reduces both the calcium carbonate and calcium sulfate scaling tendencies of the water at a given number of cycles and pH level. Sidestream lime softening is also used to lower silica levels.

Scale Inhibitors. Cooling systems can be operated at higher cycles of concentration and/or higher pH when appropriate scale inhibitors are applied. These materials interfere with crystal growth, permitting operation at "supersaturated" conditions. Organic phosphates, also called phosphonates, are commonly used to inhibit calcium carbonate scale. Phosphonates or various polymeric materials can be used to inhibit other types of scale, such as calcium sulfate and calcium phosphate.

Table 31-2 shows a relatively high-quality makeup water at various cycles of concentration.

Table 31-2. Recirculating cooling water at various cycles.

	Makeup Water	Circulating Water at 2 Cycles No Acid Feed	Circulating Water at 5 Cycles No Acid Feed	Circulating Water at 10 Cycles No Acid Feed	Circulating Water at 15 Cycles No Acid Feed	Circulating Water at 15 Cycles Acid for pH 8.7
Calcium (as $CaCO_3$), ppm	50	100	250	500	750	750
Magnesium (as $CaCO_3$), ppm	20	40	100	300	300	300
M Alkalinity (as $CaCO_3$), ppm	40	80	200	400	600	310
Sulfate (as SO_4^{2-}), ppm	40	80	200	400	600	890
Chloride (as Cl^-), ppm	10	20	50	100	150	150
Silica (as SiO_2), ppm	10	20	50	100	150	150
pH	7.0	7.6	8.3	8.9	9.2	8.7
pH_s (120 °F)	8.2	7.6	6.8	6.4	6.0	6.2
LSI	−1.2	0	+1.5	+2.5	+3.2	+2.5
RSI	9.4	7.6	5.3	3.9	2.8	3.7
$CaCO_3$ Controlled by[a]:		B	B/S	B/S	X	B/A/S

[a] B, blowdown only; B/S, blowdown plus scale inhibitor; B/A/S, blowdown plus acid plus $CaCO_3$ scale inhibitor; X, cannot operate.

With no chemical additives of any type, this water is limited to 2 cycles. At 5 cycles the pH is approximately 8.3, and the LSI is +1.5. The system can be operated without acid feed if a scale inhibitor is used. At 10 cycles with no acid feed, the LSI is +2.5 and the water is treatable with a calcium carbonate scale inhibitor. At 15 cycles and no acid feed, the theoretical pH is 9.2 and the LSI is +3.2. In this case, the water cannot be treated effectively at 15 cycles with conventional calcium carbonate inhibitors. Acid should be fed to reduce the pH to 8.7 or below so that a scale inhibitor may be used.

General Fouling Control

Species that do not form scale (iron, mud, silt, and other debris) can also cause deposition problems. Because these materials are composed of solid particles, their deposition is often more flow-related than heat-related. Suspended solids tend to drop out in low-flow areas, such as the tower sump and heat exchangers with cooling water on the shell side. In addition to serving as a water reservoir, the tower sump provides a settling basin. The accumulated solids can be removed from the sump periodically by vacuum or shoveling methods. Natural and synthetic polymers of various types can be used to minimize fouling in heat exchangers.

Organic process contaminants, such as oil and grease, can enter a system through exchanger leaks. Surfactants can be used to mitigate the effects of these materials. Fouling is addressed in further detail in Chapter 25.

Biological Fouling Control

An open recirculating cooling system provides a favorable environment for biological growth. If this growth is not controlled, severe biological fouling and accelerated corrosion can occur. Corrosion inhibitors and deposit control agents cannot function effectively in the presence of biological accumulations.

A complete discussion of microorganisms and control of biological fouling can be found in Chapter 26. Oxidizing antimicrobials (e.g., chlorine and halogen donors) are discussed in Chapter 27.

CORROSION CONTROL PROGRAMS

The addition of a single corrosion inhibitor, such as phosphate or zinc, is not sufficient for effective treatment of an open recirculating cooling system. A comprehensive treatment program that addresses corrosion and all types of deposition is required. All corrosion inhibitor programs require a good biological control program and, in some cases, supplemental deposit control agents for specific foulants.

Chromate-Based Programs

For many years, programs based on chromate provided excellent corrosion protection for cooling systems. However, it was soon recognized that chromate, as a heavy metal, had certain health and environmental hazards associated with it. Treatments employing chromate alone at 200–500 ppm rapidly gave way to programs such as "Zinc Dianodic," which incorporated zinc and phosphate to reduce chromate levels to 15–25 ppm.

Federal regulations limiting discharge of chromate to receiving streams sparked further efforts to reduce or eliminate chromate. The most recent concern relating to chromate treatment involves chromate present in cooling tower drift. When inhaled, hexavalent chrome is a suspected carcinogen. Therefore, as of May 1990, the use of chromate in comfort cooling towers was banned by the EPA. It is expected that chromate use in open recirculating cooling systems will be banned altogether by the end of 1993.

Copper Corrosion Inhibitors

Chromate is a good corrosion inhibitor for copper as well as steel. Therefore, no specific copper corrosion inhibitor was needed in most chromate-based programs. However, most other mild steel inhibitors do not effectively protect copper alloys. Therefore, nonchromate programs generally include a specific copper corrosion inhibitor when copper alloys are present in the system.

Early Phosphate/Phosphonate Programs

Many early corrosion treatment programs used polyphosphate at relatively high levels. In water, polyphosphate undergoes a process of hydrolysis, commonly called "reversion," which returns it to its orthophosphate state. In early programs, this process often resulted in calcium orthophosphate deposition.

Later improvements used combinations of ortho-, poly-, and organic phosphates. The general treatment ranges are as follows:

Orthophosphate	2–10 ppm
Polyphosphate	2–10 ppm
Phosphonate	2–10 ppm
pH	6.5–8.5

A more specific set of control limits within these ranges was developed, based on individual water characteristics and system operating conditions. Where low-calcium waters were used (i.e., less than 75 ppm), zinc was often added to provide the desired corrosion protection.

With close control of phosphate levels, pH, and cycles, it was possible to achieve satisfactory corrosion protection with minimal deposition. However, there was little room for error, and calcium phosphate deposition was frequently a problem.

Dianodic II®

The Dianodic II® concept revolutionized non-chromate treatment technology with its introduction in 1979. This program uses relatively high levels of orthophosphate to promote a protective oxide film on mild steel surfaces, providing superior corrosion inhibition. The use of high phosphate levels was made possible by the development of superior acrylate-based copolymers. These polymers are capable of keeping high levels of orthophosphate in solution under typical cooling water conditions, eliminating the problem of calcium phosphate deposition encountered with previous programs.

The general control ranges for Dianodic II are as follows:

Total inorganic phosphate	10–25 ppm
Calcium (as $CaCO_3$)	75–1200 ppm
pH	6.8–7.8

More detailed control ranges are developed for individual systems, based on water characteristics and system operating conditions.

Dianodic II programs have been successfully protecting cooling systems since their introduction. Continuing research has yielded many improvements in this treatment approach, including newer, more effective polymers, which have expanded the applicability to more diverse water chemistries. The most widely used treatment program, Dianodic II, is an industry standard in nonchromate treatment. Figure 31-12 illustrates typical results from a Dianodic II program.

Alkaline Treatment Programs

There are several advantages to operating a cooling system in an alkaline pH range of 8.0–9.2. First, the water is inherently less corrosive than at lower pH. Second, feed of sulfuric acid can be minimized or even eliminated, depending on the makeup water chemistry and desired cycles. Referring to Table 31-1, a system using this makeup could run an alkaline treatment program in the 4–10 cycle range with no acid feed. This eliminates the high cost of properly maintaining an acid feed system, along with the safety hazards and handling problems associated with acid.

Even if acid cannot be eliminated, there is still an advantage to alkaline operation. Figure 31-13 shows the approximate relationship between pH and alkalinity. A pH of 8.0–9.0 corresponds to an alkalinity range more than twice that of pH 7.0–8.0. Therefore, pH is more easily controlled at higher pH, and the higher alkalinity provides more buffering capacity in the event of acid overfeed.

A disadvantage of alkaline operation is the increased potential to form calcium carbonate and other calcium- and magnesium-based scales. This can limit cycles of concentration and necessitate the use of deposit control agents.

Alkaline Zinc Programs. One of the most effective alkaline programs relies on a combination of zinc and organic phosphate (phosphonate) for corrosion inhibition. Zinc is an excellent cathodic inhibitor that allows operation at lower calcium and alkalinity levels than other alkaline treatments. However, discharge of cooling tower blowdown containing zinc may be severely limited due to its aquatic toxicity. Zinc-based programs are most

Figure 31-12. Corrosion coupon shows results of Dianodic II treatment.

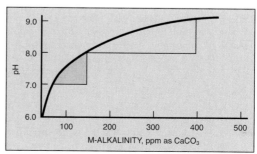

Figure 31-13. Relationship between pH and M-alkalinity shows increased buffering at higher pH.

Figure 31-14. Alkaline phosphate programs provide excellent corrosion and deposition control.

applicable in plants where zinc can be removed in the waste treatment process.

Alkaline Phosphate Programs. Combinations of organic and inorganic phosphates are also used to inhibit corrosion at alkaline pH. Superior synthetic polymer technology has been applied to eliminate many of the fouling problems encountered with early phosphate/phosphonate programs. Because of the higher pH and alkalinity, the required phosphate levels are lower than in Dianodic II treatments. General treatment ranges are as follows:

Inorganic phosphate	2–10 ppm
Organic phosphate	3–8 ppm
Calcium (as $CaCO_3$)	75–1200 ppm
pH	8.0–9.2

Typical alkaline phosphate program results are provided in Figure 31-14.

All-Organic Programs

All-organic programs use no inorganic phosphates or zinc. Corrosion protection is provided by phosphonates and organic film-forming inhibitors. These programs typically require a pH range of 8.7–9.2 to take advantage of calcium carbonate as a cathodic inhibitor.

Molybdate-Based Programs

In order to be effective, molybdate alone requires very high treatment concentrations. Therefore, it is usually applied at lower levels (e.g., 2–20 ppm) and combined with other inhibitors, such as inorganic and organic phosphates. Many investigators believe that molybdate, at the levels mentioned above, is effective in controlling pitting on mild steel. Because molybdate is more expensive than most conventional corrosion inhibitors on a parts per million basis, the benefit of molybdate addition must be weighed against the incremental

cost. Use of molybdate may be most appropriate where phosphate and/or zinc discharge is limited.

FUTURE CONSIDERATIONS

The chemical influence of cooling system blowdown on receiving streams is being closely scrutinized in the United States, where the cleanup of waterways is a high priority. Zinc and phosphate effluent limitations are in place in many states. Extensive research to develop new, more "environmentally friendly" treatment programs is underway and likely to continue. Extensive testing to determine toxicity and environmental impact of new molecules will be required. The answers are not simple, and the new programs are likely to be more expensive than current technology.

MONITORING AND CONTROL OF COOLING WATER TREATMENT

There are many factors that contribute to corrosion and fouling in cooling water systems. The choice and application of proper treatment chemicals is only a small part of the solution. Sophisticated monitoring programs are needed to identify potential problems so that treatment programs can be modified. Effective control of product feed and monitoring of chemical residuals is needed to fine-tune treatment programs. Continued monitoring is necessary to confirm treatment results and determine system trends.

Monitoring of Treatment Results

Figures 31-15 and 31-16 illustrate some of the problems that can lead to poor treatment results. Although simple monitoring tools may reveal problems, they may give no indication of the cause. The monitoring tools briefly discussed here are addressed in more detail in Chapter 36.

No monitoring tool can duplicate system conditions exactly. It is also necessary to inspect plant equipment frequently and document the results.

Corrosion. Corrosion rates can be monitored by means of corrosion coupons, instantaneous corrosion rate meters, or the Betz Monitall®, which measures the corrosion rate on heat transfer surfaces. Elevated iron or copper levels in the circulating water can also be an indication of corrosion.

Deposition. Deposition tendencies can be observed on corrosion coupons or heated apparatus, such as test heat exchangers or the Betz Monitall®. A comparison of various mineral concentrations and suspended solids levels in the makeup water to those in the blowdown may indicate the loss of some chemical species due to deposition.

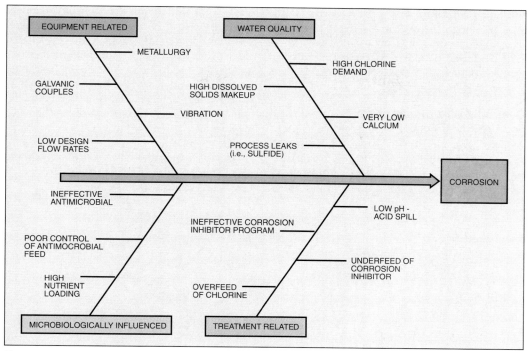

Figure 31-15. Factors affecting corrosion rate in open recirculating cooling systems.

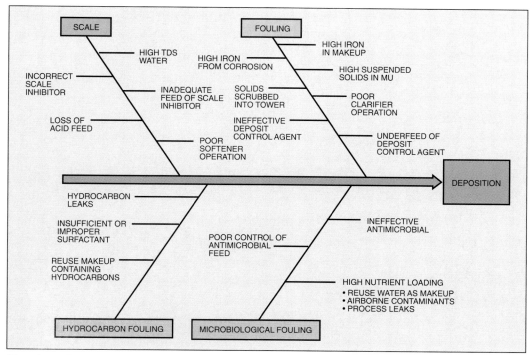

Figure 31-16. Factors contributing to deposition in open recirculating cooling systems.

Biological Fouling. Many techniques are available to monitor biological fouling. Those that monitor biological growth on actual or simulated system surfaces provide a good measure of system conditions. Bulk water counts of various species may be misleading.

Control of Water Parameters and Treatment Feed

Although some treatment programs are more forgiving than others, even the best program requires good control of cycles, pH, and treatment levels. Good control saves money. In the short term, improved control optimizes treatment levels, prevents overfeed, and minimizes chemical consumption. In the long term, cleaner heat exchanger surfaces, less frequent equipment replacement, and reduced downtime for cleaning and repair combine to improve system efficiency, contributing to higher profitability for the plant. Often, computerized feed and control systems are so effective in these areas that they soon pay for themselves.

Detailed information on system monitoring and control is provided in Chapters 35 and 36 (see also Chapters 26 and 27).

CLOSED RECIRCULATING COOLING SYSTEMS

The closed recirculating cooling water system (Figure 32-1) evolved from methods used for the cooling of early engine designs. In a closed system, water circulates in a closed cycle and is subjected to alternate cooling and heating without air contact. Heat, absorbed by the water in the closed system, is normally transferred by a water-to-water exchanger to the recirculating water of an open recirculating system, from which the heat is lost to atmosphere.

Closed recirculating cooling water systems are well suited to the cooling of gas engines and compressors. Diesel engines in stationary and locomotive service normally use radiator systems similar to the familiar automobile cooling system. Other closed recirculating cooling applications include smelt spout cooling systems on Kraft recovery boilers and lubricating oil and sample coolers in power plants. Closed systems are also widely used in air conditioning chilled water systems to transfer the refrigerant cooling to air washers, in which the air is chilled. In cold seasons, the same system can supply heat to air washers. Closed water cooling systems also provide a reliable method of industrial process temperature control.

ADVANTAGES OF CLOSED SYSTEMS

Closed recirculating systems have many advantages. They provide better control of temperatures in heat-producing equipment, and their small makeup water requirements greatly simplify control of potential waterside problems. Makeup water is needed only when leakage has occurred at pump packings or when water has been drained to allow system repair. Little, if any, evaporation occurs. Therefore, high-quality water can usually be used for makeup, and as a result, scale deposits are not a problem. The use of high-quality water also minimizes the dangers of

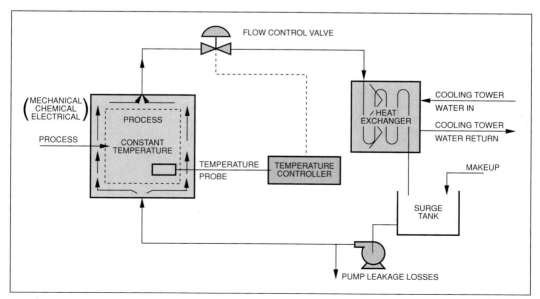

Figure 32-1. Typical closed cooling system.

cracked cylinders, broken heads, fouled exchangers, and other mechanical failures. Closed systems are also less susceptible to biological fouling from slime and algae deposits than open systems.

Closed systems also reduce corrosion problems drastically, because the recirculating water is not continuously saturated with oxygen, as in an open system. The only points of possible oxygen entry are at the surface of the surge tank or the hot well, the circulating pump packings, and the makeup water. With the small amount of makeup water required, adequate treatment can virtually eliminate corrosion and the accumulation of corrosion products.

SCALE CONTROL

Some closed systems, such as chilled water systems, operate at relatively low temperatures and require very little makeup water. Because no concentration of dissolved solids occurs, fairly hard makeup water may be used with little danger of scale formation. However, in diesel and gas engines, the high temperature of the jacket water significantly increases its tendency to deposit scale. Over a long period, the addition of even small amounts of hard makeup water causes a gradual buildup of scale in cylinders and cylinder heads. Where condensate is available, it is preferred for closed system cooling water makeup. Where condensate is not available, zeolite softening should be applied to the makeup water.

CORROSION CONTROL

An increase in water temperature causes an increase in corrosion. In a vented system, this tendency is reduced by the decreased solubility of oxygen at higher temperatures. This is the basis of mechanical deaeration.

Figure 32-2 shows corrosion rates at increasing water temperatures for two different sets of conditions. Curve A plots data from a completely closed system with no provision for the venting of oxygen to atmosphere. Curve B shows data for a vented system. At up to 170 °F (77 °C), the curves are essentially parallel. Beyond 170 °F (77 °C), curve B drops off. This occurs because the lower solubility of oxygen with increasing temperatures in a freely vented system decreases the corrosion rate faster than the rise in temperature increases it. However, in many closed systems, the dissolved oxygen entering the system in the makeup water cannot be freely vented, resulting

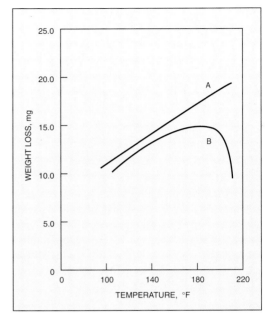

Figure 32-2. Effect of temperature on corrosion rate in closed (A) vs. open (B) systems.

in the release of oxygen at points of high heat transfer, which may cause severe corrosion.

Untreated systems can suffer serious corrosion damage from oxygen pitting, galvanic action, and crevice attack. Closed cooling systems that are shut down periodically are subjected to water temperatures that may vary from ambient to 180 °F (82 °C) or higher. During shutdown, oxygen can enter the water until its saturation limit is reached. When the system is returned to high-temperature operation, oxygen solubility drops and the released oxygen attacks metal surfaces.

The metallurgy used in constructing modern engines, compressors, and cooling systems includes cast iron, steel, copper, copper alloys, and aluminum as well as solders. Nonmetallic components, such as natural or synthetic rubber, asbestos, and carbon, are also used. If bimetallic couples are present, galvanic corrosion may develop.

The three most reliable corrosion inhibitors for closed cooling water systems are chromate, molybdate, and nitrite materials. Generally, the chromate or molybdate types have proven to be superior treatments. For mixed metallurgy systems, the molybdate inhibitors provide the best corrosion protection.

Chromate treatments in the range of 500–1000

ppm as CrO_4^{2-} are satisfactory unless bimetallic influences exist. When such bimetallic couples as steel and copper are present, chromate treatment levels should be increased to exceed 2000 ppm. Maximum inhibitor effectiveness can be achieved if the pH of these systems is kept between 7.5 and 9.5.

In a closed system, it can be quite difficult to prevent corrosion of aluminum and its alloys; the pH of the water must be maintained below 9.0. Aluminum is amphoteric—it dissolves in both acid and base, and its corrosion rate accelerates at pH levels higher than 9.0. The bimetallic couple that is most difficult to cope with is that of copper and aluminum, for which chromate concentrations even higher than 5000 ppm may not be adequate.

Where circulating pumps are equipped with certain mechanical seals, such as graphite, chromate concentrations may not exceed 250 ppm. This is due to the fact that water leaking past the seals evaporates and leaves a high concentration of abrasive salts that can damage the seal.

Another problem is encountered when chromate inhibitors are used in cooling systems serving compressors that handle sour gas. If sour gas leaks from the power cylinder into the water circuit, significant chromate reduction will occur, causing poor corrosion control and deposition of reduced chromate.

In very high heat transfer rate applications, such as continuous caster mold cooling systems, chromate levels should be maintained at 100–150 ppm maximum. Under these extreme conditions, chromate can accumulate at the grain boundaries on the mold, causing enough insulation to create equipment reliability problems.

The toxicity of high-chromate concentrations may restrict their use, particularly when a system must be drained frequently. Current legislation has significantly reduced the allowable discharge limits and the reportable quantity for the spill of chromate-based products. Depending on the type of closed system and the various factors of State/Federal laws limiting the use of chromate, a nonchromate alternative may be needed.

Molybdate treatments provide effective corrosion protection and an environmentally acceptable alternative to chromate inhibitors. Nitrite–molybdate–azole blends inhibit corrosion in steel, copper, aluminum, and mixed-metallurgy systems. Molybdates are thermally stable and can provide excellent corrosion protection in both soft and

hard water. System pH is normally controlled between 7.0 and 9.0. Recommended treatment control limits are 200–300 ppm molybdate as MoO_4^{2-}. Molybdate inhibitors should not be used with calcium levels greater than 500 ppm.

Nitrite is another widely accepted nonchromate closed cooling water inhibitor. Nitrite concentrations in the range of 600–1200 ppm as NO_2^- will suitably inhibit iron and steel corrosion when the pH is maintained above 7.0. Systems containing steel and copper couples require treatment levels in the 5000–7000 ppm range. If aluminum is also present, the corrosion problem is intensified, and a treatment level of 10,000 ppm may be required. In all cases, the pH of the circulating water should be maintained in the alkaline range, but below 9.0 when aluminum is present. When high nitrite levels are applied, acid feed may be required for pH control.

One drawback to nitrite treatments is the fact that nitrites are oxidized by microorganisms. This can lead to low inhibitor levels and biological fouling. The feed of nonoxidizing antimicrobials may be necessary to control nitrite reversion and biological fouling.

Figure 32-3 shows product performance data developed in laboratory studies simulating a mixed-metallurgy closed cooling system. The results identify steel and Admiralty corrosion rates for three closed system inhibitors at increasing treatment levels. As shown, the molybdate-based treatment provides the best overall steel and Admiralty protection. To achieve similar inhibition with chromate, higher treatment concentrations are required. Nitrite-based treatment also provides effective steel protection, with results comparable to those obtained with molybdate; however, acceptable Admiralty corrosion inhibition is not achieved.

Closed systems often require the addition of a suitable antifreeze. Nonchromate inhibitors are compatible with typical antifreeze compounds. Chromates may be used with alcohol antifreeze, but the pH of the circulating water should be maintained above 7.0 to prevent chromate reduction. Because glycol antifreezes are not compatible with chromate-based treatments, nonchromate inhibitors should be used. Molybdate treatments should not be used with brine-type antifreezes.

In closed systems that continuously run at temperatures below 32 °F (0 °C), a closed brine system is often employed. The American Society

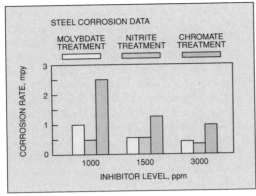
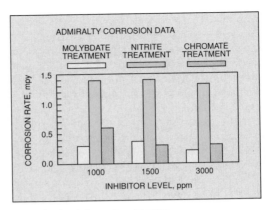

Figure 32-3. Comparison of different treatments to inhibit corrosion in a mixed-metallurgy closed system.

of Refrigeration Engineers has established chromate limits in brine treatments. Calcium brines are limited to 1250 ppm chromate, and sodium brines are limited to 2500 ppm chromate. The pH should be 7.0–8.5 with caustic adjustment only. Some success has also been recorded with nitrite-based treatment of closed brine systems at treatment levels of about 2000 ppm as NO_2^-.

PRETREATMENT OF COOLING WATER SYSTEMS

Pretreatment of cooling water systems is necessary to maximize the service life of heat exchange equipment. Usually, pretreatment consists of two phases:

1. precleaning, to remove the accumulation of foreign matter

2. prefilming, to promote the rapid formation of an inhibiting film

Precleaning is important because it prepares the surface for the prefilming phase. After the surface has been cleaned, prefilming minimizes the initial corrosion which occurs at start-up and allows the most efficient application of a corrosion inhibitor program. Economics, discharge limitations, and time requirements dictate whether pretreatment should be applied to the entire system or to individual exchangers.

PRECLEANING

All new equipment should be precleaned for removal of grease, oil, corrosion products, mill scale, and dirt. Clean surfaces enable corrosion inhibitors to promote a uniform protective film. Failure to preclean can result in increased corrosion and fouling, leading to reduced heat transfer, excessive pressure drop across critical heat exchangers, premature failures, and high maintenance costs.

Precleaning should be conducted just prior to start-up, followed immediately by rapid passivation of surfaces (as outlined in the Prefilming section of this chapter).

Normally, a cleaning solution of polyphosphate, surfactant, and antifoam is circulated though the cooling system. Precleaning is essentially a detergent treatment. The polyphosphate and surfactant in hot water help remove light rust and hydrophobic materials left by manufacturing and construction.

Polyphosphate is fed to achieve a concentration of 1–4% with the best results obtained at the higher dosages. The temperature is maintained at 120–160 °F (50–70 °C) and the pH is controlled in the range of 5.5–7.0. Higher water temperatures contribute to faster and more effective removal of unwanted materials.

Typically, cleaning is conducted over a period of 8–24 hr. When the program is completed, heavy blowdown is required for rapid removal of the precleaning chemicals.

Optimal results are attained when precleaning is performed at the recommended temperature elevations. Usually, it is difficult to achieve high temperatures unless individual exchangers are being treated with relatively small solution volumes. Lower solution temperatures can be compensated for by circulation of the precleaning solution for the maximal time period. Even at lower temperatures, precleaning significantly improves results.

Special precautions must be taken for new cooling towers built with wood that has been pretreated with a copper-based salt. The copper salts must be removed from the wood before any water is circulated through the system. Therefore, new towers should be thoroughly water-flushed and sumps should be drained and cleaned before precleaning takes place.

PREFILMING

Most methods of corrosion control involve the formation of a film to act as a barrier to corrosion. The effectiveness of the treatment depends largely on the rate at which the barrier film is formed. Materials that do not allow rapid film formation permit corrosion to take place before a complete film layer has been created. Incomplete film formation contributes to continued corrosion.

The rate at which the film forms is related to the concentration of the inhibitor.

Prefilming permits the rapid formation of a uniform film that immediately stifles the corrosion reaction (see Figure 33-1). Once the film has been established, it can be maintained through continuous, low treatment levels to deter the accumulation of corrosion products.

Serious changes in environment (e.g., severe pH depression) can destroy the film, and corrosion products can accumulate before the film is reestablished through normal treatment. When this occurs, a prefilming program may be necessary for rapid repassivation of the system.

Prefilming of equipment in cooling water systems is recommended following initial installation, chemical cleaning, and start-up after turnaround or inspection. Prefilming may also be necessitated by low pH excursions or process leaks. This depends on the severity of the condition and the length of time the cooling equipment has been exposed to severe conditions.

Phosphate and zinc are used in prefilming programs. For prefilming of copper alloys, azoles are also used.

Polyphosphates are most important because they effectively remove undesirable corrosion products as they form, while developing a protective oxide film. Generally, these materials are circulated through the system at concentration ranges of 300–600 ppm phosphate and 30–60 ppm zinc.

Figure 33-1. Inside surfaces of ³/₄ in. steel heat exchanger tubes. Top: typical steel surface, unexposed, prior to use. Middle: surface exposed to treated cooling water following pretreatment. Bottom: surface exposed to treated cooling water with no pretreatment.

A prefilming program using phosphate and zinc can be effective in rapidly providing a protective film. This program is maintained for 4–6 hr at a pH of 6.5–7 and a temperature of 120 °F. The system is then heavily blown down until the phosphate level drops below 20 ppm. If a high pH program is to be applied, the phosphate level must first be further reduced to less than 3 ppm. The pH should then be adjusted until it reaches the level of the continuous program.

As previously stated, azoles are used for the pretreatment of exchangers fabricated with copper alloy tube bundles. Typically, pretreatment is applied to individual exchangers rather than an entire system. This is especially true for cooling systems using brackish water, where it is difficult to establish or maintain effective copper corrosion protection. Under these conditions, the individual exchangers are pretreated with fresh water and a copper corrosion inhibitor. The treated water can be circulated through the exchanger for a short period of time at ambient conditions to establish a uniform protective film. After prefilming is complete and the exchanger is back in service, continuous or periodic shot feed treatment applications serve to maintain film integrity.

Another method of prefilming is an increase in inhibitor levels for 6–24 hr. Although this is somewhat easier, it is less efficient and is generally used only when the continuous inhibitor program is phosphate-based.

Other considerations and precautions involved in precleaning and pretreatment are summarized below:

■ Initial system design should include provisions for isolating critical bundles. This facilitates individual pretreatment for optimal effectiveness and efficiency (see Figure 33-2).

■ Provisions for air bumping and/or backwashing exchangers should be considered wherever feasible.

■ Quick-opening blowdown valves on the underside of an exchanger (when water is on the shell side) can minimize deposit buildup prior to baffles.

■ The benefit of prefilming is maximized only when system surfaces are clean.

Figure 33-2. Cooling system exchanger and auxiliary pump. During pretreatment, the heat exchanger is isolated from the system (valved off) and the auxiliary pump shown is used to circulate tower water or fresh water and treatment chemical to establish a protective film.

- Precleaning solution must be removed completely through blowdown prior to prefilming.

- Increasing the temperature promotes rapid film formation during prefilming.

- The application of high chemical concentrations and blowdown rates must be governed by environmental discharge limits.

- High levels of zinc can adversely affect biological oxidation waste treatment systems.

- Prefilming of individual exchangers may be the most economical procedure and may eliminate disposal problems.

- After prefilming has been completed, rapid blowdown is required immediately, and normal control must be established for maintenance.

Pretreatment is critical for cooling water systems containing steel because of the higher corrosion rates that occur. Systems containing other metals, such as Admiralty Brass or copper alloys, can also benefit from pretreatment.

In general, pretreatment, followed by ongoing treatment programs, minimizes corrosion for improved heat transfer, longer service life, and reduced plant maintenance.

AIR CONDITIONING AND REFRIGERATION SYSTEMS

For more than a century, industrial air conditioning has been used for drying, humidity control, and dust and smoke abatement. Its most familiar function is to provide a comfortable working environment, to increase the comfort and productivity of personnel in offices, commercial buildings, and industrial plants.

Air conditioning is the process of treating and distributing air to control temperature, humidity, and air quality in selected areas. For temperature and humidity control, air is moved over chilled or heated coils and/or a spray of water at a controlled temperature. Direct water sprays also remove dust and odors. Other air cleaning systems may include mechanical separation, adhesion, screening, filtration, or static attraction, depending on the type of air contaminants encountered and the required air quality.

Refrigeration is the process of lowering the temperature of a substance below that of its surroundings and includes production of chilled water for air conditioning or process applications. Chilled water for use in processes such as injection molding may be in the same temperature range as chilled water used for air conditioning. Refrigeration systems are also used to provide chilled antifreeze solutions (brines) at temperatures below the freezing point of water. Brines are used in icemaking and cold storage, in addition to a variety of chemical process applications.

Chilled water may be used in air washers, either in closed coils or as spray water. Chilled water may also be used for closed systems and for individual spray water systems.

Many methods are used to produce and distribute chilled air. In central air conditioning systems, air is passed over coils chilled by water, by brine, or by direct expansion of a volatile refrigerant. The chilled air is then distributed through ductwork.

The water systems associated with air conditioning can be classified into three general categories: open recirculating cooling, air washers, and closed or open chilled water systems. In water treatment applications, open recirculating cooling systems are similar to open chilled water systems.

The basic mechanical components of an air conditioning system are the air and water distribution systems, a refrigeration machine, and a heat rejection system. Refrigeration for air conditioning is usually provided by either absorption or compression cycles.

Absorption refrigeration (see Figure 34-2) uses low-pressure steam or high-temperature hot water as the energy source, water as the refrigerant, and lithium bromide or lithium chloride as the absorbent.

Compression refrigeration systems (see Figure 34-3) generally utilize a halocarbon compound or ammonia as the refrigerant. An internal combustion engine, turbine, or electric motor supplies the

Figure 34-1. Good water treatment practice and scheduled maintenance can help prevent unexpected shutdown of air conditioning equipment. (Courtesy of Carrier Corporation.)

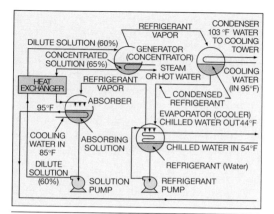

Absorption cycle based on the vaporization and condensation of refrigerant (water) under vacuum. Chilled water is cooled by the evaporation of refrigerant in the evaporator. The refrigerant vapor is absorbed by a salt solution, normally lithium bromide. The diluted absorbing solution is passed to the generator, where heat from steam or hot water is used to boil off the refrigerant vapor which was absorbed by the salt solution. The concentrated salt solution is returned to the absorber. The vapor is passed to the condenser where heat is removed by cooling water, and the vapor is condensed. The liquid refrigerant water is returned to the evaporator. Typical water temperatures are shown.

Figure 34-2. Absorption refrigeration system.

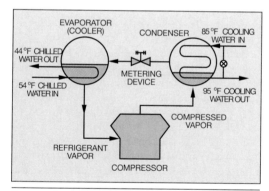

Compresssion cycle based on the vaporization and condensation of a volatile refrigerant. Water is cooled in the evaporator by evaporation of the refrigerant. The refrigerant vapor is compressed and passes to the condenser where heat is removed by cooling water and the refrigerant becomes a liquid. The liquid refrigerant is then returned to the evaporator through a metering device. Typical water temperatures are shown.

Figure 34-3. Compression refrigeration system.

power to drive a centrifugal or positive displacement compressor.

Refrigeration, or cooling, occurs when the liquid refrigerant absorbs heat by evaporation, generally at low temperature and pressure. When the refrigerant condenses, it releases heat to any available cooling medium, usually water or air.

SINGLE-STAGE REFRIGERATION CYCLE

Figure 34-3 shows the basic refrigeration cycle used for single-stage vapor compression. The four components in the system are the compressor, condenser, metering device, and evaporator. Low-pressure liquid refrigerant in an evaporator extracts heat from the fluid being cooled and evaporates. The low-pressure vapor is then compressed to a pressure at which the refrigerant vapor can be condensed by the cooling media available. The vapor then flows to the condenser, where it is cooled and condensed. The liquid refrigerant flows from the condenser to a metering device, where its pressure is reduced to that of the evaporator. The cycle is thus completed.

In industrial or commercial air conditioning systems, the heat is usually rejected to water. Once-through cooling may be used, but municipal restrictions and water costs generally dictate recirculation and evaporative cooling processes.

Evaporative condensers or cooling towers are normally used for evaporative cooling. A spray pond may be used as an alternative. Recirculation of the water in a cooling system reduces the makeup water requirement to less than 3% of the water that would be needed for once-through cooling.

Cooling capacity is measured in tons of refrigeration. A ton of refrigeration is defined as the capacity to remove heat at a rate of 12,000 Btu/hr at the evaporator or chiller.

An absorption refrigeration system that removes 12,000 Btu/hr (does 1 ton of air conditioning) requires heat energy input of approximately 18,000 Btu/hr to drive the absorption process. This means that the heat rejection at the cooling tower approximates 30,000 Btu/hr per ton of refrigeration. With a 15 °F (8 °C) temperature drop across the tower, the heat rejection of an absorption system requires circulation of approximately 4 gpm of water per ton of air conditioning. Evaporation of the recirculating water occurs at a rate of approximately 3.7 gph per ton.

Other than the solution and refrigerant pumps, there are no moving parts in an absorption system. Although this is an economical design advantage, the cost of producing the necessary low-pressure steam or high-temperature hot water (HTHW) must also be considered.

Compression systems also impose an additional heat load. This is due to the energy required to compress low-pressure, low-temperature refrigerant gas from the evaporator and deliver it to the condenser at a higher pressure. The compressor energy input is approximately 3,000 Btu/hr per ton of refrigeration. Accordingly, normal heat rejection in a compression system approximates 15,000 Btu/hr per ton of refrigeration, requiring evaporation of about 2 gal/hr of cooling water.

Compression refrigeration systems require a cooling water circulation rate of approximately 3 gpm per ton of refrigeration, with a 10 °F temperature drop across the cooling tower.

The major energy consumer in a compression refrigeration system is the compressor, which is designed to operate at a certain head pressure for a given load. This pressure equals the refrigerant pressure in the condenser. The term "high head pressure" refers to condenser pressure that is higher than it should be at a specific load condition.

High head pressure can be costly in two ways. First, it presents the danger of a system shutdown; a safety control will stop the compressor motor when the safe maximum head pressure is exceeded in the compressor. Second, an increase in power consumption results when a compressor operates at greater than design head pressure.

Fouled condenser tubes are a common cause of high head pressures (see Figure 34-4). Fouling increases the resistance to heat transfer from the refrigerant to the cooling water. In order to maintain the same heat transfer rate, the temperature

Figure 34-4. A fouled condenser can increase head pressure and waste energy.

of the refrigerant must be increased. The compressor fulfills this need by increasing the pressure at which the refrigerant is condensed. With a centrifugal chiller, a 1 °F increase in condensing temperature increases compressor energy consumption by approximately 1.7%.

Fouling and the formation of scale in absorption systems also reduce operating efficiency. Because the highest water temperatures exist in the condenser, deposition first occurs in this unit. Under extreme conditions, scale formation can also occur in the absorber.

Deposition in the condenser imposes a higher back-pressure on the generator, so that increased steam or HTHW is required to liberate the refrigerant from the absorbent. The result is an increase in refrigerant vapor pressure and a greater temperature differential between the condensing water vapor and the cooling water. Although this compensates for the resistance to heat flow, more energy is required to provide the increased heat input.

If water conditions are severe enough to cause deposition in the absorber, less refrigerant is removed by the absorber, and cooling capacity is reduced. The reduction in refrigerant circulation diminishes the ability of the equipment to satisfy cooling requirements.

If the absorption rate in the absorber is reduced while the absorbent is heated above the normal temperature in the generator, the danger of over-concentrating the brine solution also exists. This over-concentration can cause crystallization of the brine, leading to a system shutdown.

Fouling and scale formation waste energy and can ultimately cause unscheduled system shutdown. Effective water treatment can minimize the possibility of high head pressure and excessive steam consumption caused by condenser deposition.

Corrosion can cause problems in either the open recirculating or chilled water circuits. When corrosion is not properly controlled, the resulting corrosion products inhibit heat transfer, increasing energy consumption in the same manner as fouling and scale formation. Unchecked corrosion can cause heat exchanger leaks and catastrophic system failures.

In any cooling application, attention to cooling tower operation is important. Proper tower maintenance maximizes cooling efficiency, or ability to reject heat. This is critical for continuously running refrigeration machinery at full load conditions.

For best performance, the cooling tower fill should be kept clean and protected from deterioration. The water distribution system must provide uniform wetting of the fill for optimal air–water contact.

Other components, such as drift eliminators, fill supports, regulating valves, distribution decks, and tower fans, should be kept clean to maintain efficient heat rejection. Inefficient cooling or heat rejection increases the temperature of the water in the cooling tower sump and, consequently, that of the water sent to the condenser. This makes it necessary to condense the refrigerant at a higher temperature (absorption) or higher temperature and pressure (compression) to reject heat at the same rate into the warmer water. This increases the amount of energy (steam, hot water, electricity) required to operate the system.

OPEN RECIRCULATING COOLING WATER PROBLEMS

The water in open cooling systems is subject to problems of scale, corrosion, slime, and algae.

Scale

As water evaporates in a cooling tower or an evaporative condenser, pure vapor is lost and dissolved solids concentrate in the remaining water. If this concentration cycle is allowed to continue, the solubilities of various solids will eventually be exceeded. The solids will then deposit in the form of scale on hotter surfaces, such as condenser tubes. The deposit is usually calcium carbonate. Calcium sulfate, silica, and iron deposition may also occur, depending on the minerals contained in the water. Deposition inhibits heat transfer and reduces energy efficiency.

Deposition is prevented by threshold inhibitors that increase the apparent solubility of the dissolved minerals. Therefore, they do not precipitate and are removed by blowdown. Blowdown is automatically replaced by fresh water.

The ratio of dissolved solids in the circulating water to that in the makeup water is called "cycles of concentration." With correct treatment, cycles of concentration can be increased so that less makeup water and consequently less chemicals are required (Table 34-1).

The cooling capacity of a tower is affected by how finely the water is atomized into droplets. Smaller droplets lose more heat to the atmosphere; however, more of the smaller droplets are carried away by the air drawn through the tower. This "windage loss," or "drift loss," becomes part of the total blowdown from the system. Windage loss is approximately 0.1 to 0.3% of the water circulation rate.

Windage can have undesirable effects, such as the staining of buildings and the spotting and deterioration of car finishes. These problems are caused by the dissolved solids in the circulating water, which evaporate to dryness as water droplets fall on surfaces. Because water treatment chemicals produce only a slight increase in the dissolved solids content of the water, they usually do not contribute significantly to spotting problems.

Continuous blowdown, or bleed-off, is adequate for scale control in some cooling systems. The importance of continuous blowdown, as opposed to periodic complete draining, cannot be overemphasized. The volume of water in most cooling systems is small compared to the amount of water evaporated. Therefore, excess solids concentrations can develop in a short period of time. Continuous blowdown prevents excessive solids concentrations from developing in the tower water.

In order to maintain solids in solution, water that is high in alkalinity and hardness may require the feed of sulfuric acid or an acid salt in addition to blowdown. Acid feed requires careful handling and control and should only be used where the blowdown rate would otherwise be excessive.

Sodium zeolite softening of the makeup water is also an effective way to control scale. However, this process does not decrease the alkalinity of the tower water. Because the resulting low-hardness, high-alkalinity, high-pH water

Table 34-1. Maximizing cycles of concentration saves water and reduces water and sewage costs.

Tower Size (tons of refrigeration)	Daily Water Requirements (gal)		Yearly Water Costs	
	2 cycles	5 cycles	2 cycles	5 cycles
250	10,800	6,750	$972	$608
600	25,920	16,200	$2333	$1458
3000	129,600	81,000	$11,664	$7290

Based on a 12 hr/day operation, 180 days/yr, at a water cost of $0.50/1000 gal and 3 gpm per ton of refrigeration at a 10 °F temperature drop.

is particularly aggressive to copper alloys, it may necessitate acid feed in addition to softening. Also, carbon steel corrosion control is more difficult with softened water than with hard water.

Polyphosphates are of some value for scale control but must be applied cautiously, because hydrolysis of the polyphosphate results in the formation of orthophosphate ions. If this process is not properly controlled, calcium phosphate deposits may result. Chemicals are now available that inhibit scale formation without this undesirable side effect. Therefore, polyphosphates are now used primarily for corrosion inhibition.

A treatment which controls calcium carbonate particle growth and prevents deposition can permit a reasonable blowdown rate and eliminate the need for pH depression with acid. Phosphonates are particularly useful as threshold inhibitors of scale formation and as iron oxide dispersants. Certain low-molecular-weight polymers also have the ability to control calcium carbonate scale formation.

Suspended solids (airborne dust and debris from the air passing through the cooling tower) contribute to general fouling and can aggravate scale formation. The deposits may also cause localized under-deposit corrosion.

Fouling of heat transfer surfaces has an insulating effect that reduces the energy efficiency of the process. Failure to control scale formation also reduces the rate of heat transfer. Accordingly, a properly designed treatment program must include polymeric dispersants and scale control agents to minimize general fouling and inhibit scale formation.

Corrosion

Water in an open recirculating cooling system is corrosive because it is saturated with oxygen. Systems in urban areas often pick up acidic gases from the air that can be beneficial in scale reduction. However, excessive gas absorption can result in severely corrosive water.

Chromate-based corrosion inhibitors are very effective, but their use is now prohibited in comfort cooling towers. The most commonly used corrosion inhibitors are phosphate, molybdate, zinc, polyphosphate, silicate, and organic-based treatments. These inhibitors can be applied in low or alkaline pH treatment ranges.

At low pH, a high phosphate level is used to promote the passivation of steel. At high pH, a combination of various corrosion inhibitors and deposit control agents is used. These programs use organic inhibitors in combination with zinc, phosphate, or molybdate. Where these are environmentally unacceptable, silicates may be used at an alkaline pH. This type of inhibitor program also includes deposit control agents. However, silica concentration must be controlled to prevent deposition of silicate, which forms a hard and persistent scale.

Azoles, functioning as copper corrosion inhibitors, are used in most programs to improve the corrosion protection of copper and to minimize pitting of ferrous metals.

Because the heat load on many cooling tower systems varies with changing weather conditions, water evaporation rates tend to be irregular. As a result, cooling system protection can be less than desired or expected under conditions of wide load fluctuation. Automated water treatment control equipment substantially improves treatment results in systems that operate under these conditions.

Slime and Algae

Many types of antimicrobials are available for control of algae and biological slime in open cooling systems. Nonoxidizing organic materials (such as quaternary ammonium salts, other organic nitrogen compounds, and organosulfur compounds) are frequently used. Some antimicrobials can be detoxified before discharge into the environment. Microbiological programs often employ a combination of nonoxidizing and oxidizing chemicals. Oxidizing chemicals include chlorine, hypochlorites, organic chlorine donors, and bromine compounds. Chlorine gas requires chlorination equipment and controls, which are not practical for most air conditioning systems. Chlorine and hypochlorites must be applied carefully, because excessive chlorine will increase corrosion and may contribute to deterioration of cooling tower wood and reduction of heat transfer efficiency (see Figure 34-5). For more information on microbiological problems and antimicrobial use in cooling systems, see Chapter 26.

Figure 34-5. Failure to control biological slime reduces heat transfer.

AIR WASHERS

Air washers are spray chambers in which air is conditioned by direct contact with water. The chilled water is contained in an open system as shown in Figure 34-6, or circulated from a closed system as shown in Figure 34-7.

Air washers remove dust, smoke, and odors from the air. Additionally, the return air from a manufacturing process may contain unique contaminants that must be removed. Process contaminants include fiber and oil in textile plants, tobacco dust in tobacco plants, and sizing material in cloth weaving plants.

Filters remove particulate matter from the air before it passes through the spray section. Eliminator blades prevent mist or water droplets from leaving the unit with the air. In addition to cleaning, air washers usually perform other functions. Air

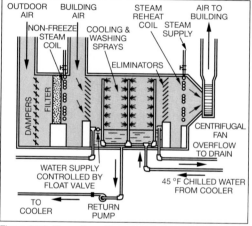

Figure 34-6. Air washer with open chilled water system.

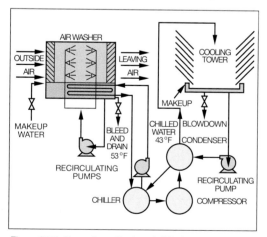

Figure 34-7. Air washer with closed chilled water system.

temperature and humidity can be controlled through adjustment of the spray water temperature.

When air must be humidified in winter, some water is evaporated. This increases the solids concentration in the remaining liquid. Generally, scale does not form, because the temperature of the water is relatively low. If the spray water temperature is below the dew point of the entering air, the air is dehumidified. In summer, dehumidification involves condensing water vapor from the air, diluting the solids in the recirculating water, and causing water to overflow from the bottom sump or pan of the air washer.

Corrosion can develop in air washers as it does in recirculating cooling water systems. The spray water is saturated with atmospheric oxygen and, when present, acidic airborne contaminants lower the pH and contribute to its corrosivity. Accordingly, the use of a corrosion inhibitor is important in air washer treatment.

Cleanliness of the air washer helps to prevent objectionable odors in the air. The volume of air in relation to the water circulation rate in air washers is much greater than it is in cooling towers. Therefore, the tendency to accumulate sludge is much greater. Sludge can cause localized corrosion or promote biological activity that produces odors. Therefore, dispersants and/or surfactants are an integral part of a water treatment program for air washers.

The air being washed also contains numerous microorganisms and materials that will feed bacteria. Therefore, biological slime is a significant problem in air washers. Nonoxidizing chemicals are used for microbiological control. However, undesirable odors may result from this treatment.

If the air washer must be sterilized, the air flow is stopped and a solution of oxidizing or nonoxidizing antimicrobial is circulated through the washer. The unit must then be hosed until the material loosened by the treatment is thoroughly flushed from the bottom of the air washer.

CLOSED WATER SYSTEMS

Closed systems are not subject to scale formation except when hard makeup water must be used. Many closed systems use zeolite-softened water or condensate as makeup to prevent scale problems.

In closed systems, the oxygen concentration is lower than that of aerated systems. Therefore, the potential for corrosion is much lower. However, some corrosion exists, and loose corrosion products can cause fouling of piping, automatic valves, and vents.

Theoretically, closed water systems should not require corrosion inhibitors. Any oxygen introduced with the initial makeup water should soon be depleted by oxidation of system metals, after which corrosion should no longer occur. However, closed systems usually lose enough water and leak enough air to require corrosion protection.

The inhibitors most commonly used are molybdate, silicate, or nitrite based. The use of chromates may be restricted because of regulations that classify them as carcinogens. The amount of inhibitor needed depends on the system water temperature and its metallurgy. Closed systems usually require little additional treatment after the initial charge. Therefore, relatively high treatment levels can be used to provide a greater margin of safety at relatively low cost.

Sulfite-based programs are also used to control corrosion. Unlike the other inhibitors, a corrosion inhibitor film is not established; the sulfite prevents oxygen corrosion by reacting with and removing dissolved oxygen. An alkaline pH is maintained to prevent acidic corrosion. In the case of air leaks, sulfite requirements are not proportional to water loss and can be very high. High sulfite consumption adds to the dissolved solids content of the circulating water and increases the cost of treatment. Therefore, air in-leakage should be minimized.

Insulating couplings are used in closed systems to control galvanic corrosion. These couplings are mostly phenolic resins, which may be attacked at high pH.

CONTROL OF WATER BALANCES

Weather changes cause solids concentration changes in open cooling water systems and particularly in air washers. Air conditioning system design does not always properly address water treatment needs. Often, water sump volumes are reduced in cooling tower designs to minimize system weight. This results in a lower ratio between volume and circulation rate, which causes a more rapid change in water solids concentration with variations in load. Also, low-capacity water pans are used in evaporative condensers and air washers in order to reduce space and weight.

A water treatment program can be complicated by any of the following factors:

■ cooling towers near smokestacks can pick up airborne dirt and acidic gases

■ cooling towers are often installed and operated in such a way that considerable water overflows from the system on shutdown

■ additional makeup water may be required during hot weather to reduce water temperature

For open systems, an effective and efficient treatment program includes continuous blowdown, continuous feed of corrosion inhibitor and dispersant, and daily water testing. Systems should not be treated and controlled solely on the basis of weekly tests. Additional chemical treatment may be needed to ensure adequate protection.

GENERAL CONSIDERATIONS

The importance of beginning chemical treatment promptly cannot be overemphasized. New installations will have mill scale on metal surfaces and will contain oil, pipe compound, brazing and welding scale, and construction debris. Systems previously operated without water treatment (or with ineffective water treatment) contain corrosion products which can slough off when protective treatment is started. Such materials can impede water flow, cause fouling, and increase galvanic corrosion potential. Suspended solids may cause automatic valves and controls to malfunction, and can shorten the life of mechanical seals on the pumps.

Such systems should be thoroughly cleaned (mechanically and chemically), drained, and flushed. Cleaning agents commonly used are organic phosphates, polyphosphates, synthetic detergents, dispersants, and combinations of these materials. The permanent protective treatment should start immediately after cleaning because clean metal surfaces in the system are particularly vulnerable to corrosion.

Air conditioning systems that do not operate all year should be protected properly during idle periods. Open cooling water systems should be drained completely. Condensers should be opened and inspected at the end of every air conditioning season. The basins of cooling towers or air washers should be thoroughly cleaned and flushed.

If the system is stored dry, the condenser should be closed tightly after it is thoroughly dried. Ideally, it should then be filled with nitrogen and sealed. If water is not removed from an idle system, additional protection is required to offset the increased corrosion potential. A higher concentration of a corrosion inhibitor suitable for long-term storage must be used.

If closed system water temperatures will be at or below freezing, antifreeze must be added for protection. When ethylene glycol antifreeze is used, chromate-treated water must be drained from the system because these materials are not compatible. However, chromate is compatible with methanol, calcium chloride, and sodium chloride brines. Molybdate, nitrite, and silicate-based inhibitors may be used with these antifreeze solutions. Molybdate treatments should not be used with brines.

Eliminator sections of cooling towers can collect salt deposits as a result of partial or intermittent wetting. Because sufficient amounts of the treated water do not reach the eliminator sections during operation, treatment chemicals added to the water cannot be expected to provide protection. Salts, dirt, and debris also accumulate on the eliminators and screens of air washers and evaporative condensers. Such areas should be hosed down at regular intervals.

Where appreciable dirt has collected, mechanical cleaning is needed. In industrial plant environments, mechanical cleaning may be required several times during an air conditioning season.

Chemical Feed and Control

Effective water purification is essential to ensure an adequate supply of high-quality water.

CHEMICAL FEED SYSTEMS

A well engineered feed system is an integral part of an effective water treatment program. If a feed system is not designed properly, chemical control will not meet specifications, program results may be inadequate, and operating costs will probably be excessive. Some of the costly problems associated with poor chemical control include:

- high chemical costs due to overfeed problems

- inconsistent product quality, reduced throughput, and higher steam and electrical costs due to waterside fouling

- high corrosion rates and resultant equipment maintenance and replacement (i.e., plugging or replacing corroded heat exchanger tubes or bundles)

- high labor costs due to an excessive requirement for operator attention

- risk of severe and widespread damage to process equipment due to poor control or spillage of acid into cooling towers

A significant investment in a chemical feed system can often be justified when compared with the high cost of these control problems. Figure 35-1 shows the distribution of control results to be expected when a chemical feed system is not properly engineered. In this situation, chemical levels are often above or below program specifications. The use of a proper feed system can prevent this situation and create a chemical control distribution similar to the one in Figure 35-2.

Chemical feed systems can be classified according to the components used, the type of material to be fed (powder or liquid), the control scheme employed, and the application.

FEED SYSTEM COMPONENTS

Chemical Storage

Treatment chemicals are usually delivered and stored in one of three ways: bulk, semibulk, and drums. The choice among these three depends on a number of factors, including usage rate, safety requirements, shipping regulations, available space, and inventory needs.

Bulk Storage. Large users often find it advantageous to handle their liquid chemical delivery and storage in bulk. Liquid treatments are delivered by vendor tank truck or common carrier. A large tank, often supplied by the water treatment company for storing the liquid treatment, is placed on the property of the user near the point of feed (Figure 35-3). Service representatives often handle all inventory management functions.

Treatment can be drawn from these storage vessels and injected directly into the water system or added to a smaller, secondary feed tank, which serves as a day tank. Day tanks are used as a

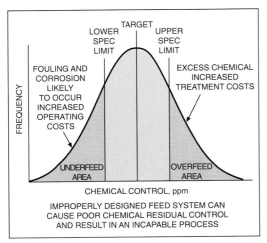

Figure 35-1. Results of an improperly designed feed system.

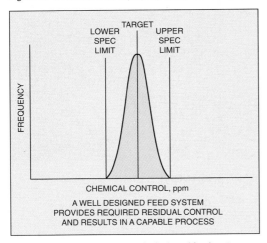

Figure 35-2. Results of a properly designed feed system.

Figure 35-3. Bulk storage tank.

Figure 35-4. Semibulk storage tank.

safeguard to prevent all of the material in the main storage tank from accidentally being emptied into the system. They also provide a convenient way to measure daily product usage rates.

Semibulk Storage. Where chemical feed rates are not large enough to justify bulk delivery and storage, chemicals can be supplied in reusable shuttle tanks (Figure 35-4). Usually, these tanks are designed in such a way that they can be stacked or placed on top of a permanent base tank for easy gravity filling of the base tank.

Drum Storage. Although 40 and 55 gallon drums were widely used for chemical delivery only a few years ago, increasing environmental concerns have sharply reduced drum usage. The restrictions on drum disposal and drum reclamation have reduced the popularity of this delivery and storage method in favor of reusable or returnable containers.

DELIVERY SYSTEMS

Delivery systems are the heart of a chemical feed system. The delivery system most often used is the chemical metering pump. Nearly 95% of all feed systems use metering pumps. However, gravity feed is gaining popularity in cooling water systems. Eductors are also used occasionally.

Metering Pumps

The most frequently used metering pumps for water treatment applications are the packed plunger, piston, and diaphragm pumps. Rotary gear and progressive cavity pumps are also used occasionally. These all fall under the general heading of "positive displacement pumps."

Design and selection of a metering pump and piping circuit are critical to ensure that pump output will match specifications. Parameters that must be considered include suction side static head, net positive suction head (NPSH), pump turndown, potential syphoning, pressure relief, and materials compatibility.

In order to ensure accurate pumping, operating conditions must be close to design specifications. For example, with a plunger pump, an increase in discharge line pressure can significantly reduce pump output. Because many factors affect pump performance, output should be checked frequently with a calibration cylinder. Some computerized chemical feed systems automatically verify metering pump output and make adjustments as necessary.

Packed Plunger Pumps. Because plunger pumps can be designed for high discharge pressures, they are often used for chemical treatment in boiler systems. Pumping action is produced by a direct-acting piston or plunger that moves back and forth in a reciprocating fashion and directly contacts the process fluid within an enclosed chamber (Figure 35-5). Motor speed and/or stroke length may be used to adjust this type of pump. The useful working range for packed plunger pumps is approximately 10–100% of rated capacity.

Packed plunger pumps use packing rings to form a seal between the plunger and the plunger bore. In some circumstances, this can necessitate periodic adjustment or replacement of the rings.

Diaphragm Pumps. Diaphragm pumps are becoming increasingly popular in water treatment applications. The diaphragm design uses the reciprocating action of a piston or plunger to transmit pressure through a hydraulic fluid to a flexible diaphragm. The diaphragm isolates and displaces the pumped fluid and is activated either mechanically or hydraulically.

Figure 35-6 shows a diaphragm pump that uses an electronic pulsing circuit to drive a solenoid, which provides the diaphragm stroke. Both stroke length and stroke frequency can be adjusted to provide a usable control range of 10–100% of capacity. Diaphragm pumps can be set up for automatic adjustment of stroke frequency based on an external signal. This capability is commonly used to control the ratio of chemical feed to water flow rate.

The diaphragm pump illustrated in Figure 35-7 uses an internal hydraulic system to operate the diaphragm in contact with the treatment solution. The pump is available in models operating at discharge pressures exceeding 1500 psig. The delivery rate of the pump is manually adjustable while the pump is running and can

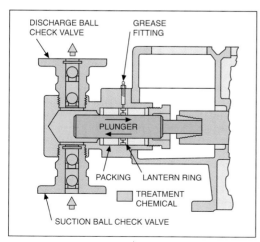

Figure 35-5. Packed plunger pump. (Reprinted from "Metering Pumps—Selection and Specification," page 8. Courtesy of Marcel Dekker, Inc.)

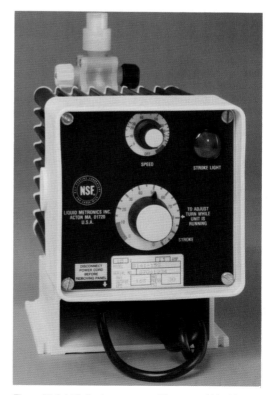

Figure 35-6. LMI diaphragm pump. (Courtesy of Liquid Metronics, Inc.)

Figure 35-7. Diaphragm pump for high-pressure applications. (Courtesy of Milton Roy Company.)

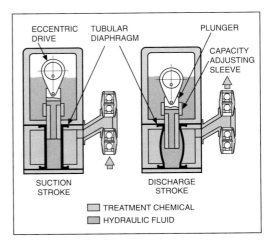

Figure 35-8. Tubular diaphragm pump. (Reprinted from "Metering Pumps—Selection and Specification," page 14. Courtesy of Marcel Dekker, Inc.)

also be adjusted automatically by a pneumatic or electric control signal. The internal hydraulic system has a built-in valve to protect against overpressure.

Some diaphragm pumps can be used to feed heavy or viscous materials, such as slurries and polymers. Figure 35-8 shows a tubular diaphragm pump that is often used in these applications. The tubular diaphragm design also uses a reciprocating plunger, but a tube-shaped diaphragm expands or contracts with pressure from the hydraulic fluid. Adjustable pumps with flow rates of up to 60 gal/hr at 100 psig are available.

An air-driven diaphragm pump operates from 1 to 200 gpm. This design is usually used for viscous products, and because of its high capacity is generally used to transfer chemical from a storage tank to a day tank. It can be used for feeding shear-sensitive polymer solutions.

The air-driven pump tolerates abrasive materials and is also used to pump sand and sludges. Discharge pressure is limited to approximately 100 psig.

Rotary Pumps. Rotary pumps have one or two rotating members to provide positive or semipositive displacement. The pump may consist of two meshing gears or a single rotating member in an eccentric housing. In the full positive displacement type, delivery rate is fixed by speed of rotation. Semipositive displacement pumps

have internal slippage, which affects delivery rate and discharge pressure. Rotary pumps generally depend on the fluid being pumped for lubrication. Most designs will not tolerate abrasive material in the fluid. They can pump highly viscous fluids and are particularly useful for polymer applications, in which low shear is desirable.

Figure 35-9 shows a rotary pump with an idler gear moving inside a rotor gear. Pumping action is achieved by the meshing of rotor and idler gear teeth and by the use of close running tolerances. With every revolution of the pump shaft, a fixed amount of liquid is drawn into the pump through the suction port. This volume of liquid fills the spaces between the teeth of the rotor, progresses through the pump, and is forced out through the discharge port.

Gravity Feed

Another commonly used delivery method, the gravity feed design uses the height difference between chemical in the tank and the point of application as the driving force. The primary advantages of gravity feed delivery are simplicity and reliability. This pumpless design eliminates moving parts and associated maintenance requirements. The elimination of check valves and their periodic failures greatly improves reliability. When feed verification methods are employed, gravity feed can provide precise chemical control.

There are several types of gravity feed systems. A shot feeder is an example of a simple but effective way to dispense premeasured chemical "shots." The shot feeder uses a measuring pot of known volume, which is filled from the bulk storage tank. A valve at the bottom of the measuring

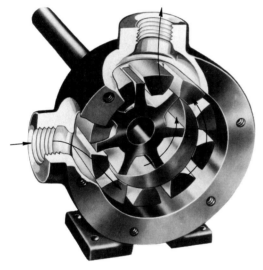

Figure 35-9. Rotary pump. (Courtesy of Viking Pump Div., Houdaille Industries, Inc.)

pot is opened and the product is allowed to flow by gravity into the system.

Feed verification can be achieved by measurement of the velocity of product flow, or volume per time. This permits accurate feed and measurement of product to a system without the traditional maintenance problems associated with metering pumps. The most sophisticated gravity feed systems combine feed verification with computerized controls to provide optimum chemical control and eliminate the need for metering pumps.

Proper size is important. An oversized system will cause spikes in chemical treatment (periodic overfeeds). If the system is undersized, it may not be able to feed enough chemical treatment. Key variables that must be considered in sizing a

gravity feed system include product viscosity, available static head, effect of fluctuating tank levels, and system friction losses.

Water-Jet Eductor

The water-jet eductor harnesses the kinetic energy of a moving liquid under pressure. An eductor entrains another liquid, gas, or gas–solid mixture, mixes it with the liquid under pressure, and discharges the mixture against a counterpressure, as shown in Figure 35-10. Application of water-jet eductors is limited by the amount of lift or suction necessary, available motive pressure, and discharge pressure. Generally, a motive-to-discharge pressure ratio of at least 3.5:1 is necessary.

Operated in conjunction with a valve, a water-jet eductor can be used for continuous injection of chemical into a water stream. It is usually used in these applications for mixing rather than proportioning. The water-jet eductor is an important component of vacuum-type chlorinators and sulfonators and is also used for transporting dry polyelectrolytes.

Eductors have many important advantages, including low cost, no moving parts, and the ability to self-prime. Moveover, because electrical power is not needed for operation, eductors are extremely well suited for use in hazardous locations where expensive explosion-proof equipment is required. Eductors can also be adapted to operate with automated control systems.

The accumulation of particulate matter in and around the eductor nozzle can cause a loss of suction. Filters and strainers can assist in reducing this problem. Eductors should be inspected and cleaned periodically in installations where deposition is likely to occur.

Accessories

Pump/Tank Packages. In most applications, a pump alone is not sufficient for chemical feed. Usually, a chemical feed system combines pump, tank, valves, gauges, mixer, strainer and relief valves (to allow chemical solution preparation), mixing (if required), and pumping.

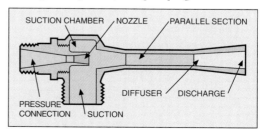

Figure 35-10. Water-jet eductor.

Mixers. A vertically mounted, shaft-driven impeller is the most common type of mixer used for chemical feed systems. If the chemical is a diluted, high molecular weight polymer, a speed reducer may be required. With certain chemicals, it is desirable to minimize air entrainment. In these cases, an electrical or air-driven recirculation pump should be used for mixing.

Timers. Timers find numerous applications in feed systems—most notably, the control of mixers and batch feeding of chemicals (particularly antimicrobials).

Alarms. Alarm systems are becoming more and more sophisticated. It is now possible to monitor and alarm as necessary based on pump status, chemical use rates, high and low tank levels, and unusual operating conditions.

Injection Nozzles. Specialized nozzles are needed to inject chemicals into pipelines. Figures 35-11 and 35-12 show typical low- and high-pressure nozzles. Low-pressure nozzles are used for injection into a liquid stream. High-pressure quill nozzles are used in vapor systems. The quill atomizes the chemical into fine droplets that are carried with the vapor stream. Care should be taken to prevent injection of liquid into steam lines immediately upstream of pipe bends, where high-velocity liquid droplets can impinge upon and erode pipe walls.

Level Gauges. The need to monitor on-site and remote tank levels has led to the development of

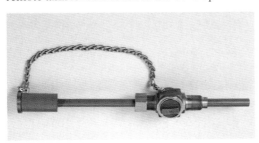

Figure 35-11. Low-pressure injection nozzle.

Figure 35-12. High-pressure injection nozzle.

many different types of level monitors. Among the more prominent methods are pressure transducers, ultrasonic monitors, capacitance sensors, pressure-sensitive linear potentiometers, and bubble tubes. Care must be used to ensure the following:

- compatibility with the materials of construction

- adequate temperature compensation

- isolation from damaging pressure shocks

CHEMICAL FEED SYSTEMS

Liquid Feed

Chemicals may be fed on a "shot" (batch) basis or on a continuous basis. The choice between these two methods depends upon the degree of control required, the application, and the system design.

Shot Feed. Chemical may be shot-fed by on–off control of a chemical feed pump or by discharge from a calibrated vessel or measuring chamber. Shot feeding may be used in cooling systems, bio-oxidation basins, and other places where the system volume to blowdown ratio is large. In these systems, the shot simply replenishes lost or consumed material. Shot feeding is also used in applications that only require periodic feed. Antimicrobials for cooling water systems are usually fed on a shot basis. Shot feed cannot be used in once-through systems, where a uniform concentration of chemical is needed.

Continuous Feed. Continuous feed systems meter chemical to the water constantly. The better systems proportion the feed according to the volume being treated and the chemical demand requirements. Continuous feed is suitable for many systems that can also be shot-fed. It is absolutely necessary in applications such as domestic water chlorination and deposit control in once-through systems, where each gallon of water must be treated and no retention vessel exists. It is also necessary when water clarification chemicals are fed to ensure that all turbidity particles encounter polymer molecules before entering the clarifier.

Continuous feed may be provided by a simple, gravity drip feed method, in which feed rate is regulated by a needle valve. Metering pumps or rotometers and control valves feeding from a recirculating pressurized line (Figure 35-13) can also be used.

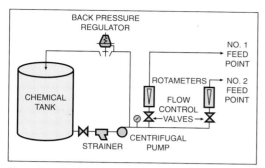

Figure 35-13. Flow diagram of constant-pressure feed using a back-pressure regulator and pump.

Dry Chemical

Large amounts of alum, lime, and soda ash are commonly applied in wastewater treatment plants and large industrial water conditioning plants. Because of the large quantities involved, these additives are usually stored and fed as dry materials. The primary advantage of dry feed is the lower shipping and storage costs. Disadvantages include dust, lack of control, high operational and maintenance costs, and increased operator attention.

Dry feed systems commonly used for water treatment applications include volumetric feeders, gravimetric feeders, and dissolving feeders.

Volumetric Feeders. Volumetric feeders accurately dispense powdered material. The material may be applied directly or used to produce a slurry that is applied to the process. Volumetric feeders are used for lime feed and lime slaking, dry polymer and clay feed, and the feed of fireside additives to boiler furnaces.

The performance and accuracy of volumetric dry feeders depend largely on the characteristics of the powder being metered. Key characteristics that affect powder feed are particle size distribution, loose and packed bulk densities, moisture content, and abrasiveness.

A typical volumetric feed system includes a bulk storage bin or silo, a feed hopper, and a metering device. The most common metering device is a helical screw or auger. The rotational speed of the screw determines the feed rate.

Some powders tend to bridge, or "rathole," causing uneven feed. To ensure even flow of powder to the helix area, auxiliary devices may be required. Among the more common are flexing hopper walls, bin vibrators, and oversized auxiliary augers positioned above the feed helix.

Gravimetric Feeders. Gravimetric feeders proportion chemicals by weight rather than by

volume, and are accurate to within 1–2%. A gravimetric feeder is a scale, balanced to ensure delivery of a desired weight of chemical to the system. The chemical discharged by a gravimetric feeder is usually put into solution or suspension.

Because gravimetric feeders are considerably more expensive than volumetric feeders, they are used only with large systems needing accurate feed or for chemicals whose flow properties prohibit the use of volumetric feeders.

Dissolving Feeders. Dissolving feeders regulate the rate at which a dry chemical is dissolved. A dissolving tank is charged with dry chemical, and a regulated flow of water is fed into the vessel. The concentration of discharged product is governed by the contact area between the dry material and water, and the rate of dissolution. Examples of dissolving feeders include solid halogen feeders and polyelectrolyte dissolving systems.

In some dissolving feeders, extra energy is required to ensure adequate dissolution (wetting) and thorough mixing. For example, in the solid halogen feeder, spray nozzles direct a high-velocity stream of water into the bed of product. In other dissolving feeders, an eductor is used for wetting and mixing.

Automatic and semiautomatic systems have been built to deliver, wet, age, transfer, and feed dry polyelectrolytes (polymers). The delivery and wetting portion of these systems is like that of a dissolving feeder. Either spray curtains (wetting chambers) or eductor devices are used to "wet" the polymer. Various tanks, controls, and pumps are used to agitate, age, transfer, and feed the made-down polymer solution (see Figure 35-14). These feed systems commonly have volumetric screw feeders for precise metering of dry polyelectrolytes. The only manual labor involved is the loading of the bin associated with the volumetric screw feeder.

CHEMICAL CONTROL SYSTEMS

Another important component of a well engineered chemical feed system is the control scheme—the method by which chemical feed rate adjustments take place. The control scheme can have a dramatic effect on chemical residual control, manpower requirements, and ultimate treatment program results. Key variables that must be considered in the selection of a control scheme include required control range, system half life, dead time, manpower availability, and economics.

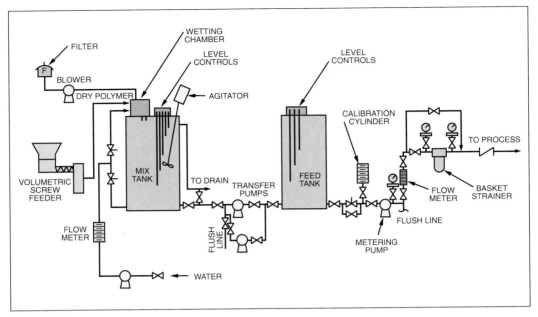

Figure 35-14. Dry polymer feeder.

Control methods may be classified according to the manner in which the final control element is regulated and the degree of sophistication of the control logic. Typical control schemes used in water treatment applications include: manual, on–off, feedforward, ratio, feedback, and feedforward/feedback combinations. The most sophisticated methods of control require the use of programmable logic controllers or computers.

Manual Control

In a manually controlled system, chemicals are added continuously and at a constant rate. Adjustments are made by plant operators at fixed time intervals—generally once per shift or once per day. These adjustments include pump stroke length or frequency, strength of chemical solution, and valve position.

Manual control is most suitable for applications in which chemical control is not critical, established control ranges are wide, and system retention time is long. In these situations, manual control maintains average chemical balances within acceptable limits.

The disadvantages of manual operation include possible lapses in control, high chemical treatment costs, increased manpower requirements, and the possibility of unacceptable results. With today's emphasis on improved quality control, there is a trend away from manual control and toward the use of more sophisticated equipment.

On–Off Constant Rate Mode

In an on–off control method, a chemical feed pump (or other constant rate feed device) is automatically cycled on and off by a control signal. This method is applicable to systems (e.g., cooling towers) that do not require continuous feed of chemical and have large volume to blowdown ratios.

An example of on–off control is an acid feed pump that turns on at a high pH setpoint and off at a low pH setpoint.

The meter-counter-timer is another on–off control system employed in cooling water systems. In this control method, a chemical pump is turned on for a fixed period of time after a preset volume of makeup water or blowdown has accumulated.

Feedforward Control

Feedforward control systems are designed to detect changes in chemical demand and compensate for them to keep the system under control. In contrast, feedback control systems react only after a system error is detected. Feedforward control is typically used to adjust corrosion inhibitor feed rate (based on changes in water temperature), chelant feed rate (based on a hardness tests), and coagulant feed rate (based on influent turbidity readings).

Ratio Control. Ratio control is a form of feedforward control in which output of the chemical pump or other metering device is automatically adjusted in proportion to a variable, such as water flow rate. Ratio control is most frequently used to maintain a fixed concentration of chemical in a water stream where the flow rate fluctuates. A common example is the feed of corrosion inhibitor in proportion to mill water supply flow rate.

The primary disadvantage of this control scheme is lack of on-line feed verification. Although the controller sends a signal to the pump, there is no guarantee that the metering pump output matches the controller signal or even that the metering pump is working. Recent advances in feedback control technology have provided a solution to this problem.

Feedback Control. With feedback control, the actual value of the controlled variable is continuously compared with the desired value. When the detected value varies from a predetermined setpoint, the controller produces a signal indicating the degree of deviation. In many water treatment applications, this signal is sent to a metering pump and the pump's output is changed.

One of the most common examples of feedback control is the feed of acid to a cooling tower based on pH. When the controller detects a difference between the setpoint and the actual pH, it changes the pump speed or valve position to adjust pH back to the setpoint.

Manual adjustment of a chemical feed pump, based on a boiler water phosphate residual test, is a simple form of feedback control. The accuracy of this method is limited only by the frequency of testing, the time required to affect a change, and the reliability of the monitoring technique.

The main disadvantage of feedback control is the fact that control action does not occur until an error develops. Another key problem with feedback control is that it is highly dependent upon the analyzer signal. In many systems, analyzer accuracy and reliability are questionable.

COOLING SYSTEM TREATMENT

Most open recirculating cooling systems require the addition of four classes of chemicals to minimize corrosion, scaling, and fouling:

■ corrosion inhibitors

■ deposit control agents or dispersants

■ antimicrobials

■ pH adjustment chemicals

Blowdown control is also an integral part of cooling water chemistry management.

Chemical Feed

Corrosion inhibitors and deposit control agents are often fed neat (undiluted) to a cooling tower basin. Common methods for chemical delivery include metering pumps and programmed gravity feed systems. Gravity feed systems may employ water-jet eductors to carry chemicals to the basin. Acids or alkalies used for pH control and some antimicrobials require dilution prior to injection into the bulk cooling water.

Feed pumps can provide accurate continuous feeding, provided that the pump rates are modulated to reflect system changes. Because of the large ratio of cooling water system volume to blowdown rate, periodic shot feed of chemical can often provide satisfactory chemical control.

Inhibitors and Dispersants. Inhibitors and dispersants may be fed neat to the cooling tower basin, where dilution in the recirculating water can easily take place. Feed systems vary from a simple continuous pump or periodic shot feed to computerized automatic control.

The two simplest feed systems in use are continuously operating chemical feed pumps and timed periodic shot feed systems. These methods provide adequate control in some cases, but are inexact if cooling system operating conditions or chemistry vary. When conditions vary, the plant operator must become an integral part of the control loop, testing chemical residuals and adjusting chemical feed rates in order to maintain proper inhibitor and dispersant levels in the water.

For improved chemical control, chemical may be fed in proportion to the flow signal from a blowdown or makeup water flowmeter. This can be done on a continuous basis with a flow signal directly controlling the pumping rate. It can also be done on a semicontinuous basis by a flow counter, which triggers a shot feed of chemical.

Additional improvement in control is possible with computerized controllers that use measured parameters to calculate cycles of concentration and combine that information with real-time flow data to calculate and feed the proper amounts of inhibitor and dispersant.

pH Control. Control of cooling water pH and alkalinity within a specified range is usually required for proper performance of the treatment program. Good pH control has become more important because chromate inhibitor treatment programs are being replaced and higher cycles are being used in cooling towers to minimize blowdown.

Commercial concentrated sulfuric acid (66° Baume) is usually used for cooling water pH control. When fed neat, it is nearly twice as dense as water and drops to the bottom of the cooling tower basin. This can damage basin concrete and cause poor pH control. For this reason, the acid should be well mixed with water prior to entering the basin. A dilution trough is used to feed acid to the cooling tower basin, using makeup water as the diluent.

Mild steel tanks are usually used to store concentrated sulfuric acid. Proper ventilation is required to prevent the buildup of explosive hydrogen gas in the storage tank. Strainers upstream of acid pumps are advisable to remove any residual corrosion products or other solids that may be present in the storage tank.

Feedback control is almost always used to control acid feed to a cooling system. The control schemes most often used are on–off and proportional-integral-derivative (PID). Metering pumps or control valves are normally used to regulate the feed of acid. The location of the pH probe is critical; in most applications, the probe should be placed close to the circulating water pumps.

Proper design is important in acid feed line installation. The lines should be installed so that slow filling and draining, which would cause excessive lag time in the control loop, are prevented. Horizontal sections should slant slightly upward in the direction of flow. Installation of an elevated loop at the discharge end of the line, higher than the acid pump, ensures a continuously filled line. In installations where the acid storage tank is higher than the feed point, an anti-siphon device can be used to provide extra protection against acid overfeed. Concentrated acid feed lines generally need to be no larger than $1/2$ in. and are usually made of 304 or 316 stainless steel tubing. Polyethylene or schedule 80 rigid PVC pipe may be used if protected from physical damage.

Other important considerations include pump/valve size, acid quality, maintenance procedures, and calibration frequency.

Sulfamic acid, hydrochloric acid, nitric acid (liquids), and sodium bisulfate (solid) may also be used for pH reduction. Sometimes, pH control is linked to chlorine gas feed, because chlorine gas combines with water to form hydrochloric acid along with the antimicrobial hypochlorous acid. This practice is not recommended, because overchlorination can result.

If increased alkalinity is needed, soda ash or caustic soda is normally used.

Blowdown Control

Cooling water dissolved solids (measured by conductivity) are maintained at a specified level by a continuous or intermittent purge (blowdown) of the recirculating water. In certain cases, it is sufficient to blow down periodically by opening a valve until the conductivity of the water in the tower reaches a certain specified level. Improved control can be obtained with an automatic blowdown controller, which opens and closes the blowdown valve based on conductivity limits or modulates a blowdown control valve to maintain a conductivity setpoint.

Even more accurate dissolved solids control can be attained when computerized control systems are used. The measured conductivity of the recirculating water divided by that of the makeup water provides an estimate of the cycles of concentration. The recirculating water conductivity setpoint is then adjusted by the on-line computer to maintain the desired number of cycles.

Computerized Chemical Feed and Control Systems

Computerized cooling water chemistry control systems can incorporate some or all of the control functions already discussed in this section, including inhibitor and dispersant feed, pH control, blowdown and cycles control, and nonoxidizing antimicrobial feed. Figure 35-15 is a schematic of a computerized cooling water chemical feed, monitoring, and control system setup.

Computerized control systems can usually be programmed to feed chemicals or adjust operating parameters according to complex customized algorithms. This allows the feed system to compensate automatically for changing operating conditions that are often highly plant-specific. For example, in some cases the makeup water may contain varying amounts of corrosion inhibitor. The corrosion inhibitor feed rate to the recirculating water must be adjusted to compensate for the inhibitor entering the system with the makeup. In other instances, the dispersant feed rate setpoint may need to be adjusted

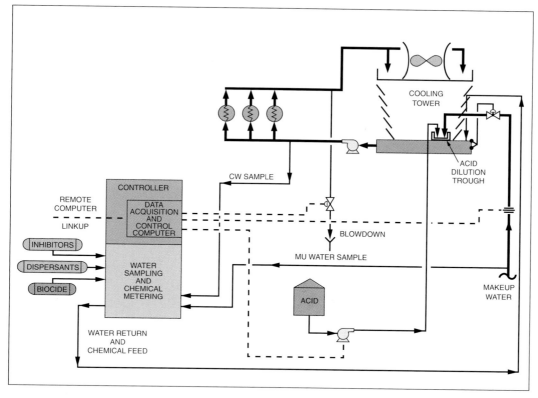

Figure 35-15. Computerized feed system for cooling towers.

according to system water chemistry (e.g., pH, conductivity, or calcium levels). In each of these cases, a computer can be used to perform the necessary calculations and implement the adjustments automatically.

Some computerized systems provide verification of chemical feed amounts. Combined with on-line chemistry monitoring and customized control algorithms, feed verification permits the most precise treatment control. The measurement system identifies the water chemistry. The computer then calculates needed chemical dosages, and the feed system verifies the quantity of chemical feed. A commonly used system is shown in Figure 35-16.

Remote computers are used to monitor and store cooling system status and program results. Parameters of interest usually include recirculating and makeup water pH and conductivity, chemical feed rates, corrosion rates, and fouling data. Following data collection, statistical techniques are used to analyze treatment program performance.

Modems are incorporated into some computerized feed systems so that alarm conditions

trigger an automatic telephone call to the proper operating personnel and advise them of the problem. This prevents minor problems from becoming serious. For example, if a valve is inadvertently left open and the contents of the acid tank begin to drain into the cooling tower basin, a low-pH alarm is activated, and a call is placed automatically to the system operator, who returns to the area and shuts off the valve. Modems are also used to allow operating personnel to make adjustments to system operating parameters and chemical feed rates from a remote location.

BOILER SYSTEM CHEMICAL FEED

For best results, all chemicals for internal treatment of a steam generating facility must be fed continuously and at proper injection points. Chemicals may be fed directly from the storage tank (neat) or may be diluted in a day tank with high-purity water. Certain chemicals may be mixed together and fed from the same day tank.

Chemical feed points are usually selected as far upstream in the boiler water circuit as possible. For chemical feed beyond the feedwater pump or

Figure 35-16. System for computerized gravity feed of cooling towers.

into the steam drum, the pump must be matched to the boiler pressure. For high-pressure boilers, proper pump selection is critical.

Product Feed Considerations

As shown in Figure 35-17, a steam generating system includes three major components for which treatment is required: the deaerator, the boiler, and the condensate system. Oxygen scavengers are usually fed to the storage section of the deaerator. The boiler internal treatment is fed to the feedwater pump suction or discharge, or to the steam drum. Condensate system feed points also vary, according to the chemical and the objective of treatment. Typical feed points include the steam header or other remote steam lines. Chemical feed may also be fed directly in combination with internal treatment chemicals or oxygen scavengers.

Chemicals

Oxygen Scavengers. Oxygen scavengers are most commonly fed from a day tank to the storage section of the deaerator. Some oxygen

scavengers have also been applied in steam headers or condensate piping to reduce oxygen-related corrosion in condensate systems. In utility systems, it is common to feed oxygen scavengers into the surface condenser hotwell. Oxygen scavenger feed rates are based on the level of oxygen in the system plus the amount of chemical additives in the system.

Sulfite. Uncatalyzed sodium sulfite may be mixed with other chemicals. The preferred location for sulfite injection is a point in the storage section of the deaerating heater where the sulfite will mix with the discharge from the deaerating section.

If sulfite is fed alone, the following equipment is needed:

- 304 stainless steel tank
- stainless steel agitator
- stainless steel relief valve
- iron piping, valves, and fittings
- a pump with machined steel or cast iron liquid end and stainless steel trim

In all cases, an injection quill should be used.

Sulfite shipped as liquid concentrate is usually acidic and, when fed neat, corrodes stainless steel tanks at the liquid level. Tanks must be polyester, Fiberglas, or polyethylene. Lines may be PVC or 316 stainless steel.

Catalyzed Sulfite. Catalyzed sulfite must be fed alone and continuously. Mixing of catalyzed sulfite with any other chemical impairs the catalyst. For the same reason, catalyzed sulfite must be diluted with only condensate or demineralized water. To protect the entire preboiler system, including any economizers, catalyzed sulfite should be fed to the storage section of the deaerating heater.

Caustic soda may be used to adjust the pH of the day tank solution; therefore, a mild steel tank cannot be used. Materials of construction for feed equipment are the same as those required for regular sulfite.

Hydrazine. Hydrazine is compatible with all boiler water treatment chemicals except organics, amines, and nitrates. However, it is good engineering practice to feed hydrazine alone. It is usually fed continuously into the storage section of the deaerating heater. Because of handling and exposure concerns associated with hydrazine, closed storage and feed systems have become standard. Materials of construction are the same as those specified for sulfite.

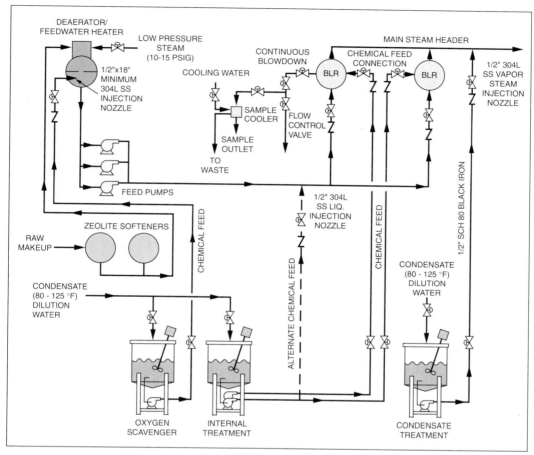

Figure 35-17. Typical piping for boiler system chemical control.

Organic Oxygen Scavengers. Many organic compounds are available, including hydroquinone and ascorbic acid. Some are catalyzed. Most should be fed alone. Like sulfite, organic oxygen scavengers are usually fed continuously into the storage section of the deaerating heater. Materials of construction are the same as those specified for sulfite.

Internal Treatment Chemicals

There are three major classifications of chemicals used in internal treatment: phosphates, chelants, and polymers. These chemicals may be fed either separately or in combination; in most balanced treatment programs, two or three chemicals are fed together. The preferred feed point varies with the chemical specified. For example, when caustic soda is used to maintain boiler water alkalinity, it is fed directly to the boiler drum. When caustic is used to adjust the feedwater pH, it is normally injected into the storage section of the deaerating heater.

Phosphates. Phosphates are usually fed directly into the steam drum of the boiler, although they may be fed to the feedwater line under certain conditions. Treatments containing orthophosphate may produce calcium phosphate feed line deposits; therefore, they should not be fed through the boiler feed line. Orthophosphate should be fed directly to the boiler steam drum through a chemical feed line. Polyphosphates must not be fed to the boiler feedwater line when economizers, heat exchangers, or stage heaters are part of the preboiler system. If the preboiler system does not include such equipment, polyphosphates may be fed to the feedwater piping provided that total hardness does not exceed 2 ppm.

In all cases, feed rates are based on feedwater hardness levels. Phosphates should be fed neat or diluted with condensate or high-purity water. Mild steel tanks, fittings, and feed lines are appropriate. If acidic phosphate solutions are fed, stainless steel and polyolefin materials are recommended.

Chelants. All chelant treatments must be fed to the boiler feedwater line by means of an injection nozzle at a point beyond the discharge of the boiler feed pumps. If heat exchangers or stage heaters are present in the boiler feed line, the injection point should be at their discharge. Care should be exercised in the selection of metals for high-temperature injection quills.

At feed solution strength and elevated temperatures, chelating agents can corrode mild steel and copper alloys; therefore, 304 or 306 stainless steel is recommended for all feed equipment. Chelant products may be fed neat or diluted with condensate. Chelant feed rates must be carefully controlled based on feedwater hardness, because misapplication can have serious consequences.

Chelants should never be fed directly into a boiler. Stainless steel chemical lines would be required, and chloride or caustic stress corrosion could cause the chelant feed line to fail inside the boiler. Localized attack of boiler metal would then occur. Chelants should not be fed if the feedwater contains a significant level of oxygen.

Polymeric Dispersants. In most applications, polymeric dispersants are provided in a combined product formulation with chelants and/or phosphates. Dilution and feed recommendations for chelants should be followed for chelant–dispersant and chelant–phosphate–dispersant programs. Dilution and feed recommendations for phosphates should be followed for phosphate–dispersant programs. These combination programs typically have the best results with respect to boiler cleanliness.

Filming Amines. All filming amines should be fed into steam headers at points that permit proper distribution. A single feed point is satisfactory for some systems. In every case, the steam distribution should be investigated and feed points established to ensure that all parts of the system receive proper treatment.

Filming amines must be mixed with condensate or demineralized water. Water containing dissolved solids cannot be used, because the solids would contaminate the steam and could produce unstable amine emulsions.

Steel tanks have been used to feed filming amines, but some corrosion can occur above the liquid level. The use of stainless steel is recommended. Equipment specifications are the same as those for regular sulfite, except that a vapor-type injection nozzle or quill is required.

Neutralizing Amines. Neutralizing amines may be fed to the storage section of the deaerating heater, directly to the boiler with the internal treatment chemicals, or into the main steam header. Some steam distribution systems may require more than one feed point to allow proper distribution. An injection quill is required for feeding into a steam distribution line.

Neutralizing amines are usually fed based on condensate system pH and measured corrosion rates. These amines may be fed neat, diluted with condensate or demineralized water, or mixed in low concentrations with the internal treatment chemicals. A standard packaged steel pump and tank can be used for feeding.

Computerized Boiler Chemical Feed Systems. Computerized boiler chemical feed systems are being used to improve program results and cut operating costs. These systems can be used to feed oxygen scavengers, amines, and internal treatment chemicals.

A typical system, as shown in Figure 35-18, incorporates a metering pump, feed verification equipment, and a microprocessor-based controller. These systems are often linked to personal computers, which are used to monitor program results, feed rates, system status, and plant operating conditions. Trend graphs and management reports can then be produced to provide documentation of program results and help in troubleshooting.

In many cases, these systems can be programmed to feed boiler treatment chemicals according to complex customized algorithms. For example, chelant feed can be adjusted automatically, based on analyzer or operator hardness test results, boiler feedwater flow, and minimum/maximum allowable product feed rates. Thus, chemical feed precisely matches system demand, virtually eliminating the possibility of underfeed or overfeed.

Figure 35-18. Computerized boiler chemical feed system.

Feed verification is another important facet of some computerized feed systems. The actual output of the pump is continuously measured and compared to a computer-calculated setpoint. If the output doesn't match the setpoint, the speed or stroke length is automatically adjusted. The benefits of this technology include the elimination of time-consuming drawdown measurements, the ability to feed most chemicals directly from bulk tanks, precise chemical residual control, and minimal manpower requirements.

WATER TREATMENT POLYMER FEED SYSTEMS

Polyelectrolytes used in water treatment systems have certain storage, handling, feeding, and dilution requirements. It is imperative that these materials be fed accurately to prevent underfeeding and overfeeding, which can result in wasted chemical treatment and poor system performance.

Polymer Types

Polymers are available as powders, liquids, and emulsions. Each form has different feeding, handling, and storage requirements.

Dry Polymers. Both cationic and anionic high molecular weight polymers are available in powdered form. These products have the advantage of being 100% polymer, which can minimize shipping and handling costs. However, it is absolutely essential that dry polymer materials be handled and diluted properly to prevent underfeeding and overfeeding.

Solution Polymers. Solution polymers are usually cationic, low molecular weight, high charge density products, and are usually used for clarification of raw water. Solution polymers are easier to dilute, handle, and feed than dry and emulsion polymers. In many cases, predilution of a solution polymer is unnecessary, and the product can be fed directly from the shipping container or bulk storage tank. Solution polymers offer the convenience of neat feed, and they can be diluted to any convenient strength consistent with chemical feed pump output.

Emulsion Polymers. Both cationic and anionic high molecular weight polymers are available as emulsions. An emulsion product allows the manufacturer to provide concentrated liquid polymer formulations that cannot be made in solution form. It is only after the emulsion polymer has "inverted" with water that the polymer is available in its active form. Therefore, these products must be diluted properly prior to use.

Storage

Dry Polymers. Dry polymers are susceptible to caking if stored under highly humid conditions. Caking is undesirable because it interferes with the polymer make-down and dilution process. Therefore, dry polymers should be kept in areas of low humidity, and opened containers of dry material should be sealed prior to restorage. In general, polymer products begin to lose their activity after 1 year of storage. Although this process is gradual, it ultimately affects the cost of chemical treatment. It is highly recommended that polymers be used before their expiration dates.

Solution Polymers. Solution polymers should be stored in an area of moderate temperature to protect them from freezing. Some solution products are susceptible to irreversible damage when frozen. Others exhibit excellent freeze–thaw recovery. In no case should solution polymers be stored at temperatures above 120 °F. As solutions, these polymers do not require periodic mixing (to prevent separation) before use. However, some solution polymers have a short shelf life, and inventory should be adjusted accordingly.

Emulsion Polymers. Because emulsion polymers are not true solutions, they separate if allowed to stand for a prolonged period of time. Therefore, emulsion polymers must be mixed prior to use with a drum mixer, tank mixer, or tank recirculation package. A bulk tank or bin recirculation package should be designed to recirculate the tank's contents at least once per day to prevent separation. Emulsion polymers contained in drums should also be mixed daily. Neat emulsion polymer must be protected from water contamination, which causes gelling of the product and can make pumping difficult or impossible. In areas of high humidity, tank vents should be outfitted with a desiccant in order to prevent water condensation within the emulsion storage tank. Even small amounts of condensation can cause significant amounts of product gelling. As with liquid products, emulsion polymers must be protected from freezing and should be stored at temperatures below 120 °F.

Dilution and Feeding

Dry Polymers. Dry polymers must be diluted with water before use. Most operations require preparation of polymer dilutions once per shift or daily. Typically, a plant operator is charged with the responsibility of measuring a correct amount of dry polymer into a container. The contents of the container are conveyed to the mixing tank through a polymer eductor. The eductor is a device that uses water pressure to create a vacuum and is designed so that dry polymer particles are wetted individually by the water as they pass through the eductor assembly (Figure 35-19). If dry polymer particles are not wetted individually before introduction into the dilution tank, "fisheyes" (undissolved globules of polymer) will form in the solution tank. Fisheyes represent wasted polymer and cause plugging in chemical feed pumps.

Dry polymer solution strengths must be limited to approximately 0.5–1% or less by weight, depending on the product used. This is necessary to keep the solution viscosity to a manageable level. The mixer employed in the solution tank should not exceed 350 rpm, and mixing should proceed only until all material is dissolved. Normally, a batch of diluted dry polymer should be used within 24 hr of preparation, because the diluted product begins to lose activity after this amount of time.

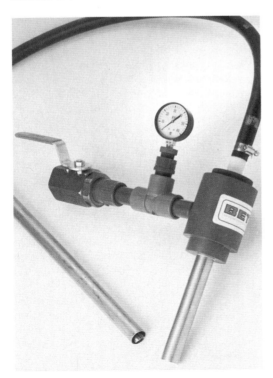

Figure 35-19. Polymer eductor.

Automatic dry polymer dilution systems can be used to perform the wetting, diluting, and mixing functions previously described; however, the system must be manually recharged with dry polymer periodically. Although costly, these systems can save appreciable time for plant personnel, and operations are usually more consistent when automatic make-down units are used.

Solution Polymers. Solution polymers may be diluted prior to use or fed neat from a shipping container, bin, or bulk storage tank. Dilution of these products becomes necessary if there is insufficient mixing available to combine the polymer with the water being treated. In-line static mixer dilution systems are acceptable for solution polymers and are the simplest method of solution polymer dilution and feed. A solution polymer can be fed through one of the many commercially available emulsion polymer dilution and make-down systems. However, in general, the use of these systems for solution polymers is not necessary. Solution polymers can be pumped most easily with gear pumps. However, many solution polymers have a viscosity low enough to be pumped by diaphragm chemical metering pumps.

Emulsion Polymers. Emulsion polymers must be diluted before use. Dilution allows the emulsion product to invert and "converts" the polymer to its active state. Proper inversion of emulsion polymers is rapid and effective. Improper inversion of the emulsion polymer can result in loss of activity due to incomplete uncoiling and dissolution of the polymer molecules.

Batch and continuous make-down systems are acceptable for emulsion polymer use. In batch preparation, a plant operator feeds a premeasured amount of neat emulsion product into the agitator vortex of a dilution tank. The product is mixed until it is homogeneous, and then the mixers are shut off. As with dry polymer products, mixer speed should always be below 350 rpm and the mixer should be shut off as soon as the product is homogeneous. This prevents excessive shearing of the polymer molecule and resultant loss of polymer activity. A batch emulsion polymer make-down system is shown in Figure 35-20.

Several manufacturers market continuous emulsion polymer make-down and feed systems. These systems pump neat polymer from the storage container into a dilution chamber, where the polymer is combined with water and fully activated. The polymer–water solution then flows

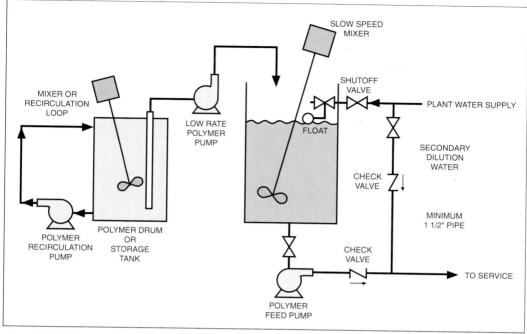

Figure 35-20. Polymer make-down system.

by water pressure to the point of application. Provision is made for secondary in-line dilution water to dilute the polymer further prior to use. These polymer feed systems are by far the easiest and best ways to feed emulsions continuously. Their manufacturers claim a superior ability to invert the polymer molecule over batch tank dilution systems. A commercially available continuous emulsion polymer makedown system is shown in Figure 35-21.

It is not acceptable to use in-line static mixing alone for dilution of emulsion polymers. However, in-line static mixing can be employed for blending secondary dilution water with diluted emulsion product prior to application. Initial dilution of emulsion polymers should be 1% or 2% by weight. This solution strength ensures proper particle-to-particle interaction during the inversion step, which aids in complete inversion.

It is usually desirable to provide secondary dilution water capabilities to emulsion polymer feed systems, because these products tend to be most effective when fed at approximately 0.1% solution strength.

General Recommendations

In addition to the above, some general guidelines apply to the feeding and handling of all water treatment polymers. In areas where the temperature routinely drops below freezing, it is

Figure 35-21. Continuous emulsion polymer make-down system. (Courtesy of Stranco.)

good practice to insulate all polymer feed lines so that feed line freezing does not occur.

For tank batches of diluted polymers, tank mixer speeds of over 350 rpm should not be used. In the preparation of diluted batches of polymer, water should always be added to the tank first. Then, the mixer should be started and the polymer added on top of the water.

Diaphragm metering pumps can be used to pump most polymer solutions. However, due to the viscosity of some products, gear pumps may be necessary. Plastic piping should be used in polymer feed systems; stainless steel is also acceptable. Most polymers are corrosive to mild steel and brass. Extra precautions should be taken to prevent spilling of polymers, because wet polymer spills can become extremely slippery and present a safety hazard. Spills should be covered with absorbent material, and the mixture should be removed promptly and disposed of properly.

MONITORING AND CONTROL OF WATER TREATMENT

Most industrial water treatment systems are dynamic. They constantly undergo changes because of seasonal variations in water chemistry, varying plant operating conditions, new environmental laws, and other factors. Because of this, proper monitoring is essential to ensure that the water treatment program applied to a boiler, cooling, wastewater or other industrial water system is satisfactorily controlled so that the desired results are achieved.

Some of the value-added benefits obtained through proper monitoring of a water treatment program include:

- reduced risks associated with chemical underfeed or overfeed

- continuing compliance with environmental regulations

- improved quality of plant operation

- increased water and energy savings

- improved plant productivity

Industrial water treatment systems may be monitored by manual methods or by continuous systems employing automatic instrumentation.

MANUAL MONITORING

Manual monitoring typically involves plant operators or technicians conducting chemical tests and comparing the results to specified chemical control limits. The testing frequency can vary from once per day to once per hour, depending on the plant resources available. The tests run can include pH, conductivity, suspended solids, alkalinity, hardness, and others. Using the test results, the plant operator manually adjusts a chemical feed pump or blowdown valve, making an estimate of the degree of change necessary.

Manual monitoring is satisfactory for noncritical water systems or systems in which water and plant operating conditions change slowly. Many systems operate with manual monitoring, using

the many tests contained in Chapters 39–71. Typical applications include the following:

- closed cooling water treatment systems

- open cooling water systems with consistent makeup water characteristics and steady load conditions

- low to medium pressure boilers

CONTINUOUS, ON-LINE MONITORING

Because of the dynamic nature of many water treatment systems and the worldwide need for improved reliability and quality, a higher degree of precision is required in the monitoring and control of water treatment programs than that obtained through manual monitoring. To achieve the degree of precision needed, continuous on-line monitoring with automatic instrumentation is required.

Because of the many technological developments in electronics and microprocessor technology over the last decade, there is a wide range of instrumentation available to monitor water treatment systems. The following sections address the systems available to monitor conductivity, pH, corrosion rate, turbidity, dissolved oxygen, sodium, fouling, biological activity, and halogens.

Specific Conductance

Dissolved solids provide a fundamental measure of water quality. In water, the specific conductance, or ability to carry an electric current, is directly related to the quantity and mobility of the dissolved solids. As such, specific conductance is widely used to monitor boiler makeup and condensate purity and to control boiler and cooling system blowdown.

Recent technological advances have improved the reliability and sophistication of conductivity controllers. Microprocessor-based instruments provide extremely reliable and accurate measurement of temperature-compensated conductivity

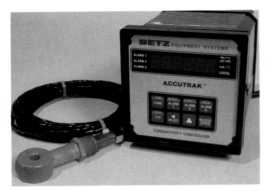

Figure 36-1. Betz Accutrak conductivity controller and toroidal probe.

combined with sophisticated control modes. The Betz Accutrak® (Figure 36-1) is an example of a conductivity controller. It offers programmable control modes, such as proportional; on/off; and proportional, integral, derivative (PID) control. It also includes self-diagnostics and a display of sensor or electronic fault conditions. Analog and standard communication protocol signals (RS-485) are provided for computer interface, facilitating data acquisition and communication.

In most conductivity measurement systems, two metallic electrodes are immersed in a liquid to complete an electrical circuit. These electrode-type sensors work well in relatively clean water, but lose accuracy if coated by dirt or fouling contaminants, which interfere with the flow of current. To reduce interference from small amounts

of contamination in the water, the probe and mounting may be designed to increase the velocity of the flowing sample, minimizing dirt buildup on the electrodes. Electrode-type probes usually show good accuracy in the specific conductance range of 50–8,000 μmho (microsiemens).

Special electrode-type probes are used for contaminant detection in high-purity water, such as steam condensate, demineralized water, and clean water rinses in metal finishing. Measurement over a range of 1–2,000 μmho is practical with these probes.

For heavy fouling conditions (found in some industrial cooling towers and boilers, waste treatment plants, and some processes, such as metal treatment baths) an electrodeless (toroidal) conductivity probe (Figure 36-1) must be used. The toroidal probe uses inductance to sense conductivity changes in a process solution.

pH

pH measurement reveals the hydrogen ion concentration in water. It is used to determine both the deposition and corrosion tendency of a water. The most widely used type of pH measurement is the electrode method. The assembly in Figure 36-2 shows the necessary elements that make up a typical pH sensor: a glass pH electrode, a reference cell, a temperature compensation element, a preamplifier, and a sensor body. Because of the difficulty of maintaining good pH control, manual systems are being replaced by continuous moni-

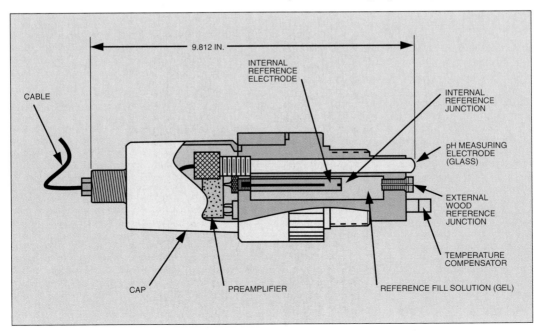

Figure 36-2. Typical pH sensor assembly schematic.

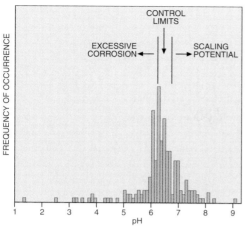

Figure 36-3. Variation of pH in a cooling tower system with manual control.

Figure 36-5. Modular pH sensor assembly.

toring and automatic control of pH in many water treatment applications. In cooling tower systems, pH has been particularly difficult to control manually because the response curve of pH to acid addition is not linear. Figure 36-3 shows the variation of pH in a cooling tower system with manually adjusted feed of sulfuric acid. Results of random plant tests were plotted to show the number of occurrences of each test value.

pH controllers use much of the same technology as the conductivity controllers discussed above. The Betz Accutrak pH controller (Figure 36-4) uses microprocessor-based electronics, programmable alarm and control modes (such as time proportional control, PID control, self-diagnostics, and a display of electronic or sensor fault conditions), and analog and RS-485 signals for computer interface for data acquisition and communication. pH sensor technology has

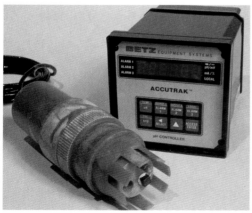

Figure 36-4. Betz Accutrak pH controller with modular pH sensor.

advanced significantly to overcome many of the problems encountered in the past, such as rapid fouling and chemical attack of the pH electrodes, contamination and rapid depletion of reference cells and electrolytes, and electrical noise and environmental interference with the low-level pH signal.

Several variations of pH sensor assemblies are available for different applications. For relatively clean water, where extensive fouling is not a problem (such as in most cooling towers), a combination pH sensor assembly is normally used. The single, molded body sensor assembly shown in Figure 36-2 combines all of the elements.

Glass electrode deterioration, reference junction plugging, and electrolyte depletion (which occurs in all pH sensor applications) proceed at approximately the same rate. This progression is slow enough in clean water to provide an acceptable economic life. When the combination sensor is worn out, it is discarded.

A rugged, modular pH assembly (Figure 36-5) is used in processes such as metal treatment baths and waste systems, where fouling or chemical attack of glass electrodes, reference junctions, and other elements is a problem. The modular assembly allows periodic maintenance and replacement of individual components.

Corrosion Rate

Corrosion rate instruments are used in many different applications to provide instantaneous corrosion rate values in mils per year. A typical package consists of an analyzer and probes, as shown in Figure 36-6. Corrosion rate instruments are used for critical cooling systems, steam condensate systems, mill water supply streams, and other applications.

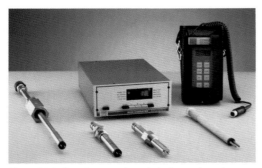

Figure 36-6. Corrater corrosion rate instruments and probes. (Courtesy of Rohrback Cosasco Systems.)

Analyzers are available for portable use or continuous operation. Portable units are generally used when several probes are installed in remote locations. The operator connects the analyzer to the probe and takes a reading, and then moves to the next probe. Continuous analyzers are used when a probe is located in a critical area that warrants continuous evaluation. They include recorder and control outputs that can be interfaced with other components such as process controllers and pumps.

Analyzers usually have internal meters and a means of checking calibrations against a standard.

The probe houses the electrodes and exposes them to the test stream. Probes are manufactured in many different configurations. Common configurations include two or three electrodes. Mild steel, stainless steel, and polyvinyl chloride (PVC) are common probe materials. Probes are available as standard and retractable assemblies and are usually provided with standard pipe connections.

Electrodes are made from many different metals, such as stainless steel, mild steel, Admiralty brass, and 90-10 copper-nickel. Electrodes fasten onto the probe, and the probe and electrode assembly are inserted into the test stream.

Turbidity

Turbidity is caused by suspended matter and can be defined as a lack of clarity in water. Turbidity measuring instruments are used to monitor and control clarifiers and lime softeners and to detect corrosion products in steam condensate.

There are presently two methods used for continuous measurement of turbidity—the nephelometric method and surface scatter technique.

Nephelometric method. In the nephelometric method, the sample flows through a cell. Near the midpoint of the cell, a light source sends a beam of light into the moving fluid. Light receivers are located at various positions in the cell. The receivers measure the amount of light scattered 90° from the incident light. The amount of light scattered increases as the turbidity in the sample increases. The instrument measures the scattered light and develops a signal that is related to nephelometer turbidity units (NTU).

Surface Scatter Technique. The surface scatter technique is similar to the nephelometric method in theory of operation. However, in this method a light source emits a beam toward the surface of a constant level reservoir. The reflected and refracted portions of the beam are discarded and the scattered portion is sensed by a photocell. The amount scattered is in direct relationship to the turbidity of the sample. Because the light transmitter and photocell are not in contact with the sample, this method eliminates fouling.

Turbidity measuring instruments usually include an analog or digital readout and a signal output that can be interfaced with a computer or chart recorder. An example of a surface scatter unit is shown in Figure 36-7, and the technique is illustrated in Figure 36-8.

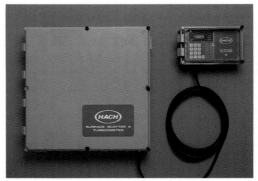

Figure 36-7. Turbidimeter. (Courtesy of Hach Company.)

Dissolved Oxygen Instrumentation

The ability to measure dissolved oxygen is very important, especially in boiler systems, where oxygen corrosion can be very damaging.

A typical dissolved oxygen measuring instrument consists of a sensor, a sensor cell, and an analyzer, as shown in Figure 36-9. The sensor measures the dissolved oxygen concentration and transmits a signal, proportional to the oxygen concentration, to the analyzer. The analyzer provides a readout in parts per billion or parts per million and an output that can be connected to a recorder or data logging device.

Dissolved oxygen is commonly measured by a membrane-isolated electrochemical cell. This cell contains a cathode, an anode, and an electrolyte

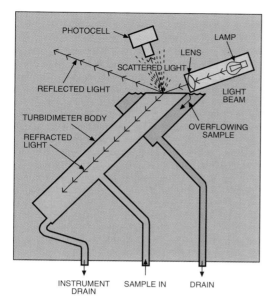

Figure 36-8. Operational diagram of surface scatter technique.

Figure 36-9. Dissolved oxygen analyzer. (Courtesy of Orbisphere Laboratories.)

solution. A gas-permeable membrane admits the dissolved oxygen from the sample to the electrodes. There, an electrochemical reaction generates an electric current with a magnitude proportional to the dissolved oxygen concentration. The reaction can be summarized by the following equation:

$$\underset{\substack{\text{dissolved}\\\text{oxygen}}}{O_2} + \underset{\text{water}}{2H_2O} + \underset{\text{electrons}}{4e^-} \rightarrow \underset{\substack{\text{hydroxide}\\\text{ions}}}{4OH^-}$$

For dissolved oxygen analyzer calibration, the sensor is exposed to humid air. The concentration of dissolved oxygen in the moisture is between 8 and 10 ppm, depending on the ambient pressure and temperature. The analyzer reading is adjusted to the correct value for the pressure and temperature. Some analyzers have an automatic calibration feature that measures the temperature and pressure at the push of a button.

Sodium

Sodium instrumentation has become very important as a means of determining steam purity. To determine the total dissolved solids concentration of the steam, the sodium level in a cooled steam sample is compared to the ratio of total solids to sodium in the boiler water.

The most common technique used to measure sodium is the specific ion electrode. The sodium specific ion electrode responds logarithmically to changes in sodium concentration. The only other factors affecting the readings are temperature and pH. Temperature is measured by an internal thermistor. A reference electrode provides the primary potential signal required for the measurement. Before the sample contacts the electrodes, the sample is circulated through a diffusion tube that is immersed in ammonia; this procedure eliminates hydrogen ion interference.

For calibration of the specific ion analyzer, both electrodes are immersed in a known standard solution. The electrodes are also immersed in another standard with a tenfold higher concentration of sodium ions for determination of the electrode slope. Modern microprocessor technology has provided advanced calibration techniques that verify electrode stability during calibration.

A typical sodium instrument is shown in Figure 36-10.

Figure 36-10. Sodium analyzer. (Courtesy of Orion Research, Inc.)

Fouling

There are several specialty systems designed to monitor the rate of fouling and corrosion in industrial equipment, including those discussed in the following sections.

MonitAll®. The Betz MonitAll® (Figure 36-11) is a portable assembly used primarily to measure the fouling and corrosion potential of cooling water streams on heated tube surfaces.

The MonitAll contains a clear flow-through assembly. Sample water flows in at the bottom and out of the top of a tube. A heat probe is inserted into the flow assembly along the axis of the tube. The heat probe generates an adjustable heat flux across a tubular metal test section. Fouling or corrosion can be accelerated if the heat flux is raised above design levels.

The test section is removable and interchangeable for other metals, which include mild steel, Admiralty brass, 304 stainless steel, 316 stainless steel, and 90-10 copper-nickel.

The heat probe includes two temperature sensors that measure probe surface temperature and bulk water temperature. The temperatures are monitored by a meter with a light-emitting diode (LED) display. The temperature meter

Figure 36-11. Betz MonitAll portable assembly.

has an analog output for a recorder or data logging device.

As the test section fouls, less heat is dissipated into the bulk water and the tube skin temperature decreases. The result is an increase in the temperature difference (ΔT) which can be related to fouling factor R_f by the following equation:

$$R_f = \frac{\Delta T_{final} - \Delta T_{initial}}{\text{Heat Flux (Btu/hr/ft}^2)}$$

The MonitAll is equipped with flow control valves to maintain a constant flow rate and insertion tubes to increase velocity in the clear flow cell.

Model Condenser. The Betz Model Condenser, Figure 36-12, is a test device used primarily to simulate surface condenser fouling and corrosion. It consists of a horizontal, cylindrical stainless steel shell with one, two, or four removable tubes. The tubes run the length of the shell and terminate at the tube sheets. An electric heater is located in the bottom of the shell to generate a constant heat flux. Temperature sensors are located in the shell and tube discharges to monitor temperature difference.

The principle of operation is very similar to that of a standard surface condenser. The test water flows through the tube(s) and discharges to drain. The shell is filled with distilled water, which covers the electric heater but is below the tubes. A vacuum of 27 inches of mercury is applied to the shell to simulate condenser conditions. Heat is applied to the distilled water with the electric heater. The distilled water boils and the vapor rises to the tube surfaces. Cool water flowing through the tubes condenses

Figure 36-12. Betz model condenser.

steam on the tube surfaces. The condensation falls to the reservoir of distilled water and the cycle repeats.

Condenser operating conditions, such as heat flux and tubeside velocity, are simulated by the model condenser. Shell temperature and tube discharge temperature are monitored continuously. As foulant accumulates on the internal tube surfaces, less heat is transferred through the tube wall. As a result, the shell temperature increases and the tube discharge temperature decreases. At a constant flow rate, the increase in the temperature difference can be related to a fouling factor by the same equation given for the MonitAll. Typically, tubes are removed and sent to a laboratory for further analysis.

Test Heat Exchanger. A test heat exchanger (Figure 36-13) is used to monitor the fouling and corrosion tendencies of a particular cooling water stream. Cooling water passes through two removable tubes contained in a cylindrical shell. The tubes, which are available in many different materials, can be arranged for two single-pass or one two-pass operation. Steam or hot condensate flows into the shell and heats the water flowing through the tubes. The condensate exits the shell through a flowmeter that is used to monitor heat input.

Figure 36-13. Test heat exchanger.

COSMOS™ Cooling System Monitoring Station.

Monitoring and analysis of key operating parameters are important tools in the development of an effective cooling water treatment program. The Betz Cooling System Monitoring Station (COSMOS) is a versatile tool that can be used for this purpose. It monitors pH, conductivity, and corrosion rates. In addition, a MonitAll hot tester can be included for evaluation of heat flux, water velocity, and fouling factors. A variety of metals can be evaluated.

The monitoring station consists of two units: a data acquisition cabinet and a piping and

Figure 36-14. Betz cooling system monitoring station (COSMOS).

instrumentation cabinet. Figure 36-14 shows the data acquisition cabinet, with the panel door open, connected to the piping and instrumentation cabinet.

The piping and instrumentation cabinet (wet side) includes the MonitAll hot tester, a flow sensor, two corrosion probes, a conductivity sensor, a pH sensor, coupon holders, stainless steel piping, and a drain line. A small electrical enclosure within the cabinet supplies the electrical power for the MonitAll and the space heater provided for climate control.

The data acquisition cabinet contains the microprocessor-based controller, which manages all of the data acquisition, storage, and display. It also controls a printer, the floppy disk drive, automatic corrosion probe switching, automatic shutdown safeguards and alarms, and the climate controls. The controller has a keypad and a display window for operator interface. A personal computer can be used to generate reports, graphs, and statistical analysis from the acquired data.

Biological Activity in Cooling Systems

A biofilm fouling monitor (Figure 36-15) is used to determine levels of microorganisms attached to surfaces in a cooling system. The monitor consists of a holder that is threaded on both ends. Each half of the holder contains a screen that secures glass beads to the sampling surfaces.

The biofilm monitor can be attached at any suitable location in the hot water return where the flow through the monitor is at least 1–2 gpm. The monitor is normally on line at least 1 week before the sampling starts. The time required to develop a steady-state biofilm on the beads varies depending on system conditions. Steady-state is reached when the amount of biological material

Figure 36-15. Biofilm fouling monitor.

removed by turbulent flow is equal to the amount of new biofilm produced by microbial growth. After steady-state is achieved, changes in levels of biofilm reflect changes in the system environment; for example, increased nutrient levels lead to greater amounts of biofilm, while the addition of toxic materials causes a reduction in biofilm levels. Individual systems must be monitored to determine what level of fixed microorganisms is acceptable.

Macrofouling monitors (Figure 36-16) are used to monitor the growth rate of zebra mussels, Asiatic clams, and other mollusks. The problems caused by these animals are described in Chapter 28. The strategic placement of macrofouling control monitors helps to quantify growth and settlement cycles in a particular area. They also provide quantification of kill rates following chemical treatment.

A macrofouling unit contains a set of fouling plates. Water flows upward through the unit. The mussel or clam larvae attach themselves to the fouling plates. Their rate of growth is monitored visually by regular examination of the plates.

Halogen Residuals

Continuous on-line measurement devices used to monitor halogen residuals fall into two categories: amperometric and colorimetric.

Amperometric analyzers, depending on the mode of use, measure free or total halogen concentrations in water samples. Changing halogen concentrations in the sample produce a corresponding change in the electrical current that flows from the cathode to the anode of a sensor. Some amperometric analyzers also correct for variations in sample temperature and pH.

Figure 36-17 shows a colorimetric analyzer, which changes color intensity depending on the chlorine concentration of the sample. Small

Figure 36-16. Macrofouling monitor.

volumes of sample, an indicator agent, and a buffer solution are precisely metered and mixed. During a development interval, the indicator oxidizes and produces a magenta or red compound which is photometrically measured. The color intensity is compared to a reference and the difference is used to characterize the chlorine concentration of the sample. Measurement accuracy can be affected by the presence of chromates, chloramines, nitrite, iron, manganese, and other strong oxidants in the sample. Careful selection of the chlorine analyzer and proper installation should help to minimize these measurement interferences.

Continuous monitoring is an important part of many chlorine applications:

■ to control feed rate in potable water supplies

■ to prevent damage to ion exchange demineralizers or reverse osmosis systems in municipal and industrial water supplies

Figure 36-17. Colorimetric-type chlorine analyzer. (Courtesy of Hach Company.)

- as an antimicrobial in cooling tower applications
- verifying regulatory discharge requirements for wastewater or industrial process streams

OTHER MONITORING TECHNIQUES
Visual Inspection

Visual inspection equipment is often useful for the inspection of internal surfaces in boiler tubes, condenser tubes, heat exchangers, and turbines. Visual inspection is used to determine failure potential due to deposit accumulation or corrosion.

Fiber Optics. A fiber optics device (Figure 36-18) is commonly used for equipment inspection. A lens on each end of the fiber optics bundle provides a clear, undistorted, color image. Video equipment and 35 mm cameras may be used with a fiber optics system.

Video Inspection. Television camera inspection equipment provides an alternative to fiber optics.

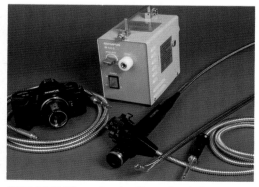

Figure 36-18. Fiber optics visual inspection device.

The typical package consists of a miniature camera, lights, a rotating mirror for radial viewing, and a monitor.

Additional Monitoring Technologies

Ion Chromatography. Ion chromatography (IC) is widely used in laboratories and is finding a place in some continuous process analysis applications. It is used to detect low-level contamination of otherwise high-purity streams. The detection capability of IC permits routine analysis in the parts per billion range and, in some cases, in the parts per trillion range. The advantages of IC are its selectivity, sensitivity, and speed in the analysis of anions and cations.

An ion chromatograph consists of an anion or cation separation column and an anion or cation suppressor column. In the separation process, metals are eluted from a separating resin with a strong acid, such as HCl. The metals are then exposed to the suppressor column, which is a strong anion exchanger in the hydroxide form. The chloride (Cl) is removed by the anion resin and the eluted hydroxide reacts with the acid proton to form H_2O. Thus, the metals elute in very dilute water solution as metal hydroxides, and the conductivity is measured. For alkali metals and many other metals, the conductivity imparted to pure water is a simple function of species concentration. Anions are separated in an analogous process.

Flow Injection Analysis. Air-segmented continuous process analyzers have been the foundation of automated industrial water testing for over 30 years. The technology has evolved to a point at which such systems are cost-effective and productive for a wide variety of industrial process monitoring applications. However, in the 1990's, nonsegmented flow injection analysis (FIA) was introduced as an alternative method for these applications.

In the FIA method, small quantities of sample are transported through a narrow bore tube and then mixed with reagents to develop a color that is monitored by a detector. In this new technique, air bubbles are not used to separate individual samples. Samples are injected into a flowing, continuous stream of reagent. To maintain sample integrity, injection intervals must be long enough to prevent cross-contamination. FIA is highly reproducible due to the elimination of air bubbles, the use of precise injection techniques, constant flow rates, and exact timing of analytical reaction from injection to detection.

Wastewater and Gas Cleaning Systems

The finite amount of water on the planet participates in a complicated recycling scheme that provides for its reuse.

WASTEWATER TREATMENT

Many industries use large volumes of water in their manufacturing operations. Because some of this water becomes contaminated, it requires treatment before discharge (Figure 37-1).

Improvements in determining the effects of industrial waste discharges have led to the adoption of stringent environmental laws, which define the degree of treatment necessary to protect water quality. Discharge permits, issued under the National Pollutant Discharge Elimination System (NPDES), regulate the amount of pollutants that an industry can return to the water source. The permitted quantities are designed to ensure that other users of the water will have a source that meets their needs, whether these needs are for municipal water supply, industrial or agricultural uses, or fishing and recreation. Consideration is given to the feasibility of removing a pollutant, as well as the natural assimilative capacity of the receiving stream. This assimilative capacity varies with the type and amount of pollutant.

Wastewater treatment plants are designed to convert liquid wastes into an acceptable final effluent and to dispose of solids removed or generated during the process. In most cases, treatment is required for both suspended and dissolved contaminants. Special processes are required for the removal of certain pollutants, such as phosphorus or heavy metals.

Wastewater can be recycled for reuse in plant processes to reduce disposal requirements (Figure 37-2). This practice also reduces water consumption.

POLLUTANTS
Organic Compounds

The amount of organic material that can be discharged safely is defined by the effect of the material on the dissolved oxygen level in the water. Organisms in the water use the organic matter as a food source. In a biochemical reaction,

dissolved oxygen is consumed as the end products of water and carbon dioxide are formed. Atmospheric oxygen can replenish the dissolved oxygen supply, but only at a slow rate. When the organic load causes oxygen consumption to exceed this resupply, the dissolved oxygen level drops, leading to the death of fish and other aquatic life. Under extreme conditions, when the dissolved oxygen concentration reaches zero, the water may turn black and produce foul odors, such as the "rotten egg" smell of hydrogen sulfide. Organic compounds are normally measured as chemical oxygen demand (COD) or biochemical oxygen demand (BOD).

Nutrients

Nitrogen and phosphorus are essential to the growth of plants and other organisms. However, nitrogen compounds can have the same effect on a water source as carbon-containing organic compounds. Certain organisms use nitrogen as a food source and consume oxygen.

Figure 37-1. Wastewater requires proper treatment before it is discharged from a plant.

Figure 37-2. Water consumption can be reduced through recycling and reuse of wastewater.

Phosphorus is a concern because of algae blooms that occur in surface waters due to its presence. During the day, algae produce oxygen through photosynthesis, but at night they consume oxygen.

Solids

Solids discharged with a waste stream may settle immediately at the discharge point or may remain suspended in the water. Settled solids cover the bottom-dwelling organisms, causing disruptions in population and building a reservoir of oxygen-consuming materials. Suspended solids increase the turbidity of the water, thereby inhibiting light transmittance. Deprived of a light source, photosynthetic organisms die. Some solids can coat fish gills and cause suffocation.

Acids and Alkalies

The natural buffering system of a water source is exhausted by the discharge of acids and alkalies. Aquatic life is affected by the wide swings in pH as well as the destruction of bicarbonate alkalinity levels.

Metals

Certain metals are toxic and affect industrial, agricultural, and municipal users of the water source. Metals can cause product quality problems for industrial users. Large quantities of discharged salts necessitate expensive removal by downstream industries using the receiving stream for boiler makeup water.

REMOVAL OF INSOLUBLE CONTAMINANTS

Various physical methods may be used for the removal of wastewater contaminants that are insoluble in water, such as suspended solids, oil, and grease. Ordinarily, water-soluble contami-

nants are chemically converted to an insoluble form to allow removal by physical methods. Essentially, biological waste treatment is this conversion of soluble contaminants to insoluble forms.

Gravity Separation

Most waste treatment systems employ a gravity separation step for suspended particle or oil removal.

The settling rate of a particle is defined in terms of "free" versus "hindered" settling. A free settling particle's motion is not affected by that of other particles, the vessel's walls, or turbulent currents. A particle has a hindered settling rate if there is any interference from these effects.

The free settling of a discrete particle in a rising fluid can be described as the resolution of several forces—gravity, the drag exerted on the particle, and the buoyant force as described by Archimedes' principle. The particle's velocity increases until it reaches a terminal velocity as determined by these forces. The terminal velocity is then:

$$v = \sqrt{\frac{2gm_p(\rho_p - \rho_f)}{A_c C_d \rho_p \rho_f}}$$

where:

v = velocity, ft/sec
g = gravitation constant, ft/sec^2
m_p = mass of the particle, lb
ρ_p = density of the particle, lb/ft^3
ρ_f = density of the fluid, lb/ft^3
A_c = cross sectional area of the particle exposed to the direction of motion, ft^2
C_d = drag coefficient, a function of particle geometry

Gravity settling is employed primarily for removal of inorganic suspended solids, such as grit and sand. Therefore, in the approximation of the drag coefficient, it is assumed that particles are spherical. Further, if a Reynolds number of less than 2.0 is assumed, the settling velocity of a discrete particle can be described by Stokes' settling equation:

$$v = \frac{gD_p^{\;2}(\rho_p - \rho_f)}{18\mu}$$

where:

D_p = particle diameter, ft
μ = fluid viscosity, lb/ft-sec

The terminal velocity of a particle in the "free" settling zone is a function of its diameter, the density difference between the particle and the fluid, and the fluid viscosity.

The equipment employed for gravity separation for waste treatment is normally either a rectangular basin with moving bottom scrapers for solids removal or a circular tank with a rotating bottom scraper. Rectangular tanks are normally sized to decrease horizontal fluid velocity to approximately 1 ft/min. Their lengths are three to five times their width, and their depths are 3–8 ft.

Circular clarifiers (see Figure 37-3) are ordinarily sized according to surface area, because velocity must be reduced below the design particle's terminal velocity. The typical design provides a rise rate of 600–800 gpd/ft².

When wastewater contains appreciable amounts of hydrocarbons, removal of these contaminants becomes a problem. Oil is commonly lower in density than water; therefore, if it is not emulsified, it can be floated in a separate removal stage or in a dual-purpose vessel that allows sedimentation of solids. For example, the refining industry uses a rectangular clarifier with a surface skimmer for oil and a bottom rake for solids as standard equipment. This design, specified by the American Petroleum Institute, is designated as an API separator. The basic principles governing the separation of oil from water by gravity differential are also expressed by Stokes' Law.

Air Flotation

Where the density differential is not sufficient to separate oil and oil-wetted solids, air flotation may be used to enhance oil removal. In this method, air bubbles are attached to the contaminant particles, and thus the apparent density difference between the particles is increased.

Dissolved air flotation (DAF) is a method of introducing air to a side stream or recycle stream at elevated pressures in order to create a supersaturated stream. When this stream is introduced into the waste stream, the pressure is reduced to atmospheric, and the air is released as small bubbles. These bubbles attach to contaminants in the waste, decreasing their effective density and aiding in their separation.

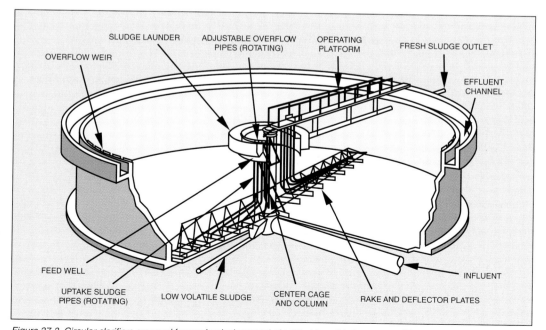

Figure 37-3. Circular clarifiers are used for mechanical removal of settleable solids from waste. (Reprinted with permission from Power.)

The most important operational parameters for contaminant removal by dissolved air flotation are:

■ air pressure

■ recycle or slip stream flow rate

■ influent total suspended solids (TSS) including oil and grease

■ bubble size

■ dispersion

Air pressure, recycle, and influent TSS are normally related in an air-to-solids (A/S) ratio expressed as:

$$A/S = \frac{K S_a (fP - 1)R}{SSQ}$$

where:

K   =   a constant, approximately 1.3
S_a =   the solubility of air at standard conditions, mL/L
f   =   air dissolved/S_a, usually 0.5–0.8
P   =   operating pressure, atm
R   =   recycle rate, gpm
SS =   influent suspended solids, mg/L
Q   =   wastewater flow, gpm

The A/S ratio is most important in determining effluent TSS. Recycle flow and pressure can be varied to maintain an optimal A/S ratio. Typical values are 0.02–0.06.

In a DAF system, the supersaturated stream may be the entire influent, a slip stream, fresh water, or a recycle stream. Recycle streams are most common, because pressurization of a high-solids stream through a pump stabilizes and disperses oil and oil-wetted solids.

As in gravity settling, air flotation units are designed for a surface loading rate that is a function of the waste flow and rise velocity of the contaminants floated by air bubbles. The retention time is a function of tank depth.

DAF units can be rectangular in design but are usually circular, resembling a primary clarifier or thickener. They are often single-stage units.

Induced air flotation (IAF) is another method of decreasing particle density by attaching air bubbles to the particles; however, the method of generating the air bubbles differs. A mechanical action is employed to create the air bubbles and their contact with the waste contaminants. The most common methods use high-speed agitators or recycle a slip stream through venturi nozzles to entrain air into the wastewater.

In contrast to DAF units, IAF units are usually rectangular and incorporate four or more air flotation stages in series. The retention time per stage is significantly less than in DAF circular tanks.

As in gravity settling, the diameter of the particle plays an important role in separation. Polyelectrolytes may be used to increase effective particle diameters. Polymers are also used to destabilize oil–water emulsions, thereby allowing the free oil to be separated from the water. Polymers do this by charge neutralization, which destabilizes an oil globule surface and allows it to contact other oil globules and air bubbles. Emulsion breakers, surfactants, or surface-active agents are also used in air flotation to destabilize emulsions and increase the effectiveness of the air bubbles.

Filtration

Filtration is employed in waste treatment wherever suspended solids must be removed. In practice, filtration is most often used to polish wastewater following treatment. In primary waste treatment, filters are often employed to remove oil and suspended solids prior to biological treatment. More commonly, filters are used following biological treatment, prior to final discharge or reuse.

Filters used for waste treatment may be designed with single-, dual-, or multimedia and may be of the pressure or gravity type.

REMOVAL OF INSOLUBLE CONTAMINANTS

pH Adjustment—Chemical Precipitation

Often, industrial wastewaters contain high concentrations of metals, many of which are soluble at a low pH.

Adjustment of pH precipitates these metals as metal oxides or metal hydroxides. The pH must be carefully controlled to minimize the solubility of the contaminant. As shown in Figure 37-4, some compounds, such as zinc, are amphoteric and redissolve at a high pH. Chemicals used for pH adjustment include lime, sodium hydroxide, and soda ash.

Chemical precipitation of soluble ions often occurs as the result of pH adjustment. Contaminants are removed either by chemical reaction leading to precipitation or by adsorption of ions on an already formed precipitate.

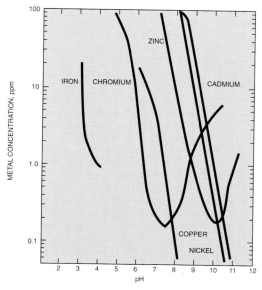

Figure 37-4. Proper pH adjustment is critical for optimum precipitation of metals.

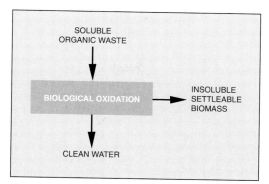

Figure 37-5. Biological oxidation converts soluble waste into clean water and an insoluble biomass.

Biological Oxidation— Biochemical Reactions

One of the most common ways to convert soluble organic matter to insoluble matter is through biological oxidation. Soluble organics metabolized by bacteria are converted to carbon dioxide and bacterial floc, which can be settled from solution.

Various microorganisms feed on dissolved and suspended organic compounds. This natural biodegradation can occur in streams and lakes. If the assimilative capacity of the stream is surpassed, the reduced oxygen content can cause asphyxiation of fish and other higher life forms. This natural ability of microorganisms to break down complex organics can be harnessed to remove materials within the confines of the waste plant, making wastewater safe for discharge.

The biodegradable contaminants in water are usually measured in terms of biochemical oxygen demand (BOD). BOD is actually a measure of the oxygen consumed by microorganisms as they assimilate organics.

Bacteria metabolize oxygen along with certain nutrients and trace metals to form cellular matter, energy, carbon dioxide, water, and more bacteria. This process may be represented in the form of a chemical reaction:

Food (organic compounds) + Microorganisms + Oxygen + Nutrients	$\rightarrow$	Cellular matter + Microorganisms + Carbon dioxide + Water + Energy

The purity of the water depends on minimizing the amount of "food" (organic compounds) that remains after treatment. Therefore, biological waste treatment facilities are operated to provide an environment that will maximize the health and metabolism of microorganisms. An integral part of the biological process is the conversion of soluble organic material into insoluble materials for subsequent removal (Figure 37-5). An overview of factors involved in biological oxidation is given in Table 37-1.

Open Lagoon Biological Oxidation

Where organic loads are low and sufficient land area is available, open lagoons may be used for biological treatment. Lagoons provide an ideal habitat for microorganisms. Natural infiltration of oxygen is sufficient for biological oxidation if the organic loading is not too high. However,

Table 37-1. Factors affecting biological oxidation.

Food, BOD	To maintain control with efficient BOD removal, the proper amount of food must be supplied.
Dissolved oxygen	Insufficient oxygen levels inhibit BOD removal.
pH/toxicants	With time, bacteria adapt to changes in conditions. Rapid changes in pH or type of waste organic inhibit the process.
Time	The degree of degradation varies with time.
Nutrients	Bacteria require trace amounts of nitrogen and phosphorus for cell maintenance.
Temperature	Low temperatures slow reaction rates; higher temperatures kill many strains of bacteria.

Figure 37-6. Mechanical aeration provides oxygen for increased bacterial metabolism of dissolved organic pollutants in an activated sludge plant.

mechanical aeration (Figure 37-6) is often used to increase the ability to handle a higher loading.

Lagoons are nothing more than long-term retention basins. Ordinarily shallow in depth, they depend on surface area, wind, and wave action for oxygen transfer from the atmosphere. Depending on the influent BOD loading and oxygen transfer, lagoons may be aerobic or anaerobic. Lagoons are used primarily for low BOD wastes or as polishing units after other biological operations.

Aerated Lagoons. As BOD loading increases, naturally occurring surface oxygen transfer becomes insufficient to sustain aerobic bacteria. It then becomes necessary to control the environment artificially by supplying supplemental oxygen. Oxygen, as air, is introduced either by mechanical agitators or by blowers and subsurface aerators. Because energy must be expended, the efficiency of the oxygen transfer is a consideration. Therefore, although unaerated lagoons are typically 3–5 ft deep, allowing large surface areas for natural transfer, aerated lagoons are usually 10–15 ft deep in order to provide a longer, more difficult path for oxygen to escape unconsumed. Aerated lagoons also operate with higher dissolved oxygen content.

Facultative Lagoons. Lagoons without mechanical aeration are usually populated by facultative organisms. These organisms have the ability to survive with or without oxygen. A lagoon designed specifically to be facultative is slightly deeper than an unaerated lagoon. Influent suspended

solids and solids created by the metabolism of the aerobic bacteria settle to the bottom of the lagoon where they undergo further decomposition in an anaerobic environment.

Activated Sludge Oxidation

According to the reaction presented previously, control of contaminant oxidation at high BOD loadings requires a bacteria population that is equal to the level of food. This need is the basis for the activated sludge process.

In the activated sludge process, reactants, food, and microorganisms are mixed in a controlled environment to optimize BOD removal. The process incorporates the return of concentrated microorganisms to the influent waste.

When bacteria are separated from wastewater leaving an aeration basin and reintroduced to the influent, they continue to thrive. The recirculated bacteria continue to oxidize wastewater contaminants, and if present in sufficient quantity, produce a relatively low BOD effluent water.

Because the activated sludge process incorporates the return of concentrated microorganisms, it must include a process for microorganism concentration and removal. This process includes an aeration stage and a sedimentation stage (Figure 37-7). Because suspended solids are considered wastewater contaminants, the sedimentation stage accomplishes two functions: concentration of bacteria and removal of solids.

The operating parameters that affect the performance of any activated sludge process are BOD, microorganisms, dissolved oxygen, retention time, nutrient concentration, and the external influences of temperature and pH. In order to understand the various activated sludge designs, it is necessary to examine the relationship between available food and bacteria population.

If a seed culture of bacteria is introduced into a fixed amount of food, the conditions shown in Figure 37-8 are created. Initially, excess food is present; therefore, the bacteria reproduce in a geometric fashion. This is termed the "log growth phase." As the population increases and food

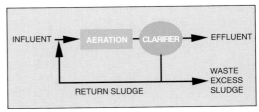

Figure 37-7. Activated sludge process returns active biomass to enhance waste removal.

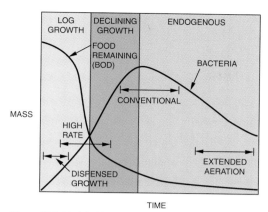

Figure 37-8. Model of bacterial population as a function of time and amount of food.

oxidation time provided. The point of operation determines the remaining bacteria population and BOD of the effluent.

Optimization of an activated sludge plant requires the integration of mechanical, operational, and chemical approaches for the most practical overall program. Mechanical problems can include excessive hydraulic loading, insufficient aeration, and short-circuiting. Operational problems may include spills and shock loads, pH shocks, failure to maintain correct mixed liquor concentration, and excessive sludge retention in the clarifier.

Various chemical treatment programs are described below. Table 37-2 presents a comparison of various treatment schemes.

decreases, a plateau is reached in population. From the inflection point on the curve to the plateau, population is increasing but at a decreasing rate. This is called the "declining growth phase." Once the plateau is crossed, the bacteria are actively competing for the remaining food. The bacteria begin to metabolize stored materials, and the population decreases. This area of the curve is termed "endogenous respiration." Eventually, the bacteria population and BOD are at a minimum.

Because activated sludge is a continuous, steady-state process, each plant operates at some specific point on this curve, as determined by the

Sedimentation. Because activated sludge depends on microorganism recirculation, sedimentation is a key stage. The settleability of the biomass is a crucial factor. As bacteria multiply and generate colonies, they excrete natural biopolymers. These polymers and the slime layer that encapsulates the bacteria influence the flocculation and settling characteristics of bacteria colonies. It has been determined empirically that the natural settleability of bacteria colonies is also a function of their position on the time chart represented in Figure 37-8. Newly formed colonies in the log growth phase are relatively nonsettleable. At the end of the declining growth phase and the first part of the endogenous phase,

Table 37-2. Typical removal efficiencies for oil refinery treatment processes.

Process	Process Influent	Removal Efficiency, %							
		BOD	COD	TOC	SS	Oil	Phenol	Ammonia	Sulfide
API separator	raw waste	5–40	5–30	NA[a]	10–50	60–99	0–50	NA[a]	NA[a]
Primary clarifier	API effluent	30–60	20–50	NA[a]	50–80	60–95	0–50	NA[a]	NA[a]
Dissolved air flotation	separator effluent	20–70	10–60	NA[a]	50–85	70–85	10–75	NA[a]	NA[a]
Filter	API effluent	40–70	20–55	NA[a]	75–95	65–90	5–20	NA[a]	NA[a]
Secondary oxidation pond	API effluent	40–95	30–65	60	20–70	50–90	60–99	0–15	70–100
Aerated lagoon	primary effluent	75–95	60–85	NA[a]	40–65	70–90	90–99	10–45	95–100
Activated sludge	primary effluent	80–99	50–95	40–90	60–85	80–99	95–99+	33–99	97–100
Trickling filter	API effluent	60–85	30–70	NA[a]	60–85	50–80	70–98	15–90	70–100
Cooling tower	primary effluent	50–95	40–90	10–70	50–85	60–75	75–99+	60–95	NA[a]
Activated carbon	primary effluent	70–95	70–90	50–80	60–90	75–95	90–100	7–33	NA[a]
Tertiary filter granular media	secondary effluent	NA[a]	NA[a]	50–65	75–95	65–95	5–20	NA[a]	NA[a]
Activated carbon	secondary + filter effluent	91–98	86–94	50–80	60–90	70–95	90–99	33–87	NA[a]

[a] NA = data not available.

natural flocculation is at an optimum. As the endogenous phase continues, colonies break up and floc particles are dispersed, decreasing the biomass settleability.

Although microbes are eventually able to break down most complex organics and can tolerate very poor environments, they are very intolerant of sudden changes in pH, dissolved oxygen, and the organic compounds that normally upset an activated sludge system. These upsets normally result in poor BOD removal and excessive carry-over of suspended solids (unsettled microorganisms) in the final effluent.

Aeration. Aeration is a critical stage in the activated sludge process. Several methods of aeration are used:

High Rate Aeration. High rate aeration operates in the log growth phase. Excess food is provided, by recirculation, to the biomass population. Therefore, the effluent from this design contains appreciable levels of BOD (i.e., the oxidation process is not carried to completion). Further, the settling characteristics of the biomass produced are poor. High sludge return rates are necessary to offset poor settling and to maintain the relatively high biomass population. Poor settling increases the suspended solids content of the effluent. The relatively poor effluent produced limits this design to facilities which need only pretreatment before discharge to a municipal system. The advantage of high rate aeration is low capital investment (i.e., smaller tanks and basins due to the short oxidation time).

Conventional Aeration. The most common activated sludge design used by municipalities and

industry operates in the endogenous phase, in order to produce an acceptable effluent in BOD and TSS levels. Conventional aeration represents a "middle of the road" approach because its capital and operating costs are higher than those of the high rate process, but lower than those of the extended aeration plants. As shown in Figure 37-8, the conventional plant operates in the area of the BOD curve where further oxidation time produces little reduction in BOD. Natural flocculation is optimum, so the required sedimentation time for removal of suspended solids from the effluent is minimized.

Extended Aeration. Extended aeration plants operate in the endogenous phase, but use longer periods of oxidation to reduce effluent BOD levels. This necessitates higher capital and operating costs (i.e., larger basins and more air). In conjunction with lower BOD, extended aeration produces a relatively high suspended solids effluent when optimum natural settling ranges are exceeded.

Extended aeration designs may be necessary to meet effluent BOD requirements when the influent is relatively concentrated in BOD or the wastes are difficult to biodegrade. Because extended aeration operates on the declining side of the biomass population curve, net production of excess solids is minimized, as shown in Table 37-3. Therefore, savings in sludge handling and disposal costs may offset the higher plant capital and operating costs required for extended aeration.

Step Aeration/Tapered Aeration. In a plug flow basin, the head of the basin receives the waste in its most concentrated form. Therefore, metabolism

Table 37-3. Activated sludge.

	Aeration Retention Time, hr	MLSS[a], ppm	Aeration D.O., ppm	Sludge Recycle, %	BOD Loading, lb/Mft³	F/M lb BOD/lb MLVSS[b]	Sludge Production, lb/lb of BOD	BOD Removal, %
High rate	1/2–3	300–1,000	0.5–2.0	5–15	2.5	1.5–5.0	0.65–0.85	75–85
Conventional activated sludge	6–8 (diffused)	1,000–3,000	0.5–2.0	20–30	20–40	0.2–0.5	0.35–0.55	85–90
	9–12 (mechanical)	500–1,500	0.5–2.0	10–20	20–40	0.2–0.5	0.35–0.55	80–95
Extended aeration	18–36	3,000–6,000	0.5–2.0	75–100	10–25	0.03–0.15	0.15–0.20	90–95
Step aeration	3–5	2,000–3,500	0.5–2.0	25–75	40–60	0.2–0.5	0.35–0.55	85–90
Contact stabilization	3–6	1,000–3,000 (aeration)	0.5–2.0	25–100	60–75	0.2–0.6	0.35–0.55	85–90
		4,000–10,000 (contact basin)	0.5–2.0	25–100	60–75	0.2–0.6	0.35–0.55	85–90
Pure oxygen	1–3	3,000–8,000	2–6	25–50	100–250	0.25–1.0	0.35–0.55	95–98
Complete mix	3–5	3,000–6,000	0.5–2.0	25–100	50–120	0.2–0.6	0.35–0.55	85–95

[a] Mixed liquor suspended solids.

[b] Mixed liquor volatile suspended solids.

and oxygen demand are greatest at that point. As the waste proceeds through the basin, the rate of oxygen uptake (respiration rate) decreases, reflecting the advanced stage of oxidation.

Tapered aeration and step aeration reduce this inherent disadvantage. Tapered aeration provides more oxygen at the head of the basin and slowly reduces oxygen supply to match demand as the waste flows through the basin. This results in better control of the oxidation process and reduced air costs.

Step aeration modifies the introduction of influent waste. The basin is divided into several stages, and raw influent is introduced to each stage proportionately. All return microorganisms (sludge) are introduced at the head of the basin. This design reduces aeration time to 3–5 hr, while BOD removal efficiency is maintained. The shorter aeration time reduces capital expenses because a smaller basin can be used. Operating costs are similar to those of a conventional plant.

Contact Stabilization. Due to the highly efficient sorptive capabilities of activated biomass, the time necessary for the biomass to "capture" the colloidal and soluble BOD is approximately 30 min to 1 hr. Oxidation of fresh food requires the normal aeration time of 4–8 hr. In the contact stabilization design, relatively quick sorption time reduces aeration tank volume requirements. The influent waste is mixed with return biomass in the initial aeration tank (or contact tank) for 30–90 min. The entire flow goes to sedimentation, where the biomass and its captured organics are separated and returned to a reaeration tank. In the reaeration tank the wastes undergo metabolism at a high biomass population. The system is designed to reduce tank volume by containing the large majority of flow for a short period of time.

This process is not generally as efficient in BOD removal as the conventional plant process, due to mixing limitations in the contact basin. Operating costs are equivalent. Due to the unstabilized state of the biomass at sedimentation, flocculation is inferior. Suspended solids in the effluent are problematic.

Because this design exposes only a portion of the active biomass to the raw effluent at a time, it is less susceptible to feed variations and toxicants. For this reason it can be beneficial for treatment of industrial wastes.

Pure Oxygen Sludge Processes. Oxygen supply and transfer often become limiting factors in industrial waste treatment. As the name implies, pure oxygen activated sludge processes supply oxygen (90–99% O_2) to the biomass instead of air. The increased partial pressure increases transfer rates and provides several advantages. Comparable or higher BOD removal efficiencies are maintained at higher BOD influent loadings and shorter retention times. Generally, aeration time is 2–3 hr. A further advantage is the production of lower net solids per pound of BOD removed. Therefore, sludge disposal costs are reduced.

The units are usually enclosed. Normally, three or four concrete box stages in series are provided for aeration. The raw wastewater, return biomass, and pure oxygen enter the first stage. Wastewater passes from stage to stage in the underflow.

The atmosphere flows over the open surface of each stage to the last stage, from which it is vented to control the oxygen content. Oxygen purity and the demand for oxygen decline through the stages. Each stage contains a mechanical agitator for mixing and oxygen transfer. By design, each stage is completely mixed. After aeration, the waste flows to a conventional sedimentation stage. BOD and TSS removals are usually somewhat better than in a conventional aeration system.

Chemical Treatment Programs. The following additives represent a variety of chemical programs that may be used to address problems and improve system efficiency.

Essential Nutrients. Nutrients, particularly nitrogen and phosphorus, may be added to ensure complete digestion of organic contaminants.

Polymers. Polymer feeding improves the settling of suspended solids. Cationic polymers can increase the settling rate of bacterial floc and improve capture of dispersed floc and cell fragments. This more rapid concentration of solids minimizes the volume of recycle flow so that the oxygen content of the sludge is not depleted. Further, the wasted sludge is usually more concentrated and requires less treatment for eventual dewatering. Polymers may also be used on a temporary basis to improve the removal of undesirable organisms, such as filamentous bacteria or yeast infestations, that cause sludge bulking or carryover of floating clumps of sludge.

Oxidizing Agents. Peroxide, chlorine, or other agents may be used for the selective oxidation of troublesome filamentous bacteria.

Antifoam Agents. Antifoam agents may be used to control excessive foam.

Coagulants. In addition to antifoam agents, coagulants may be fed continuously to improve efficiency, or to address particularly difficult conditions. They may also be used intermittently to compensate for hydraulic peak loads or upset conditions.

Fixed Media Biological Oxidation

In contrast to activated sludge, in which the biomass is in a fluid state, fixed media oxidation passes influent wastewater across a substructure laden with fixed biomass. The parameters for healthy microorganisms remain the same, except the manner in which food and microorganisms come into contact.

Fixed media designs allow a biological slime layer to grow on a substructure continually exposed to raw wastewater. As the slime layer grows in thickness, oxygen transfer to the innermost layers is impeded. Therefore, mixed media designs develop aerobic, facultative, and anaerobic bacteria as a function of the thickness of the slime layer. Eventually, either because of size and wastewater shear or the death of the microorganisms, some of the slime layer sloughs off. In a continuous process, this constantly sloughing material is carried to a sedimentation stage, where it is removed. There are no provisions to recycle the microorganisms, because return sludge would plug the fixed media structure. In fact, media plugging and lack of oxygen transfer are the primary difficulties encountered with fixed media designs. Plugging problems can be alleviated by increased wastewater shear. This is normally accomplished by recycling of a portion of the effluent wastewater.

Trickling Filters. Trickling filters are not really filters but a filter-like form of fixed media oxidation. Wastewater is sprayed over a bed of stones, 3–5 in. in diameter. Bed depths range from 5 to 7 ft. Because air contact is the sole means of oxygen transfer, microorganisms become more oxygen deficient as depth increases.

Trickling filters can be classified by hydraulic loading as low-rate, high-rate, or roughing. Due to inherent oxygen transfer difficulties, even low rate filters cannot achieve the BOD removal possible in conventional activated sludge systems. Industrial trickling filters are usually followed by an activated sludge unit. They may be used as a pretreatment step before discharge to a municipal sewage system.

Biological Towers. Another form of fixed media filter uses synthetic materials in grid fashion as a substructure for biological growth. The high porosity available with artificially designed media alleviates the oxygen transfer problems of trickling filters and allows greater bed depths. Bed depths of up to 20 ft with adequate oxygen allow longer contact and consequently better BOD removal.

Biodiscs. Biodiscs are a recently developed form of fixed media oxidation. The media is fixed to a rotating shaft that exposes the media alternately to food (wastewater) and oxygen (atmosphere). Design parameters include speed of rotation, depth of the wastewater pool, porosity of the synthetic media, and number of series and parallel stages. These units circumvent the oxygen limitations of the trickling filter and therefore provide BOD removal comparable to conventional activated sludge systems. Solids produced are easily settled in the sedimentation stage, providing acceptable TSS levels in the effluent. Little operational attention is required.

SOLID WASTE HANDLING

Wastewater treatment is a concentration process in which waterborne contaminants are removed from the larger wastewater stream and concentrated in a smaller side stream. The side stream is too large to be disposed of directly, so further concentration processes are required. These processes are called "solid waste handling" operations.

Stabilization/Digestion

Sludge stabilization is a treatment technique applied to biological sludge to reduce its odor-causing or toxic properties. This treatment often reduces the amount of solids as a side effect. Anaerobic and aerobic digestion, lime treatment, chlorine oxidation, heat treatment, and composting fall into this category.

Anaerobic Digestion. Anaerobic digestion takes place in an enclosed tank, as depicted in Figure 37-9. The biochemical reactions take place in the following phases:

$$\text{Organics} + \text{Acid-forming} \rightarrow \text{Volatile acids}$$
$$\text{organisms}$$

$$\text{Volatile acids} + \text{Methane} \rightarrow \text{Methane} + \text{Carbon}$$
$$\text{formers} \quad\quad\quad\quad\quad \text{dioxide}$$

Sludge solids are decreased due to the conversion of biomass to methane and carbon dioxide. The methane can be recovered for its heating value.

Aerobic Digestion. Aerobic digestion is the separate aeration of sludge in an open tank.

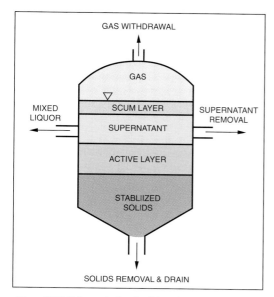

Figure 37-9. Odor control and solids reduction are accomplished in a digester.

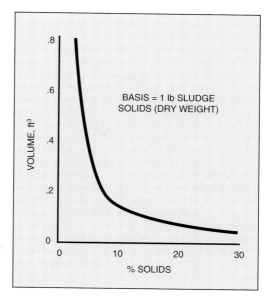

Figure 37-10. Solid waste volume is drastically reduced when water is removed.

Oxidation of biodegradable matter, including cell mass, occurs under these conditions. As in anaerobic digestion, there is a decrease in sludge solids, and the sludge is well stabilized with respect to odor formation. Capital costs are lower than those of anaerobic digestion, but operating costs are higher, and there is no by-product methane production.

Lime Treatment. Stabilization by lime treatment does not result in a reduction of organic matter. Addition of sufficient lime to maintain the pH of the sludge above 11.0 for 1–14 days is considered sufficient to destroy most bacteria.

Composting. A natural digestion process, composting usually incorporates sludge material that later will be applied to farmland. Sludge is combined with a bulking material, such as other solid wastes or wood chips, and piled in windrows. Aeration is provided by periodic turning of the sludge mass or by mechanical aerators. The energy produced by the decomposition reaction can bring the waste temperature to 140–160 °F, destroying pathogenic bacteria. At the end of the composting period, the bulking material is separated, and the stabilized sludge is applied to land or sent to a landfill.

Sludge Conditioning

Typically, sludge from a final liquid–solids separation unit may contain from 1 to 5% total suspended solids. Figure 37-10 shows the rela-

tionship between the volume of sludge to be handled and the solids content in the sludge. Because of the cost savings associated with handling smaller volumes of sludge, there is an economic incentive to remove additional water. Dewatering equipment is designed to remove water in a much shorter time span than nature would by gravity. Usually, an energy gradient is used to promote rapid drainage. This requires frequent conditioning of the sludge prior to the dewatering step.

Conditioning is necessary due to the nature of the sludge particles. Both inorganic and organic sludge consist of colloidal (less than 1 μm), intermediate, and large particles (greater than 200 μm). The large particles, or flocs, are usually compressible. Under an energy gradient, these large flocs compress and prevent water from escaping. The small particles also participate in this mechanism, plugging the pores of the sludge cake, as shown in Figure 37-11. The pressure drop through the sludge cake, due to the decrease in porosity and pore sizing, exceeds available energy, and dewatering ceases.

The purpose of sludge conditioning is to provide a rigid sludge structure of a porosity and pore size sufficient to allow drainage. Biological sludges are conditioned with $FeCl_3$, lime, and synthetic cationic polymers, either separately or in combination. Heat conditioning and low-pressure oxidation are also used for biological

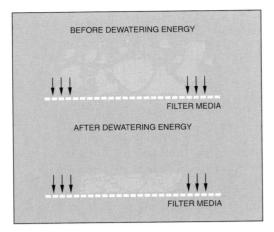

Figure 37-11. Unconditioned sludge can be difficult to dewater.

sludges. Inorganic sludges are conditioned with $FeCl_3$, lime, and either cationic or anionic polymers.

Dewatering

Belt Filter Press. Belt filter presses have been used in Europe since the 1960's and in the United States since the early 1970's. They were initially designed to dewater paper pulp and were subsequently modified to dewater sewage sludge.

Belt filter presses are designed on the basis of a very simple concept. Sludge is sandwiched between two tensioned porous belts and passed over and under rollers of various diameters. At a constant belt tension, rollers of decreasing dia-

meters exert increasing pressure on the sludge, thus squeezing out water. Although many different designs of belt filter presses are available, they all incorporate a polymer conditioning unit, a gravity drainage zone, a compression (low-pressure) zone, and a shear (high-pressure) zone. Figure 37-12 shows these zones in a simplified schematic of a belt filter press.

Polymer Conditioning Unit. Polymer conditioning can take place in a small tank, in a rotating drum attached to the top of the press, or in the sludge line. Usually, the press manufacturer supplies a polymer conditioning unit with the belt filter press.

Gravity Drainage Zone. The gravity drainage zone is a flat or slightly inclined belt, which is unique to each press model. In this section, sludge is dewatered by the gravity drainage of free water. The gravity drainage zone should increase the solids concentration of the sludge by 5–10%. If the sludge does not drain well in this zone, the sludge can squeeze out from between the belts or the belt mesh can become blinded. The effectiveness of the gravity drainage zone is a function of sludge type, quality, and conditioning, along with the screen mesh and the design of the drainage zone.

Compression (Low-Pressure) Area. The compression, or low-pressure, area is the point at which the sludge is "sandwiched" between the

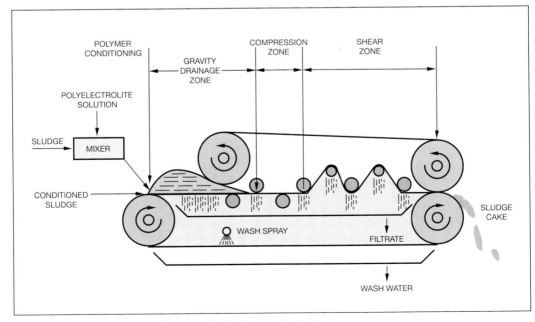

Figure 37-12. Gravity drainage is an important step in belt press dewatering.

upper and lower belts. A firm sludge cake is formed in this zone in preparation for the shear forces encountered in the high-pressure zone.

Shear (High-Pressure) Zone. In the shear, or high-pressure, zone, forces are exerted on the sludge by the movement of the upper and lower belts, relative to each other, as they go over and under a series of rollers with decreasing diameters. Some manufacturers have an independent high-pressure zone which uses belts or hydraulic cylinders to increase the pressure on the sludge, producing a drier cake. A dry cake is particularly important for plants that use incineration as the final disposal.

Dewatering belts are usually woven from monofilament polyester fibers. Various weave combinations, air permeabilities, and particle retention capabilities are available. These parameters greatly influence press performance.

Usually, cationic polymers are used for sludge conditioning. A two-polymer system is often used on a belt filter press to improve cake release from the upper dewatering belt. The polymer must be selected carefully to ensure optimum performance.

Odors are controlled by proper ventilation, by ensuring that the sludge does not turn septic, and by the use of added chemicals, such as potassium permanganate or ferric sulfate, to neutralize the odor-causing chemicals.

Screw Press. A recent development in sludge dewatering equipment, used primarily in the pulp and paper industry, is the screw press. Screw presses are most effective for primary sludges, producing cake solids of 50–55%, but are also appropriate for primary and secondary blended sludges.

Sludge is conditioned and thickened prior to dewatering. The conditioned sludge enters one end of the machine, as shown in Figure 37-13. A slowly rotating screw, analogous to a solid bowl centrifuge, conveys and compresses the solids.

The screw has the same outer diameter and pitch for the entire length of the press. In some models, the diameter of the screw shaft increases toward the discharge end of the screw press to enhance dewatering. The compression ratio (the ratio of free space at the inlet to the space at the discharge end of the screw) is selected according to the nature of the material to be dewatered and the dewatering requirement. Dewatered cake is discharged as it is pressed against the spring or hydraulically loaded cone mounted at the end of the screw press.

The drum of the screw press consists of a fine strainer screen, a thicker punched holding plate, and a reinforcement rib.

Filtrate is collected in the collecting pan located under the screw press, and the cake is transported to the next stage.

Vacuum Filters. Vacuum filtration uses various porous materials as filter media, including cloth, steel mesh, and tightly wound coil springs. Under an applied vacuum, the porous medium retains

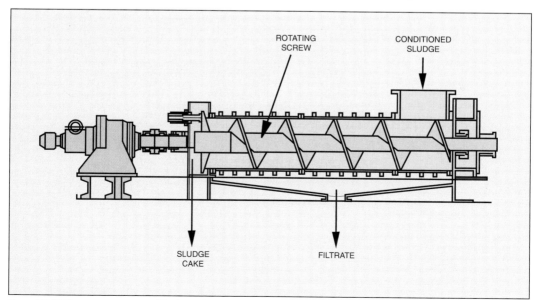

Figure 37-13. Screw presses are making inroads in the pulp and paper industry.

the solids, but allows water to pass through. The relative importance of cake dryness, filtrate quality, and filter cake yield can vary from one system to another.

A decrease in drum speed allows more time for drying of the sludge to increase cake dryness. However, this also decreases the filter cake yield, defined as pounds of dry solids per hour per square foot of filter area. Polymers can help produce a drier cake without the problem of a lower filter cake yield. Synthetic polymers improve cake dryness by agglomerating sludge particles that may hinder the removal of water. This agglomeration also increases the solids capture across the unit, which results in a higher-quality filtrate.

Centrifuges. Centrifugal force, 3500–6000 times the force of gravity, is used to increase the sedimentation rate of solid sludge particles.

The most common centrifuge found in waste treatment dewatering applications is the continuous bowl centrifuge (Figure 37-14). The two principal elements of a continuous solid bowl centrifuge are the rotating bowl and the inner screw conveyor. The bowl acts as a settling vessel; the solids settle due to centrifugal force from its rotating motion. The screw conveyor picks up the solids and conveys them to the discharge port.

Often, operation of centrifugal dewatering equipment is a compromise between centrate quality, cake dryness, and sludge throughput. For example, an increase in solids throughput reduces clarification capacity, causing a decrease in solids capture. At the same time, the cake is drier due to the elimination of fine particles that become entrained in the centrate. The addition of polymers, with their ability to agglomerate fine particles, can result in increased production rates without a loss in centrate quality.

Polymers are usually fed inside the bowl because shear forces may destroy flocs if they are formed prior to entry. Also, large particles settle rapidly in the first stage of the bowl. Thus, economical solids recovery can be achieved through internal feeding of polymers after the large particles have settled.

Plate and Frame Press. A plate and frame filter press is a batch operation consisting of vertical plates held in a frame. A filter cloth is mounted on both sides of each plate. Sludge pumped into the unit is subjected to pressures of up to 25 psig as the plates are pressed together. As the sludge fills the chamber between individual plates, the filtrate flow ceases, and the dewatering cycle is completed. This cycle usually lasts from ½ to 2 hr.

Because of the high pressures, blinding of the filter cloth by small sludge particles can occur. A

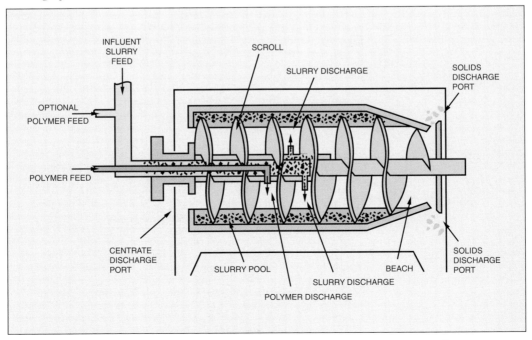

Figure 37-14. Proper adjustments for cake dryness and centrate quality are key to the efficient operation of a continuous bowl centrifuge.

filter precoat (e.g., diatomaceous earth) can be used to prevent filter blinding. Proper chemical conditioning of the sludge reduces or eliminates the need for precoat materials. At 5–10 psig, polymers can produce a rigid floc and eliminate fine particles. At greater pressures, the effectiveness of synthetic polymers is reduced; therefore, inorganic chemicals, such as ferric chloride and lime, are often used instead of polymers.

Sludge Drying Beds. Sludge drying beds consist of a layer of sand over a gravel bed. Underdrains spaced throughout the system collect the filtrate, which usually is returned to the wastewater plant.

Water is drained from the sludge cake by gravity through the sand and gravel bed. This process is complete within the first 2 days. All additional drying occurs by evaporation, which takes from 2 to 6 weeks. For this reason, climatic conditions, such as frequency and rate of precipitation, wind velocity, temperature, and relative humidity, play an important role in the operation of sludge drying beds. Often, these beds are enclosed to aid in dewatering. Chemical conditioning also reduces the time necessary to achieve the desired cake solids.

Sludge Disposal

Disposal of the sludge generated by wastewater treatment plants is dependent on government regulations (such as the Resource Conservation and Recovery Act), geographical location, and sludge characteristics, among other things. Final disposal methods include reclamation, incineration, land application, and landfill.

Reclamation. Because of costs associated with the disposal of wastewater sludge, each waste stream should be evaluated for its reclamation potential. Energy value, mineral content, raw material makeup, and by-product markets for each sludge should be evaluated. Examples include burning of digester gas to run compressors, recalcination of lime sludge to recover CaO, return of steel mill thickener sludge to the sinter plant, and marketing of by-product metallic salts for wastewater treatment use.

Incineration. Biological sludge can be disposed of by incineration; the carbon, nitrogen, and sulfur are removed as gaseous by-products, and the inorganic portion is removed as ash. Old landfill sites are filling up and new ones are becoming increasingly difficult to obtain. Therefore, waste reduction through incineration is becoming a favored disposal practice.

Several combustion methods are available, including hogged fuel boilers, wet air oxidation and kiln, multiple hearth furnace, and fluidized bed combustion processes.

Sludge incineration is a two-step process involving drying and combustion. Incineration of waste sludge usually requires auxiliary fuel to maintain temperature and evaporate the water contained in the sludge. It is critically important to maintain a low and relatively constant sludge moisture.

Land Application. Sludge produced from biological oxidation of industrial wastes can be used for land application as a fertilizer or soil conditioner. A detailed analysis of the sludge is important in order to evaluate toxic compound and heavy metal content, leachate quality, and nitrogen concentration.

Soil, geology, and climate characteristics are all important considerations in determining the suitability of land application, along with the type of crops to be grown on the sludge-amended soil. Sludge application rates vary according to all of these factors.

Landfill. Landfill is the most common method of industrial wastewater treatment plant sludge disposal.

Care must be taken to avoid pollution of groundwater. The movement and consequent recharge of groundwater is a slow process, so contamination that would be very small for a stream or river can result in irreversible long-term pollution of the groundwater. Many states require impermeable liners, defined as having a permeability of 10^{-7} cm/sec, in landfill disposal sites. This requirement limits liners to a few natural clays and commercial plastic liners. In addition to impermeable liners, leachate collection and treatment systems are typically required for new and remediated landfills.

Steps can be taken to reduce leachate and leachate contamination. Decreasing the moisture in the sludge removes water that would eventually be available as leachate. Proper consideration of the hydraulics of the landfill site can capture more rainfall as runoff and eliminate ponding and its contribution to leachate.

ENVIRONMENTAL REGULATIONS

Many governmental regulations have been established in recent years for the protection of the environment. The Clean Water Act and the Resource Conservation and Recovery Act are among the most significant.

Clean Water Act

The Clean Water Act (CWA) of 1972 established regulations for wastewater discharge, provided funding for Publicly Owned Treatment Works (municipal waste treatment plants), and authorized the National Pollutant Discharge Elimination Systems (NPDES) to regulate and establish wastewater discharge permits for industrial and municipal plants.

Resource Conservation and Recovery Act (RCRA)

The Resource Conservation and Recovery Act (RCRA) of 1976 provided regulations for management of hazardous solid wastes, cleanup of hazardous waste sites, waste minimization, underground storage, and groundwater monitoring.

GAS CLEANING SYSTEMS

Gas cleaning and process systems characterized by high solids content present a unique challenge in industrial water treatment. In comparison with "cleaner" systems (e.g., cooling systems, boiler systems, potable water systems) these systems can develop severe deposition and/or corrosion problems in a very short period of time. The deposits are often of an unusual composition, which can vary widely throughout a single system. The deposit and corrosion problems can cause sudden losses of production time, increased labor costs, and frequent equipment replacement.

High solids water systems are used in most industries, usually as part of manufacturing processes that use water for solids transport, chemical conversion and reaction, gas stripping, or solids separation (see Figure 38-1). The following processes are typical high-solids applications:

■ wet scrubbers, wet electrostatic precipitators, and dry scrubbers used in air pollution control systems

■ waste treatment and disposal processes, from acid mine drainage streams to ash sluicing systems

■ reagent makedown and storage systems (e.g., lime slaking)

Many plants with high solids water systems have addressed deposit and corrosion problems with a successful combination of operational control, mechanical cleaning, and chemical treatment.

High solids systems can experience a combination of classic deposition mechanisms. High levels of dissolved solids often cause scaling problems. As concentrations increase, the solubility of certain ions is exceeded and precipitation occurs. This precipitation can take place on system surfaces, such as piping, pumps, and spray nozzles. Precipitation can also occur on the surfaces of suspended matter, creating even larger particles. Precipitation can be triggered

Figure 38-1. Gas cleaning systems are now found in almost all industries.

by minor changes in system chemistry and mechanical factors such as velocity and turbulence.

An increase in suspended solids concentration increases the deposition potential in two ways. Significant concentrations of suspended solids (greater than 25 ppm) contribute to mechanical fouling, which usually occurs in low-velocity areas, restricted areas, or areas where large particle size creates problems (e.g., strainers, filters). There is no chemical reaction involved in this mechanism as there is in scaling or precipitation. The physical characteristics of the suspended particles are not altered—they simply settle in low flow areas or become entrapped by adhering to other particles.

The presence of suspended solids also contributes to precipitation and scaling. By increasing the available surface area, suspended solids increase the potential for precipitation. Experiments have shown that solutions of calcium and

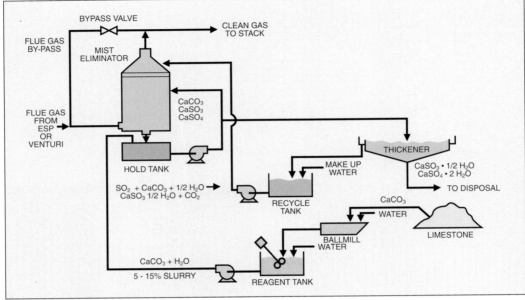

Figure 38-2. Limestone scrubbing is the most common method of SO_2 removal.

sulfate ions, at concentrations greater than saturation, can be prepared without precipitation occurring. However, when a small amount of suspended solids is added to the solution, calcium sulfate crystals precipitate onto the solids and drop out of solution. A change in solution "mechanics" can make a substantial difference in solids behavior.

Most high solids systems are subject to both scaling and fouling. Certain modifications can be made to the chemistry or operation to minimize these problems. However, many systems (especially process streams and wet scrubbers) are designed to promote precipitation and separation of solids. Their purpose is to collect gaseous contaminants from the gas stream and precipitate them so that they can be removed and disposed of safely. In these systems, deposition must be controlled without interfering with the reactions. Figure 38-2 shows a sulfur dioxide wet scrubber that uses limestone or lime as a reagent. The scrubber is designed to circulate a 3–12% solids slurry at flow rates as high as 30,000–40,000 gpm. The goal is to create a reaction between the $CaCO_3$ or $Ca(OH)_2$ slurry and the SO_2 gas to generate $CaSO_3$ and $CaSO_4$ solids. This can produce tons of material a day. These SO_2 scrubbers can experience severe scaling (with $CaSO_3$ and $CaSO_4$); some systems only run 3–4 days before being shut down for cleaning. Standard deposit control programs cannot be utilized because they would also retard or stop the reaction of SO_2 gas with the lime or limestone slurry.

Deposition can occur in high solids systems despite effective control of system chemistry and operation. Under these conditions, the realistic goal is to extend the time between scheduled shutdowns and eliminate forced downtime.

SOLIDS CONTROL

Reduction of suspended solids is often a large step in controlling deposition problems. Many systems have solids removal as part of the circulating loop, but this practice is not always effective.

Clarifiers/Thickeners

Clarifiers and thickeners (Figures 38-3 and 38-4) are frequently included in high solids systems. They are designed to receive the entire high solids stream and return the overflow to the process or discharge it. These units can also be designed to receive the blowdown from a high

Figure 38-3. Solids control from process thickeners and clarifiers is often critical to optimum operation of the system.

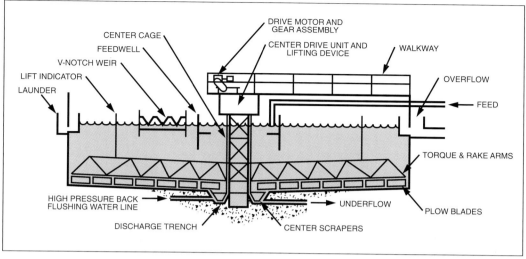

Figure 38-4. Cross section of a typical high solids thickener.

solids system. In such cases, the overflow is either returned as makeup to the primary system or discharged. Underflow from a clarifier or thickener is usually sent to a dewatering unit, such as a vacuum filter, filter press, or centrifuge. It may also be sent directly to a disposal pond or landfill.

It is especially important to maximize the settling of solids when trying to improve the operation of a high solids system. This goal involves several key considerations:

- hydraulic design of the unit
- recirculation of underflow
- extraneous streams
- microbiological contamination
- foaming
- height of sludge blanket

Hydraulic design of the unit. The unit may simply be overloaded by flow and/or solids. Design information should be consulted for maximum design levels.

Recirculation of underflow. Operation can be improved by a reduction in recirculation. Most inorganic sludges do not require the "seeding" of incoming solids by conditioned underflow solids. Recirculation of underflow may increase the solids loading. When recirculation is required, the underflow recirculation line should discharge into a splitter box or centerwell.

Extraneous streams. Many high solids systems are a "catch-all" for various streams in a plant. Some of these streams contain solids, organic materials, oils, and chemicals that disrupt settling. All flows to the clarifier or thickener should be

analyzed (even if the flow is intermittent) to determine their impact on clarifier operation. If necessary, they should be diverted elsewhere.

Microbiological contamination. Some clarifiers and thickeners operate with only periodic draw-off of underflow solids. This can lead to septic conditions in the settled sludge, causing odor problems and generating gases that can disrupt settling. Major microbiological populations in the sludge can also cause dewatering problems, such as filter cloth blinding. Periodic treatment of the system or settling unit with an antimicrobial usually eliminates the problem. Sludge should be sampled and examined for identification of the organism.

Foaming. Many high solids systems experience frequent problems with foaming. The settling unit develops a foam layer that traps incoming particles, adds solids to the overflow, and even creates housekeeping and safety problems. Foam can be caused by surfactant or hydrocarbon materials, microbiological contamination, organics in the makeup water, and a variety of other factors. Mechanical solutions include water sprays located on top of the settling unit and a boom arrangement at the overflow weir to prevent foam from entering the overflow. Antifoams can be fed to the influent or to the water sprays on top of the unit.

Height of sludge blanket. With insufficient underflow solids removal, the sludge blanket level can become too high. This limits "clean water" depth and residence time and often causes high solids in the overflow. A low sludge depth can also be detrimental. If the settling unit is designed so that the incoming solids travel up through the sludge blanket to improve contact and removal, a low sludge depth decreases the contact, causing high solids in the overflow.

Most settling units can benefit from a properly designed polymer addition program. In many cases, the feed of an anionic polyelectrolyte, at approximately 0.5–3 ppm, greatly enhances solids settling. This feed often results in improved dewatering, although a supplemental feed of polymer just upstream of the dewatering unit may be required. In some applications, a cationic polymer fed prior to the anionic polymer feed produces optimum results. The need for a cationic polymer can easily be determined by the standard jar testing procedure described in Chapter 5. If foam is present, the addition of an antifoam to the high solids system or just prior to the settling unit can minimize or eliminate foaming.

Solids monitoring in the overflow provides a direct indication of settling unit efficiency. A realistic goal for most systems is maintenance of less than 100 ppm suspended solids in the overflow; maintenance of less than 25 ppm ensures minimal impact on the receiving system.

Hydrocyclones

Some high solids systems are designed with hydrocyclones to improve suspended solids removal. A side stream is taken from the primary process stream, sent to the hydrocyclone for solids reduction, and returned to the system. Theoretically, this process is similar to side stream filtration in a cooling system. However, hydrocyclones can be difficult to operate properly, and plugging often becomes a problem.

Hydrocyclones (Figure 38-5) are designed to perform a "gross" separation of solids by density. The process uses centrifugal force to separate and remove large particles from the incoming water or slurry. The water and remaining suspended particles are returned to the system. There are no moving parts and no parts that should require periodic cleaning.

In practice, operational problems are caused by fluctuations in dissolved solids concentration (which influences densities) and suspended solids loading. The degree of separation is satisfactory as long as the influent stream characteristics do not vary significantly from design. However, most high solids systems do not operate smoothly, and the hydrocyclone is often misapplied. The result is a shift in "separation" so that too many solids return to the system or too many solids are removed, plugging the removal piping. Usually,

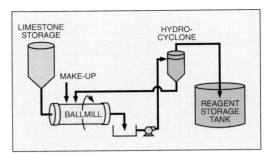

Figure 38-5. Hydrocyclones are often incorporated in ball mill systems to improve solids separation.

chemical treatment cannot solve this problem. Sometimes, a redesign of the hydrocyclone or the addition of cyclones for a multicyclone arrangement improves the control of solids separation.

Settling Ponds

Many high solids systems use settling ponds as part of a recirculation loop or as a final disposal site. Ash sluicing systems usually have ponds that operate in a recirculation mode.

Slurry is pumped or gravity-fed to the pond. Usually, sufficient retention time (several days) is provided for most solids to settle. Baffles may be included to direct the flow. After settling, the liquid is pumped back to the system as makeup or recirculating flow.

Due to accumulation of solids and/or low water levels, actual retention times may be insufficient for solids settling. Short-circuiting of flows in the pond (often due to solids accumulation) can also be a contributing factor.

Periodic removal of accumulated solids maximizes the capacity of settling ponds. Polymers can be added to enhance settling.

Even with effective solids removal, scaling can occur in the return line. Often, in systems with high dissolved solids, scaling does not develop in lines going to the settling pond. This is due to the abrasive nature of the high suspended solids in the slurry; any scale is immediately scoured off. Also, the surface area provided by the suspended solids causes most precipitation to occur preferentially on the suspended particles, not on pipe walls or pumps. However, water leaving the pond is often saturated or even supersaturated with dissolved species. In the absence of suspended particles, precipitation can occur on system surfaces. Deposit control is often required to protect the "clean water" return lines.

SOLIDS BLOWDOWN

Some high solids systems are operated to maintain a certain level of suspended solids in the system.

As described previously, maintenance of the required concentration can be critical to the efficiency of the systems, as in an SO_2 scrubber. However, in any system, excess solids can cause catastrophic problems due to plugging.

It is important to establish a blowdown procedure that includes solids sampling. In critical systems, density may be monitored continuously. In other systems, sampling one or more times a day is sufficient for setting or checking blowdown rates. Solids levels can be determined through gravimetric analysis, hydrometers, settling cones, or measurement of centrifuged samples. A Marcy meter can be used for day-to-day comparison by operators.

Turbidity is not always an accurate indication of suspended solids. Dissolved solids that contribute "color" to a sample can affect turbidity readings. To compensate for this interference, suspended solids may be measured gravimetrically and compared to turbidity readings for a large number of samples to develop a correlation for a specific system. Turbidity can then be used as a measure of suspended solids for daily plant operation, provided that the correlation to suspended solids is checked periodically.

Conductivity is usually a reliable measurement of dissolved solids when used on a comparative basis.

DEPOSIT CONTROL

Deposit control is the most difficult aspect of operating most high solids systems. The prospect of high dissolved and suspended solids concentrations leads most equipment designers to anticipate severe erosion/corrosion rates. The low pH of some systems also contributes to corrosion problems. Many systems are designed with exotic metals or coatings and rubber linings or plastic internals.

In operation, most systems develop substantial deposition problems (Figure 38-6). The formation of calcium sulfate, calcium fluoride, and silicate deposits is not significantly influenced by pH, so even low-pH systems can experience heavy deposition. Many plastic internal parts (such as mist eliminators, Figures 38-7 and 38-8) cannot withstand even light accumulations of deposits, due to their structural strength limitations. Many systems experience cracking, breaking, and loss of internal parts due to solids accumulation. This immediately affects efficiency and maintenance costs.

Figure 38-6. Deposits can develop rapidly in high solids systems.

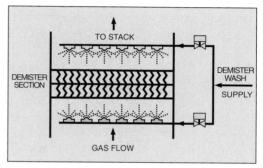

Figure 38-7. Most demister sections incorporate wash sprays to help keep surfaces clean.

Figure 38-8. Demister sections can develop serious deposition problems due to carryover and the use of recycled water for wash water.

The following sections address operational procedures that affect deposition.

pH Control

Calcium carbonate scale formation is greatly affected by pH. Any high solids system with a pH over 8.2 has a potential for calcium carbonate scaling. In many applications, pH is maintained above 7.0 to prevent corrosion problems. However, if pH control is not closely maintained, system pH can approach the calcium carbonate scaling range. A pH in the range of 6.0–7.0 can reduce corrosion to acceptable limits and eliminate any potential for calcium carbonate scale.

Some processes require a specific pH level for the most efficient chemical reaction or adsorption (Figure 38-9). In these cases, pH control is critical. The choice of instrumentation and reagent chemical are important for successful operation. Feedforward control loops are relatively simple but often do not maintain the optimum system pH. Caustic is often used as a pH control reagent. However, it is expensive, and it is a strongly alkaline agent. This means that a small amount of caustic can swing the pH drastically, and the control loop may not be responsive enough to maintain pH in the desired range. This results in wide pH variations—often into the calcium carbonate scaling range. Lime and limestone are less expensive than caustic, and their lower basicity often improves control of pH. However, the addition of calcium can contribute significantly to deposition problems. Therefore, several factors must be considered in the choice of a reagent: cost, instrumentation, scaling potential without reagent addition, and solids generation.

High pH does not always mean high deposition potential. Above a pH of approximately 10.2, the potential for calcium carbonate ($CaCO_3$) deposition decreases rapidly as calcium hydroxide

($Ca(OH)_2$) becomes the predominant calcium species. $Ca(OH)_2$ is much more soluble than $CaCO_3$ and consequently less likely to deposit. In some plants, acid is fed to reduce pH, based on the assumption that high pH is always undesirable. This acid addition usually lowers the pH into the 7.5–10 range, promoting $CaCO_3$ deposition.

Good pH control is dependent on good instrument maintenance. Probes and transmitters must be checked and calibrated regularly to ensure reliable pH control.

Density/Solids Control

The control of solids concentrations is important to maintaining efficient operation. A system that is prone to deposition problems may benefit from an evaluation of dissolved and suspended solids levels. Tight control of blowdown minimizes these levels and reduces deposition.

Flushing Procedures

Proper shutdown procedures prevent deposition in idle system components. Many systems have redundant equipment, such as spare pumps and alternative piping systems or bypass lines. Pumps and lines holding high solids slurries must be flushed with "clean" water prior to shutdown. Even if only a few hours will elapse between shutdown and start-up, these sections of the system should be liberally flushed. Fresh water connections should be installed throughout a high solids system to facilitate flushing.

Location of Makeup and Reagent Feed Points

Makeup and reagent feed points are a primary consideration in systems with high concentrations of dissolved solids. Severe deposition can occur at addition points, caused by the use of makeup water whose chemistry is dissimilar to that of the system water, or due to the use of a strong reagent that causes localized pH swings. Therefore, feed points should be located in areas where precipitation of solids can be controlled (e.g., thickeners or clarifiers, sumps, and large reaction tanks with extended residence time) or in areas with good mixing, distribution, and turbulence.

In many high solids systems, significant deposition occurs despite all efforts to control it mechanically and operationally. Because of the low solubility of many compounds, it is often impossible to eliminate deposition problems completely. However, it is possible to reduce

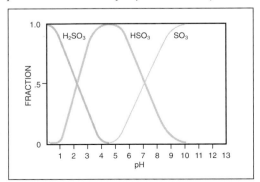

Figure 38-9. Distribution of sulfite and sulfate species based on system pH.

deposition and alter its characteristics to ensure satisfactory performance with a predictable maintenance schedule.

Many of the deposit control agents used in cooling systems are not suitable for high solids systems. Most phosphonates and polymers react readily with suspended solids and newly precipitated crystals. However, some of these materials exhibit low solubility with certain ions. For example, the incorrect application of some phosphonates in high-calcium waters can result in the precipitation of a calcium–phosphonate complex that can aggravate deposition.

Several phosphonates and polymeric dispersants work very effectively in the presence of high dissolved and suspended solids and in demanding environments. Some work well at both low and high pH and at moderately high temperatures. In some systems, the deposit control agent must be effective for only a relatively short period of time, to control deposition until the flow reaches a solids removal unit. That is why the feed of the deposit control agent can usually be below stoichiometric levels. For example, precipitation and fouling need to be controlled only for a matter of minutes in some systems, although they must be controlled for hours or even days in boilers and cooling systems.

The effect of suspended solids on a deposit control program is an important consideration. Most deposit control agents adsorb onto the surfaces of solids. This must be considered in determination of feed rates and feed points. For example, if a deposit control agent is added to a reagent feed tank hours before application to the system, its efficacy may be severely reduced due to exposure to solids.

Treatment chemicals are removed along with solids when the stream passes through solids removal equipment, such as clarifiers and hydrocyclones. Therefore, water leaving a solids removal unit is essentially untreated for deposit control. Because this water can be highly saturated with scaling ions, a deposit control treatment may be required at this point to prevent scaling of lines.

MONITORING

High solids systems can be difficult to monitor. High dissolved and suspended solids can plug sampling lines and continuous monitors. They can also interfere with chemical tests.

Many high solids systems have an in-line pH probe to track system pH and control acid or base feed. Regular calibration and maintenance is required to prevent losses in efficiency, reagent waste, plugging, and accelerated corrosion. In many cases, a side stream "pot" arrangement for the pH probe provides more representative results than a probe placed directly in the process stream.

Evaluating the rate of deposit accumulation in these systems may require a creative approach to monitoring and results analysis. Pressure drop across a system may be compared with design specifications to indicate accumulation of deposition. Gas velocities, process stream flows, and system efficiency levels (in the case of air pollution control equipment) can all be used to assess the operating condition of the system. Reliable instrumentation and constant maintenance is needed to track these parameters. Depending on the needs of the application, optional monitoring configurations such as the following may be employed:

- A flanged "spool" piece can be inserted in the system piping. This line can be valved to a bypass loop, or the system can be intermittently shut down to permit inspection of the spool piece to evaluate deposition.

- Bypass racks with coupons are usually not effective in high solids applications because they are too susceptible to plugging. Deposit coupons placed directly in the system with a retractable assembly can give a comparative measure of deposit accumulation (Figure 38-10). However, they are susceptible to severe erosion/corrosion and total metal loss due to the impingement of solids.

- Portable heat exchangers can be used to generate representative deposits. These exchangers can also be used to evaluate various deposit control programs by feeding the program directly to the portable heat exchangers.

Figure 38-10. Use of a retractable deposit coupon assembly can provide realistic information on system deposits.

Control Water Analyses and Their Interpretation

Because of the increasing demands for fresh water, there is a continuing need to share resources and return quality water to the environment.

ANALYTICAL METHODS AND EQUIPMENT

In industrial water conditioning, chemical analyses are needed to govern the treatment processes. Analysis should be conducted promptly after sample collection so that the chemical nature of the sample does not change. On-site testing may be supplemented by plant central laboratories or Betz Customer Service Laboratories.

The methods included in this handbook are suitable for on-site analysis. They involve the use of apparatus and chemicals that are evaluated, approved, and supplied by Betz Laboratories. Lists of required materials are provided with each titrimetric, spectrophotometric, and colorometric procedure. In some cases, other appropriate equipment may be substituted for the apparatus listed. Substitution of reagents, unless otherwise noted, is not recommended. Microbiological tests and tests which are not suitable for on-site analysis have been excluded from this text.

Several authoritative sources have been referenced in the development of these procedures. These sources include *The Annual Book of ASTM Standards* and the APHA-AWWA *Standard Methods for the Examination of Water and Wastewater*, along with other well known and widely accepted analytical methodology. The procedures are not for EPA or governmental reporting purposes and are not to be used where litigation may be involved.

SAMPLE COLLECTION AND PREPARATION

In order to ensure that results obtained from an analysis are useful, it is necessary to secure a representative sample from the system to be tested. Sample lines must be flushed before samples are taken, and all sampling locations and procedures must be well defined.

For most tests, the samples should be cooled to room temperature (21-26 °C, 70-80 °F) prior to

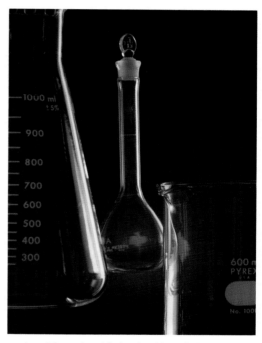

testing. They should also be filtered through 0.2-2.5 μm filters, if required.

METHODS OF ANALYSIS
Titration Methods

Historically, titration has been the most common method of plant control analysis. Titration is based on the use of a buret, from which a standard solution is added to the sample until an "end point" is reached. The end point is generally indicated by a color change or detected by potentiometric device (e.g., pH meter).

Several types of burets are available for plant use:

■ semimicro burets (2.0 or 3.0 mL capacity) are used to titrate low concentrations of species in the sample

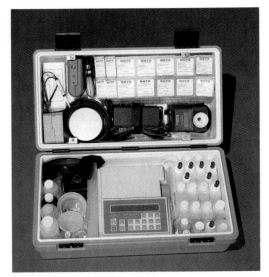

Figure 39-1. Water analysis field test kit.

- large burets (25 or 50 mL capacity) are used to titrate species found in higher concentrations

- automatic burets feature a reservoir for "automatic" filling of the buret and an overflow and reset to 0 mL

Digital titrators provide a more portable approach to titration in the field. These hand-held units are widely accepted because they are rugged and easily carried from one location to another. The digital titrator is equivalent to a buret in the conventional titration methods. The titrator acts as a plunger and forces concentrated titrant from an attached plastic cartridge. Each cartridge can perform the same amount of testing as one quart of titrant in conventional tests. However, in most plant laboratories and testing locations, automatic burets are still being used.

Photometric Methods

Photometers or spectrophotometers provide the most accurate means of measuring the color of a reacted sample. In field analysis applications, simple filter photometers have been replaced by monochromator-based spectrophotometers. The essential components of a spectrophotometer (Figure 39-2) include the following:

- a stable source of radiant energy

- a system of lenses, mirrors, and slits that define, collimate (make parallel), and focus the beam

- a monochromator, to resolve the radiation into component wavelengths or "bands" of wavelengths

- a transparent container to hold the sample

- a radiation detector with an associated read-out system

Light from a tungsten bulb is reflected off of a parabolic mirror and dispersed with a double pass through a high-dispersion prism. The selected wavelength is imaged onto a movable slit, ensuring a uniform band width.

Colorimetric Methods

Colorimetric comparator tests are not as accurate as the photometric or spectrophotometric methods. Color comparisons may be used as a backup to on-line or other optical instrumentation. Colorimetric methods have become popular because of their simplicity and relatively low cost. However, tight control of most industrial water systems should not be entrusted to this technique alone.

In a comparator test, a color is developed that is proportional to the concentration of the substance being determined. The concentration present in the sample is determined by comparison with sealed color standards. The color standards are made of colored plastic or glass, or liquids sealed in air-tight containers.

INSTRUMENTAL METHODS USED IN THE LABORATORY

Common new methods of water analysis often involve highly sophisticated electronic instrumentation not generally used on-site for plant control.

- Ion Chromatography is used to measure trace levels of anions in feedwater, steam, condensate, and boiler water.

- Atomic Absorption Spectroscopy (AA), Inductively Coupled Ion Spectroscopy (ICP), X-ray Fluorescence Spectroscopy, and other laboratory procedures are used routinely to measure many elements at trace levels in a fraction of the time required for wet chemical methods. Some instruments can provide concurrent readouts of over 40 elements in ppb measurements.

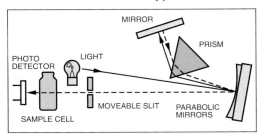

Figure 39-2. Schematic diagram for DR 2000 spectrophotometers.

■ Gas Chromatography (GC), or Gas Chromatography and Mass Spectroscopy (GC/MS), quantitatively separates and detects volatile components (e.g., neutralizing amines) in boiler condensate.

■ High-Pressure Liquid Chromatography (HPLC) permits the separation and detection of trace organic compounds in antimicrobial applications.

■ Total Organic Carbon (TOC) measurements are used to determine the amount of organic compounds present in water as a result of water treatments or process leaks. This process is also useful for measuring organic fouling of resins in demineralizer systems.

■ Nuclear Magnetic Resonance Spectroscopy (NMR) provides an analytical tool to aid in determining the structure of organic polymers and other organic water treatment chemicals.

■ Fourier Transform Infrared Analysis (FT-IR) permits the qualitative and quantitative determination of the composition of boiler and cooling system deposits.

■ Specific ion electrode detection is an electrometric method that can measure trace amounts of both anions and cations in water and is within the reach of most laboratories and testing sites.

The field-testing methods presented in this handbook are often supplemented by these instrumental methods to optimize treatment effectiveness.

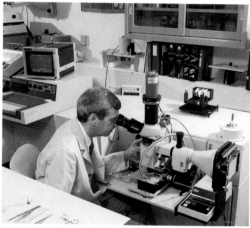

Figure 39-3. FT-IR microscopy.

EXPRESSION OF ANALYTICAL RESULTS

40

PARTS PER MILLION (PPM)

Water analysis involves the detection of minute amounts of a variety of substances. The expression of results in percentage would require the use of cumbersome figures. For this reason, the results of a water analysis are usually expressed in parts per million (ppm) instead of percentage. One part per million equals one ten-thousandth of one percent (0.0001%), or one part (by weight) in a million parts—for example, 1 oz in 1,000,000 oz of water, or 1 lb in 1,000,000 lb of water. It makes no difference what units are used as long as both weights are expressed in the same units.

When elements are present in minute or trace quantities, the use of parts per million results in small decimal values. Therefore, it is more convenient to use parts per billion (ppb) in these cases. One part per billion is equal to one-thousandth of one part per million (0.001 ppm). For example, in studies of steam purity using a specific ion electrode to measure sodium content, values as low as 0.001 ppm are not uncommon. This is more conveniently reported as 1.0 ppb.

In recent times, the convention for reporting analytical results has been shifting toward the use of milligrams per liter (mg/L) as a replacement for parts per million and micrograms per liter (µg/L) as a replacement for parts per billion.

Test procedures and calculations of results are based on the milliliter (mL) rather than the more common cubic centimeter (cc or cm³). The distinction between the two terms is very slight. By definition, a milliliter is the volume occupied by 1 g of water at 4 °C, whereas a cubic centimeter is the volume enclosed within a cube 1 cm on each edge (1 mL = 1.000028 cm³).

MILLIGRAMS PER LITER (mg/L)

The milligrams per liter (mg/L) convention is closely related to parts per million (ppm). This relationship is given by:

$$ppm \times \text{solution density} = mg/L$$

Thus, if the solution density is close or equal to 1, then ppm = mg/L. This is normally the case in dilute, aqueous solutions of the type typically found in industrial water systems. Control testing is usually conducted without measurement of a solution's density. For common water samples, this poses no great inaccuracy, because the density of the sample is approximately 1. Milligrams per liter (mg/L) and parts per million (ppm) begin to diverge as the solution density varies from 1. Examples of this are a dense sludge from a clarifier underflow (density greater than 1) or closed cooling system water with high concentrations of organic compounds (density less than 1). All of the analytical methods discussed in this text contain calculations required to obtain the results in milligrams or micrograms per liter.

EQUIVALENTS PER MILLION (EPM)

In reporting water analyses on an ion basis, results are also expressed in equivalents per million (epm). Closely allied to the use of parts per million, this approach reduces all constituents to a common denominator—the chemical equivalent weight.

The use of equivalents per million is not recommended for normal plant control. Parts per million is a simpler form of expressing results and is accepted as the common standard basis of reporting a water analysis. However, whenever extensive calculations must be performed, the use of equivalents per million greatly simplifies the mathematics, because all constituents are on a chemical equivalent weight basis. The remainder of this section provides a discussion of parts per million and equivalents per million for those who desire a working knowledge of these methods of expression for purposes of calculations.

The units of ppm and epm are commonly combined in normal reporting of water analyses, and many different constituents are frequently reported on a common unit weight basis. For example, calcium (equivalent weight 20.0) is reported in terms of "calcium as $CaCO_3$" (equivalent weight 50.0). The test for calcium is calibrated in terms of $CaCO_3$, so the conversion factor 2.5 (50/20) is not needed. Hardness, magnesium, alkalinity, and free mineral acid are often reported in terms of $CaCO_3$; the value reported is the weight of $CaCO_3$ that is chemically equivalent to the amount of material present. Among these substances, ionic balances may be calculated. When constituents are of the same unit weight basis, they can be added or subtracted directly. For example, ppm total hardness as $CaCO_3$ minus ppm calcium as $CaCO_3$ equals ppm magnesium as $CaCO_3$. However, ppm magnesium as Mg^{2+} equals 12.2 (magnesium equivalent weight) divided by 50.0 ($CaCO_3$ equivalent weight) times the ppm magnesium as $CaCO_3$.

In every case, it is necessary to define the unit weight basis of the results—"ppm alkalinity as $CaCO_3$" or "ppm sulfate as SO_4^{2-}" or "ppm silica as SiO_2." Where the unit weight basis is different, calculations must be based on the use of chemical equations.

The following rules outline where epm can be used and where ppm must be used. In general, either may be used where an exact chemical formula is known. When such knowledge is lacking, ppm must be used.

■ The concentration of all dissolved salts of the individually determined ions must be in ppm.

■ Two or more ions of similar properties whose joint effect is measured by a single determination (e.g., total hardness, acidity, or alkalinity) may be reported in either ppm or epm.

■ The concentration of undissolved or suspended solids should be reported in ppm only.

■ The concentration of organic matter should be reported in ppm only.

■ The concentration of dissolved solids (by evaporation) should be expressed as ppm only.

■ Total dissolved solids by calculation may be expressed in either ppm or epm.

■ Concentration of individual gases dissolved in water should be reported in ppm. The total concentration of each gas when combined in water may be calculated to its respective ionic concentration in either ppm or epm.

CALCULATION OF TOTAL DISSOLVED SOLIDS BY EPM

Starting with a reasonably complete water analysis, total dissolved solids may be calculated as epm. In a complete water analysis, the negative ion epm should equal the positive ion epm. Where there is an excess of negative ion epm, the remaining positive ion epm is likely to be sodium or potassium (or both). For the sake of convenience, it is generally assumed to be sodium. Where there is an excess of positive epm, the remaining negative epm usually is assumed to be nitrate.

To calculate dissolved solids, convert the various constituents from ppm to epm and total the various cations (positively charged ions) and anions (negative ions). The cations should equal the anions. If not, add either sodium (plus) or nitrate (minus) ions to balance the columns. Convert each component ionic epm to ppm and total to obtain ppm dissolved solids. For example, to convert 150 ppm calcium as $CaCO_3$ to epm (Table 40-1) divide by 50 (the equivalent weight of calcium carbonate) and obtain 3.0 epm. To convert 96 ppm sulfate as SO_4^{2-} to epm, divide by 48 (the equivalent weight of sulfate) and obtain 2.0 epm. After balancing the cations and anions by adding sodium, convert to ionic ppm by multiplying the epm by the particular ionic equivalent of weight. For example, to convert 3.0 epm calcium to ppm calcium as Ca^{2+}, multiply by 20 (the equivalent weight of calcium) and obtain 60 ppm calcium as Ca^{2+}. To obtain the ppm dissolved solids, total the ppm of the individual ions.

Table 40-1. Calculation of dissolved solids.

		epm (+) Cations	epm (−) Anions	Ionic ppm
	ppm			
Calcium as $CaCO_3$	150 =	3.0		= 60 as Ca
Magnesium as $CaCO_3$	50 =	1.0		= 12 as Mg
Sulfate as SO_4	96 =		2.0	= 96 as SO_4
Chloride as Cl⁻	18 =		0.5	= 18 as Cl
Bicarbonate as $CaCO_3$	120 =		2.4	= 146 as HCO_3
Sodium (difference) as Na		0.9		= 21 as Na
Total dissolved solids		4.9	4.9	353

Table 40-2. Conversion table.

Formula		Number of Equivalents	Equivalent Weight
POSITIVE IONS			
Aluminum	Al^{3+}	3	9.0
Ammonium	NH_4^+	1	18.0
Calcium	Ca^{2+}	2	20.0
Copper	Cu^{2+}	2	31.8
Hydrogen	H^+	1	1.0
Ferrous Ion	Fe^{2+}	2	27.9
Ferric Ion	Fe^{3+}	3	18.6
Magnesium	Mg^{2+}	2	12.2
Manganese	Mn^{2+}	2	27.5
Potassium	K^+	1	39.1
Sodium	Na^+	1	23.0
NEGATIVE IONS			
Bicarbonate	HCO_3^-	1	61.0
Carbonate	CO_3^{2-}	2	30.0
Chloride	Cl^-	1	35.5
Fluoride	F^-	1	19.0
Iodide	I^-	1	126.9
Hydroxide	OH^-	1	17.0
Nitrate	NO_3^-	1	62.0
Phosphate (tribasic)	PO_4^{3-}	3	31.7
Phosphate (dibasic)	HPO_4^{2-}	2	48.0
Phosphate (monobasic)	$H_2PO_4^-$	1	97.0
Sulfate	SO_4^{2-}	2	48.0
Bisulfate	HSO_4^-	1	97.1
Sulfite	SO_3^{2-}	2	40.0
Bisulfite	HSO_3^-	1	81.1
Sulfide	S^{2-}	2	16.0
COMPOUNDS			
Alum	$Al_2(SO_4)_3 \cdot 18H_2O$	6	111.0
Aluminum Sulfate (anhydrous)	$Al_2(SO_4)_3$	6	57.0
Aluminum Hydroxide	$Al(OH)_3$	3	26.0
Aluminum Oxide	Al_2O_3	6	17.0
Ammonia	NH_3	1	17.0
Sodium Aluminate	$Na_2Al_2O_4$	6	27.3
Calcium Bicarbonate	$Ca(HCO_3)_2$	2	81.1
Calcium Carbonate	$CaCO_3$	2	50.0
Calcium Chloride	$CaCl_2$	2	55.5
Calcium Hydroxide	$Ca(OH)_2$	2	37.0
Calcium Oxide	CaO	2	28.0
Calcium Sulfate (anhydrous)	$CaSO_4$	2	68.1

Formula		Number of Equivalents	Equivalent Weight
Calcium Sulfate (gypsum)	$CaSO_4 \cdot 2H_2O$	2	86.1
Calcium Phosphate	$Ca_3(PO_4)_2$	6	51.7
Carbon Dioxide	CO_2	2	22.0
Chlorine	Cl_2	2	35.5
Ferrous Sulfate (anhydrous)	$FeSO_4$	2	76.0
Ferric Sulfate	$Fe_2(SO_4)_3$	6	66.6
Magnesium Oxide	MgO	2	20.2
Magnesium Bicarbonate	$Mg(HCO_3)_2$	2	73.2
Magnesium Carbonate	$MgCO_3$	2	42.2
Magnesium Chloride	$MgCl_2$	2	47.6
Magnesium Hydroxide	$Mg(OH)_2$	2	29.2
Magnesium Phosphate	$Mg_3(PO_4)_2$	6	43.8
Magnesium Sulfate (anhydrous)	$MgSO_4$	2	60.2
Magnesium Sulfate (Epsom Salts)	$MgSO_4 \cdot 7H_2O$	2	123.2
Manganese Hydroxide	$Mn(OH)_2$	2	44.5
Silica	SiO_2	2	30.0
Sodium Bicarbonate	$NaHCO_3$	1	84.0
Sodium Carbonate	Na_2CO_3	2	53.0
Sodium Chloride	$NaCl$	1	58.4
Sodium Hydroxide	$NaOH$	1	40.0
Sodium Nitrate	$NaNO_3$	1	85.0
Trisodium Phosphate	$Na_3PO_4 \cdot 12H_2O$	3	126.7
Trisodium Phosphate (anhydrous)	Na_3PO_4	3	54.7
Disodium Phosphate	$Na_2HPO_4 \cdot 12H_2O$	2	179.1
Disodium Phosphate (anhydrous)	Na_2HPO_4	2	71.0
Monosodium Phosphate	$NaH_2PO_4 \cdot H_2O$	1	138.0
Monosodium Phosphate (anhydrous)	NaH_2PO_4	1	120.0
Sodium Silicate	Na_2SiO_3	2	61.0
Sulfuric Acid	H_2SO_4	2	49.0
Sodium Metaphosphate	$NaPO_3$	1	102.0
Sodium Sulfate	Na_2SO_4	2	71.0
Sodium Sulfite	Na_2SO_3	2	63.0

ACIDITY

45

TITRATION METHOD

Free mineral acidity (FMA) is present in any sample with a pH of less than 4.3. Although FMA is not usually encountered in raw water, it is very significant when acidity is present because acids contribute to corrosiveness and influence certain chemical and biological processes. FMA is caused by the presence of acids, such as sulfuric, nitric, and hydrochloric, and does not include carbonic acid formed by the combination of carbon dioxide and water. The effluent from a hydrogen zeolite exchange unit contains FMA in proportion to the chloride, sulfate, and nitrate present in the raw water. In nonpolluted waters, the total acidity of the water is mainly caused by dissolved carbon dioxide and the pH is between 4.3 and 8.3. For measurement of total acidity, the sample is titrated to a pH of 8.3.

SUMMARY OF METHOD

For the determination of free mineral acidity, the sample is titrated with a standard solution of sodium hydroxide to a methyl purple end point (pH 4.3). For total acidity, the sample is titrated with the same solution to a phenolphthalein end point (pH 8.3).

APPARATUS REQUIRED

Apparatus	Betz Code
Buret, automatic, 25 mL	103
Casserole, porcelain, 210 mL	156
Cylinder, graduated, 100 mL	121
Stirring Rod, glass	114

CHEMICALS REQUIRED

Chemical	Betz Code
Methyl Purple Indicator	297
Phenolphthalein Indicator	212
Sodium Hydroxide, 1 N	255

INTERFERENCES

Colored or turbid samples may obscure the color change at the end point; a pH meter should be used in such situations.

Free residual chlorine in the sample may bleach the indicator. Eliminate this source of interference by treating the sample with 1 drop of 0.1 N sodium thiosulfate.

PROCEDURE FOR TEST

Free Acidity (FMA)

1. Measure a 100 mL sample of the water in the graduated cylinder and transfer to the casserole.

2. Add 2 drops of methyl purple indicator.

3. Titrate with 0.02 N sodium hydroxide (see Notes section for preparation information), adding from the buret drop-by-drop while stirring constantly, until 1 drop changes the color of the sample from purple to blue (pH 4.3).

4. Record the number of milliliters of 0.02 N sodium hydroxide required for the free acidity titration.

Total Acidity

1. Measure a 100 mL sample of the water in a graduated cylinder and transfer to the casserole.

2. Add 3 drops of phenolphthalein indicator.

3. Titrate with 0.02 N sodium hydroxide, adding from the buret drop-by-drop while stirring constantly, until 1 drop turns the sample color to faint pink (pH 8.3).

4. Record the number of milliliters of 0.02 N sodium hydroxide required for the total acidity titration.

CALCULATION OF RESULTS

For a 100 mL sample, the free acidity in milligrams per liter as $CaCO_3$ is equal to the milliliters of 0.02 N sodium hydroxide required for the methyl purple end point multiplied by 10. The total acidity is equal to the milliliters of 0.02 N sodium hydroxide required for the phenolphthalein end point multiplied by 10.

NOTES

The titrating solution, 0.02 N sodium hydroxide, has limited stability and must be freshly prepared once per week. To prepare, measure 20.0 mL of 1 N sodium hydroxide into a 1000 mL volumetric flask and dilute to 1 L with carbon dioxide-free distilled water. For maximum accuracy, the titrating solution should be standardized and protected in keeping with standard methods.

ALKALINITY

TITRATION METHOD

The presence of alkalinity in a water sample may be caused by any of a number of substances. However, in most natural waters the majority of the alkalinity comes from the presence of bicarbonate and carbonate ions.

Very high alkalinity values can be undesirable in an industrial water supply. For example, the presence of a high total M-alkalinity in boiler feedwater results in a high level of carbon dioxide in the steam and condensate. Carbon dioxide is often responsible for the corrosion of condensate systems. High boiler water alkalinity is also undesirable because the presence of high hydroxide ion concentration is a potential contributor to caustic metal embrittlement. High boiler alkalinity can also lead to steam contamination. On the other hand, the alkalinity of boiler water must be sufficiently high to protect the boiler metal from acidic corrosion and to ensure precipitation of any scale-forming salts present in the feedwater. In typical treatment approaches, both a minimum and a maximum level are set for boiler water alkalinity.

Phenolphthalein and methyl orange indicators, which change color at pH 8.3 and pH 4.3 respectively, provide standard reference points which are almost universally used to express alkalinity (see Chapter 41). Methyl purple is often used in place of methyl orange because the end point is easier to see. In order to determine hydroxide alkalinity, barium chloride may be added prior to the titration to precipitate the carbonate ion from solution. Subsequent titration to the phenolphthalein end point provides a direct measure of the hydroxide.

SUMMARY OF METHOD

The sample is titrated with a standard solution of sulfuric acid to the phenolphthalein end point (pH 8.3) and then to the methyl purple end point (4.3). The phenolphthalein end point is the point at which the hydroxide ion and half of the carbonate ion have been neutralized. At the methyl purple end point, the other half of the carbonate ion and any bicarbonate ion initially present in the sample have been neutralized.

APPARATUS REQUIRED

Apparatus	Betz Code
Buret, automatic, 25 mL	103
Casserole, porcelain, 210 mL	156
Cylinder, graduated, 50 mL	109
Stirring Rod, glass	114

CHEMICALS REQUIRED

Chemical	Betz Code
Methyl Purple Indicator	297
Phenolphthalein Indicator	212
Sulfuric Acid, 0.02 N	202

PROCEDURE FOR TEST

1. Use the graduated cylinder to transfer a clear 50 mL sample of water to the casserole (see Note 1).

2. Add 4–5 drops of phenolphthalein indicator.

3. Add 0.02 N sulfuric acid from the buret, drop-by-drop, to the sample in the casserole. Stir constantly until 1 drop removes the last trace of red and the sample becomes colorless.

4. Record the total milliliters of acid used to this point as the phenolphthalein (P) reading.

5. Add 4 drops of methyl purple indicator.

6. Add the sulfuric acid, drop-by-drop, until 1 drop changes the color from blue to purple.

7. Record the final buret reading as the total (M) reading.

CALCULATION OF RESULTS

For a 50 mL sample, phenolphthalein alkalinity in milligrams per liter as $CaCO_3$ is equal to the milliliters of 0.02 N sulfuric acid required for the phenolphthalein (P) end point multiplied by 20.

The total alkalinity is equal to the total (M) end point multiplied by 20 (see Note 2).

NOTES

1. If the water sample does not settle to clear, the total (M) reading can be determined on a filtered sample. However, the phenolphthalein (P) reading must always be determined on an unfiltered sample.

2. Example of Total (M) reading:

Titration to (P) reading	25.0 mL
Additional titration to (M) reading	+20.0 mL
Total (M) reading	45.0 mL

The total alkalinity is equal to:

$$20 \times 45.0 = 900 \text{ mg/L as } CaCO_3$$

ALUMINUM
(0.03–0.80 mg/L)

Aluminum is one of the most abundant metals in the earth's crust. It is present in natural waters because of the waters' contact with rock, soil, and clay. However, the amount of aluminum added to groundwater by these natural processes is small in comparison to that which can appear as residual from a water treatment clarifier.

Many clarifier pretreatments use some form of alum coagulation, and any upset or excess of alum results in high aluminum concentrations in the treated water. Therefore, the determination of aluminum in industrial waters is made principally to check clarifier performance.

SUMMARY OF METHOD

Aluminon indicator combines with aluminum in the sample to form a red-orange color. The intensity of color is proportional to the aluminum concentration. Ascorbate reagent powder is added to remove iron interference. The AluVer® 3 aluminum reagent, packaged in powder form, shows exceptional stability and is applicable for fresh water samples. This method was adapted from the APHA-AWWA *Standard Methods for the Examination of Water and Wastewater.*

APPARATUS REQUIRED

Apparatus	Betz Code
Cell, sample, 25 mL, 2.5 cm (2 required)	2601
Clippers, large	2635
Cylinder, graduated, mixing, 50 mL	2631
Pipet, graduated, plastic, 1 mL	371
Safety Bulb, rubber	1575
Spectrophotometer, DR 2000	2776

CHEMICALS REQUIRED

Chemical	Betz Code
Aluminum Standard Solution, 20 mg/L as Al	672
AluVer® 3 Reagent Powder Pillows	2014
Ascorbate Reagent Powder Pillows	2016
Bleaching 3 Reagent Powder Pillows	2015

INTERFERENCES

The following do not interfere up to the indicated concentrations:

Alkalinity	1000 mg/L as $CaCO_3$
Iron	20 mg/L
Phosphate	50 mg/L

Polyphosphate, at all levels, causes negative interference in the test. Before running the test,

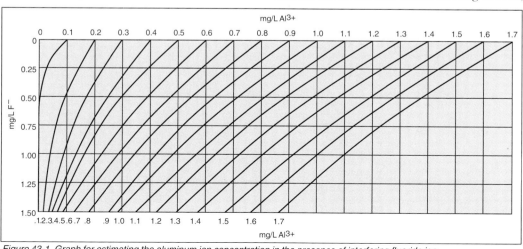

Figure 43-1. Graph for estimating the aluminum ion concentration in the presence of interfering fluoride ion.

polyphosphate must be removed or converted to orthophosphate by means of the Total Inorganic Phosphate test procedure.

Acidity interferes at greater than 300 mg/L as $CaCO_3$.

Calcium does not interfere.

Fluoride interferes at all levels by complexing with aluminum. When the fluoride concentration is known, the actual aluminum concentration can be determined according to Figure 43-1. To use the fluoride interference graph, select the vertical grid line along the top of the graph that represents the aluminum reading obtained in the calculation of results. Locate the point on the line where it intersects with the horizontal grid line appropriate for the fluoride present in the sample.

Extrapolate the true aluminum concentration from the curves on either side of the intersect point. For example, if the aluminum test result was 0.7 mg/L Al^{3+} and the fluoride present in the sample was 1 mg/L F^-, the point where the 0.7 grid line intersects with the 1 mg/L F^- grid line falls between the 1.2 and 1.3 mg/L Al^{3+} curves. In this case, the true aluminum content would be 1.27 mg/L.

PROCEDURE FOR TEST

1. Fill the 50 mL graduated mixing cylinder to the 50 mL mark with sample.

2. Add the contents of one ascorbate reagent powder pillow. Stopper and invert several times to dissolve powder.

3. Add the contents of one AluVer® 3 aluminum reagent powder pillow. Stopper and invert repeatedly for 1 min to dissolve.

4. Pour 25 mL of mixture into a 25 mL sample cell (the prepared sample).

5. Add the contents of one bleaching 3 reagent powder pillow to the remaining 25 mL in the mixing graduated cylinder. Stopper and shake vigorously for 30 sec.

6. Pour the remaining 25 mL of mixture into a second 25 mL sample cell (the blank).

7. Wait 15 min.

8. Place the blank into the cell holder. Close the light shield. Zero the photometer at 522 nm.

9. Immediately, place the prepared sample into the cell holder. Close the light shield.

10. Read the absorbance.

CALCULATION OF RESULTS

The aluminum in milligrams per liter is read off a calibration curve of absorbance versus concentration from 0.03 to 0.80 mg/L aluminum.

NOTES

AluVer® 3 is a registered trademark of Hach Co.

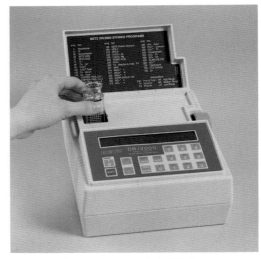

Figure 43-2. Spectrophotometer for field testing.

AMMONIA
(0.04–0.61 mg/L NH₃)

44

SALICYLATE METHOD

Nitrogen exists in many forms in aqueous systems: nitrate, nitrite, ammonia, and organic nitrogen. Organic nitrogen, sometimes referred to analytically as albuminoid or fixed nitrogen, occurs naturally in surface water from the decomposition of organic matter. Ammonia results from the breakdown of these nitrogenous organic compounds by microbiological organisms or boiler conditions.

Ammonia nitrogen is of importance in connection with the determination of the specific conductance of steam condensate. Ammonia possesses a high specific conductance, and when it is present in steam condensate, conductivity measurements must be corrected. Otherwise, results would indicate an erroneously high solids content. Where effective degassing equipment is used to remove ammonia and carbon dioxide prior to the measurement of steam purity, ammonia can be reduced to such a low value in the test sample that it may require no correction for its conductance effect.

Another significant effect of ammonia is its high corrosivity to copper and copper-bearing alloys. Ammonia dissolves these metals by forming the copper-ammonium ion. The ammonia may be removed through break point chlorination or cation exchange by hydrogen zeolite.

Measurement of organic nitrogen can also be used in tracing process chemical leaks. In these instances, organic nitrogen is converted to ammonia by Kjeldahl digestion. The ammonia generated is detected by Nesslerization. This digestion procedure should be performed in a qualified, fully equipped laboratory. Samples may be preserved by the addition of a strong acid to yield a pH of less than 2 in the sample.

SUMMARY OF METHOD

Ammonia compounds combine with chlorine to form monochloramine. Monochloramine reacts with salicylate to form 5-aminosalicylate. The 5-aminosalicylate is oxidized in the presence of a sodium nitroprusside catalyst to form a blue-colored compound. The blue color is masked by the yellow color from the excess reagent present to give a final green-colored solution. This method was adapted from *Clin. Chim. Acta.*, **14**, 403 (1966).

APPARATUS REQUIRED

Apparatus	Betz Code
Cell, sample, 25 mL, 2.5 cm (2 required)	2601
Cylinder, graduated, plastic, 25 mL (2 required)	495
Clippers, large	2635
Stopper, polyethylene	2601A
Spectrophotometer, DR 2000	2776

CHEMICALS REQUIRED

Chemical	Betz Code
Alkaline Cyanurate Powder Pillows	2348
Ammonia Salicylate Reagent Powder Pillows	2347
Deionized Water	243

INTERFERENCES

The following ions may interfere when present in concentrations exceeding those listed:

Calcium	1000 mg/L as $CaCO_3$
Magnesium	6000 mg/L as $CaCO_3$
Nitrite	40 mg/L as NO_2^-
Nitrate	440 mg/L as NO_3^-
Orthophosphate	300 mg/L as PO_4^{3-}
Sulfate	300 mg/L as SO_4^{2-}

Sulfide will intensify the color.

Iron will interfere with the test. Eliminate iron interference as follows:

1. Determine the amount of iron present in the sample following the Iron, Reactive procedure.

2. Add the same iron concentration to the deionized water sample in step 5.

The interference from iron in the sample will then be successfully blanked out in step 11.

Extremely acidic or alkaline samples should be adjusted to a pH of approximately 7. Use 1 N sodium hydroxide standard solution (Betz Code 255) for acidic samples or 1 N sulfuric acid standard solution (Betz Code 2066) for basic samples.

Less common interferences such as hydrazine and glycine result in intensified colors in the prepared sample. Turbidity and sample color will give erroneously high values. Samples with severe interferences require distillation. Albuminoid nitrogen samples also require distillation.

PROCEDURE FOR TEST

1. Pour 25 mL of sample into a 25 mL mixing graduated cylinder.

2. Pour 25 mL of deionized water into a second cylinder.

3. Add the contents of one ammonia salicylate reagent powder pillow to each cylinder. Stopper and shake to dissolve.

4. Wait 3 min and add the contents of one ammonia cyanurate powder pillow to each cylinder. Stopper and shake to dissolve.

5. Wait 15 min.

6. Use the deionized water sample (blank) to zero the photometer at 655 nm.

7. Place the reacted sample into the photometer and read the absorbance.

CALCULATION OF RESULTS

The ammonia concentration in milligrams per liter as NH_3 is read off a prepared calibration curve made from anhydrous ammonium chloride dissolved in deionized water in the range of 0.04–0.61 mg/L as NH_3.

BETZ POLYMER INHIBITORS
(10–80 mg/L as product)

TURBIDIMETRIC METHOD

Betz polymer inhibitors are principally used to control calcium phosphate scale in cooling water systems. Their use allows higher orthophosphate control limits than were possible prior to their development. This provides improved corrosion protection without the tendency for calcium phosphate deposition.

SUMMARY OF METHOD

A sample containing Betz polymer inhibitor becomes turbid when the Polymer Reagent II (a cationic surfactant) is added to the sample after buffering with Polymer Buffer II. The amount of turbidity is proportional to the concentration of Betz polymer inhibitor in the sample.

APPARATUS REQUIRED

Apparatus	Betz Code
Cell, sample, 25 mL, 2.5 cm	2601
Cylinder, graduated, mixing, 50 mL	2631
Filter Paper, prefolded, 12.5 cm*	2623
Flask, Erlenmeyer, 125 mL	126
Funnel, plastic, 3 in.	366
Pipet, graduated, plastic, 1 mL	371
Pipet, graduated, plastic, 5 mL	380
Safety Bulb, rubber	1575
Spectrophotometer, DR 2000	2776

*or an equivalent filter paper (2.5 μm retention).

CHEMICALS REQUIRED

Chemical	Betz Code
Polymer Buffer II	1254
Polymer Reagent II	1253

INTERFERENCES

Suspended solids and cationic polymers will combine with Betz polymer inhibitor to yield a lower test result. The presence of a polyacrylamide can yield a higher test result.

High hardness (greater than 1000 mg/L) in the sample can cause lower results. To minimize this interference, the sample can be diluted with deionized water (1:1 dilution). The sample must still read in the working range of the test (i.e., not less than 10 mg/L after the proper dilution).

PROCEDURE FOR TEST

1. Filter the sample.
2. Transfer 50 mL of the sample to a 125 mL Erlenmeyer flask.
3. Add 5 mL of Polymer Buffer II by pipet to the filtered sample and swirl for 10 sec.
4. Fill a spectrophotometer cell with a portion of the buffered sample and zero the spectrophotometer at 415 nm.
5. Pour the buffered sample back into the Erlenmeyer flask. Add 1 mL of Polymer Reagent II to the entire 50 mL of buffered sample.
6. Begin a timed 5-min reaction period.
7. After 4–4 1/2 min, fill a cell with the prepared sample.
8. Place the cell into the spectrophotometer and read the absorbance of the reacted sample at 5 min.

CALCULATION OF RESULTS

The Betz polymer inhibitor concentration is read on a calibration curve showing product in the range of 10–80 mg/L versus absorbance.

Figure 45-1. Liquid chromatograph for polymer and monomer analysis.

CALCIUM
(2–500 mg/L as CaCO₃)

TITRATION METHOD

The determination of calcium in a water analysis is closely allied to the determination of hardness, because hardness is caused primarily by the presence of calcium and magnesium.

Under certain circumstances, determination of calcium is made so as to subdivide the total hardness into calcium hardness and magnesium hardness. The total hardness minus the calcium hardness equals the magnesium hardness.

The determination of calcium usually is not made as a control test. However, this determination is necessary for making chemical calculations for treatment chemicals. For example, when using internal boiler water treatment of the phosphate type, calcium hardness is precipitated by the phosphate whereas magnesium hardness is precipitated as magnesium silicate or magnesium hydroxide. Therefore, in calculating the quantity of materials required, it is necessary to subdivide the hardness into calcium hardness and magnesium hardness.

In the treatment of cooling water, the determination of calcium can also play an important role. Langelier's Saturation Index is often used to predict the tendency of a water to form calcium carbonate scale. Use of the Langelier Index requires the determination of calcium.

SUMMARY OF METHOD

The test is based on the titration of calcium ions with EDTA in the presence of an indicator which changes from a salmon-pink to an orchid purple at the end point.

Because magnesium ions are precipitated as magnesium hydroxide prior to the titration, they are not measured. This precipitation occurs when the pH is elevated above 12.

APPARATUS REQUIRED

Apparatus	Betz Code
Buret, automatic, 25 mL	103
Casserole, porcelain, 210 mL	156
Cylinder, graduated, 50 mL	109
Measuring Dipper, brass	113
Pipet, 2 mL	146
Stirring Rod, glass	114
Safety Bulb, rubber	1575

CHEMICALS REQUIRED

Chemical	Betz Code
Calcium Indicator	293
Hardness Titrating Solution,	
1 mL = 1 mg CaCO₃	292
Sodium Hydroxide, 1.0 N	255

INTERFERENCES

Some transition and heavy metals (copper, zinc, cobalt, and nickel) can complex the indicator and prevent a sharp end point. Dilution of the sample with deionized water is recommended as a means of reducing or eliminating these as well as other interferences.

PROCEDURE FOR TEST

1. Measure 50 mL of sample and transfer to a casserole.
2. Add 2.0 mL of 1.0 N sodium hydroxide to the sample and stir.
3. Add 1 brass dipper (0.2 g) of calcium indicator and stir.
4. If calcium is present, the solution will turn salmon-pink.
5. Titrate with hardness titrating solution until the end point is signaled by an orchid purple color.

CALCULATION OF RESULTS

To calculate the calcium hardness in milligrams per liter as $CaCO_3$, first divide the milliliters of the water sample into 1000. Then multiply this number by the milliliters of the hardness titrating solution.

For a 50 mL sample,

$$\frac{1000 \text{ mL}}{50 \text{ mL}} = 20$$

$$\text{mg/L calcium as } CaCO_3 = \frac{\text{mL hardness}}{\text{titrating solution}} \times 20$$

CARBON DIOXIDE

Carbon dioxide exists in aqueous solutions as "free" carbon dioxide and the combined forms of carbonate and bicarbonate ions. "Free" carbon dioxide refers to the uncombined dissolved gas. Its presence in solution is due either to absorption from the atmosphere or to the decay of organic matter. Because the carbon dioxide content of the normal atmosphere is quite low (less than 0.04%), surface waters typically contain less than 10 mg/L free carbon dioxide. On the other hand, well waters may contain several hundred milligrams per liter of carbon dioxide gas.

Corrosion is the principal effect of dissolved carbon dioxide. The gas will dissolve in water, producing corrosive carbonic acid.

$$H_2O \ + \ CO_2 \ \rightleftharpoons \ H_2CO_3 \ \rightleftharpoons \ H^+ + HCO_3^-$$

The low pH resulting from this reaction also enhances the corrosive effects of oxygen.

In boiler systems, corrosion resulting from carbon dioxide is most often encountered in the condensate system. Because feedwater deaeration normally removes carbon dioxide from the boiler feedwater, the presence of the gas in condensate is typically due to carbonate and bicarbonate decomposition under boiler conditions. For an approximation of the carbon dioxide content of the steam, the feedwater total (M) alkalinity may be multiplied by a factor of 0.79, which takes into account 80% sodium carbonate decomposition. For example, feedwater with a total (M) alkalinity of 100 mg/L as calcium carbonate could be expected to generate a carbon dioxide level of 79 mg/L in the steam. Such a high carbon dioxide level would create a very corrosive condensate.

Free carbon dioxide is the term used to designate carbon dioxide gas dissolved in water as opposed to combined CO_2 present as bicarbonate and carbonate ions.

SUMMARY OF METHOD

Free carbon dioxide reacts with sodium carbonate to form sodium bicarbonate. At the equivalence point (pH 8.3), the phenolphthalein indicator turns pink. If free mineral acidity is present (e.g., due to sulfuric or hydrochloric acids), it is measured along with the free carbon dioxide. To correct for this, a second sample is titrated with the sodium carbonate to a pH of 4.3. Subtraction of the two titrations provides a corrected free carbon dioxide concentration.

APPARATUS REQUIRED

Apparatus	Betz Code
Buret, automatic, 25 mL	103
Cylinder, graduated, 100 mL	121
Stirring Rod, plastic	488
Tubing, sulfur-free rubber, 10 ft	130

CHEMICALS REQUIRED

Chemical	Betz Code
Phenolphthalein Indicator	212
Methyl Orange Indicator	211
Sodium Carbonate, N/22	232

Reliable results are obtained only when a fresh water sample that has been secured under careful sampling conditions is tested.

INTERFERENCES

Free mineral acids, aluminum, chromium, copper, amines, ammonia, borate, nitrite, sulfide, high total dissolved solids, and excess indicator can interfere. If color obscures the indicator end points, a pH meter should be used.

PROCEDURE FOR TEST

1. Obtain the sample by running the water through sulfur-free rubber tubing that discharges into the bottom of a 100 mL graduated

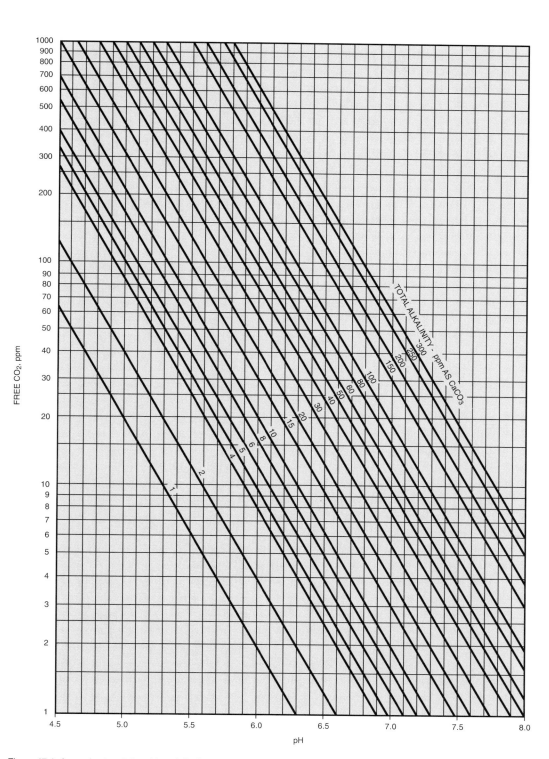

Figure 47-1. Approximate relationships of alkalinity, pH, and free carbon dioxide in raw waters.

cylinder. Allow the sample to overflow for a few minutes, then withdraw the tubing while the sample is still flowing. Flick the graduated cylinder to throw off any excess sample lodged above the 100 mL mark.

2. Immediately add 5 drops of phenolphthalein indicator. If the sample turns pink or red, no carbon dioxide or free mineral acidity is present and the pH is greater than 8.3.

3. If the sample remains colorless, the pH is less than 8.3. It must be titrated to 8.3 with N/22 sodium carbonate. Use a pH meter or titrate until the pink color stabilizes for 15 sec. Record the milliliters of N/22 sodium carbonate required. Multiply this number by 10. This is a value for CO_2 (in milligrams per liter) that has not been corrected for free mineral acidity (FMA).

4. To correct the CO_2 for FMA, obtain a second 100 mL sample and measure the initial pH using a pH meter or methyl orange indicator.

A salmon-pink color indicates a pH of less than 4.3 and a yellow color indicates a pH of greater than 4.3.

5. If the sample pH is below 4.3, titrate the sample with N/22 sodium carbonate to 4.3 to obtain a salmon-pink to yellow end point. Record the milliliters of N/22 sodium carbonate required.

CALCULATION OF RESULTS

To find the corrected CO_2 content in milligrams per liter (for a 100 mL sample), use the following formula:

$$mg/L\ CO_2 = 10\ (T_1 - 2T_2)$$

where

T_1 = milliliters required to titrate the first sample to a pH of 8.3 (the phenolphthalein end point from step 3)

T_2 = milliliters required to titrate the second sample to a pH of 4.3 (the methyl orange end point from step 6)

CHLORIDE
(5–1000 mg/L as Cl⁻)

TITRIMETRIC METHOD

The chlorides of calcium, magnesium, sodium, iron, and other cations normally found in water are extremely soluble. Because no precipitation occurs, the chlorides present in boiler and cooling waters are usually proportional to the cycles of concentration.

In industrial water conditioning, the principal application of the chloride test is in the control of blowdown from boiler and cooling systems. The chloride test may also be used to calculate the rate of blowdown, provided that the chloride concentration of the feedwater is of sufficient magnitude for accurate determination. A slight analytical error in determining feedwater chloride can cause an appreciable error in the calculated rate of blowdown. Chloride determination is also used to estimate the percent of makeup present in boiler feedwater when sodium zeolite softened makeup is used.

In the control of ion exchange softeners, the chloride test helps determine whether or not the regeneration salt has been washed out completely. It is also useful in detecting condenser leakage, especially aboard ships and in other places where seawater is used for condensing purposes. When leakage is occurring, the condensate shows increased chloride.

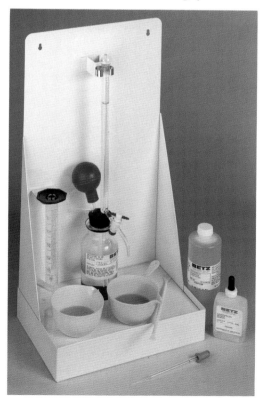

Figure 48-1. Automatic buret titration assembly.

SUMMARY OF METHOD

In the titrimetric method for chloride, a standard solution of silver nitrate is used as the titrant and chromate ion is used as the indicator. The chloride in the sample reacts with the silver ions in the titrant to form insoluble silver chloride. When all of the chloride has been precipitated out of solution as silver chloride, the excess silver ions react with the chromate ions to form insoluble silver chromate, which is a brick-red color.

APPARATUS REQUIRED

Apparatus	Betz Code
Buret, automatic, 25 mL	103T
Casserole, porcelain, 210 mL	156
Cylinder, graduated, 50 mL	109
Flask, Erlenmeyer, 250 mL	111
Funnel (3 in. dia.), glass, fluted	112
Stirring Rod, plastic	488

CHEMICALS REQUIRED

Chemical	Betz Code
Hydrogen Peroxide, 3%	241
Phenolphthalein Indicator	212
Potassium Chromate Indicator	213
Silver Nitrate (1 mL = 1 mg Cl⁻)	207

PROCEDURE FOR TEST

1. If the phenolphthalein (P) or methyl orange (M) alkalinity has been determined, the same sample may be used directly for the chloride test without further neutralization. If not, the sample must be properly neutralized with sulfuric acid.

2. Take a 50 mL sample of the water to be tested and place the sample into the 250 mL Erlenmeyer flask. Add 4–5 drops of phenolphthalein indicator and add sufficient 0.02 N sulfuric acid to neutralize to the colorless side of phenolphthalein (pH < 8.3).

3. Add 5 drops of potassium chromate indicator to the neutralized sample.

4. Add silver nitrate (1 mL = 1 mg Cl⁻) from the buret, stirring constantly, until 1 drop produces a rust-like color that does not disappear.

CALCULATION OF RESULTS

In a 50 mL sample, chloride in milligrams per liter as Cl⁻ is equal to the milliliters of silver nitrate required to reach the end point, minus a blank of 0.2 mL, multiplied by 20.

NOTES

1. Where brackish waters are encountered, a stronger solution of silver nitrate should be used (1 mL = 5 mg Cl⁻, Betz Code 226). No blank is necessary with this stronger solution. With a 50 mL sample and (1 mL = 5 mg Cl⁻) silver nitrate, chloride in milligrams per liter as Cl⁻ is equal to the milliliters of silver nitrate required, multiplied by 100.

2. When the sulfite content of the water tested exceeds 10 mg/L, the sulfite should be oxidized to prevent interference with the chloride test. After neutralization (step 2), add 2 mL of 3% hydrogen peroxide and stir well. Then continue with steps 3 and 4 of the procedure.

3. When the EDTA content of the water tested exceeds 50 mg/L, the EDTA should be oxidized to prevent interference with the chloride test. After completing step 2, add 25 mL of 3% hydrogen peroxide and stir well. Insert the stem of the fluted funnel into the flask; place the flask on a hot plate and boil the solution until 1–2 mL remains. Add 50 mL of deionized water. Continue the test procedure, repeating step 2.

CHLORINE, FREE
(0.1–2.0 mg/L)

DPD-2 METHOD

The test for chlorine is a test for the hydrolysis products of chlorine gas in water. Chlorine gas is soluble in water, forming hypochlorite ion and hypochlorous acid. It is in the latter form that chlorine exerts its disinfecting and oxidizing properties.

In industrial water conditioning, chlorination is used primarily for the control of slime and algae, although chlorine may also be used to assist in coagulation, taste, odor, color, and iron removal.

There is a marked difference in the germ control properties of a chlorine residual, depending on the form in which chlorine exists in the treated water. The free available chlorine residual is defined as that portion of the total residual chlorine that will react chemically and biologically as hypochlorous acid. It is in this form that chlorine exerts the most potent antibacterial effect.

Figure 49-1. Scintillation counter for radioisotope analysis.

SUMMARY OF METHOD

Chlorine in the sample as hypochlorous acid or hypochlorite ion (free chlorine or free available chlorine) reacts immediately with DPD-2 (*N,N*-diethyl-*p*-phenylenediamine) indicator to form a red color proportional in strength to the chlorine concentration. This method was adapted from the APHA-AWWA *Standard Methods for the Examination of Water and Wastewater.*

APPARATUS REQUIRED

Apparatus	Betz Code
Cell, sample, 25 mL, 2.5 cm (2 required)	2601
Clippers, large	2635
Pipet, graduated, plastic, 1 mL	371
Safety Bulb, rubber	1575
Spectrophotometer, DR 2000	2776

CHEMICALS REQUIRED

Chemical	Betz Code
Chlorine Standard Solution Ampules	1219
DPD-2 Free Chlorine Reagent Powder Pillows	2317
Free Chlorine Arsenite Reagent	2095

INTERFERENCES

In samples containing more than 250 mg/L alkalinity or 150 mg/L acidity as $CaCO_3$, colors may not develop fully, or they may fade instantly. Neutralize these samples to a pH of 6–7 with 1 N H_2SO_4 or 1 N NaOH. Determine the amount required on a separate 25 mL sample; then add the same amount to the sample to be tested.

Samples containing monochloramine will show a gradual drift to higher chlorine readings. When read within 30 sec of reagent addition, 3.0

mg/L monochloramine will cause an increase of at least 0.1 mg/L in the free chlorine reading.

Bromine, iodine, ozone, and oxidized forms of manganese and chromium also react as chlorine.

PROCEDURE FOR TEST

1. Fill a clean 25 mL sample cell to the mark with 25 mL of the water sample. Place the cell into the cell holder of the spectrophotometer and zero the meter at 530 nm.

2. Fill a second sample cell with the original water sample. Add the contents of one DPD-2 free chlorine reagent powder pillow and swirl several times to mix.

3. Place the prepared sample into the cell holder and record the absorbance of the sample 30 sec after adding the powder pillow.

4. Fill another sample cell with 25 mL of the original water sample. This will be the correction sample. Add 1 mL of free chlorine arsenite reagent to the correction sample and swirl to mix. Wait 1 min. This will eliminate all of the free chlorine present in the sample.

5. Add the contents of one DPD-2 free chlorine reagent powder pillow to the correction sample and swirl to mix.

6. Place the correction sample into the cell holder and record the absorbance after 30 sec. This reading is related only to interferences.

CALCULATION OF RESULTS

In order to determine the concentration of free chlorine in the sample, first subtract the absorbance obtained in step 6 from the absorbance in step 3. The free chlorine concentration is read off a calibration curve of absorbance versus free chlorine concentration (0.1–2.0 mg/L).

CHLORINE, TOTAL
(0.1–2.0 mg/L)

DPD-2 METHOD

Combined available chlorine residual is defined as that portion of the total residual chlorine which will react chemically and biologically as chloramine or organic chloramine. In these forms, chlorine is a relatively mild antibacterial and oxidizing agent.

SUMMARY OF METHOD

Chlorine can be present in water as free available chlorine and as combined available chlorine. Both forms can exist in the same water and be determined together as the total available chlorine. Free chlorine is present as hypochlorous acid and/or hypochlorite ion. Combined chlorine exists as monochloramine, dichloramine, nitrogen trichloride, and other chloro derivatives. In the presence of iodide ion, the chloramines and free

chlorine in the sample react with DPD-2 (*N,N*-diethyl-*p*-phenylenediamine) indicator to form a red color proportional in strength to the total chlorine concentration. To determine the concentration of combined chlorine, run a free chlorine test. Subtract the results of the free chlorine test from the results of the total chlorine test to obtain combined chlorine. This method was adapted from the APHA-AWWA *Standard Methods for the Examination of Water and Wastewater.*

APPARATUS REQUIRED

Apparatus	Betz Code
Cell, sample, 25 mL, 2.5 cm (2 required)	2601
Clippers, large	2635
Pipet, graduated, plastic, 1 mL	371
Safety Bulb, rubber	1575
Spectrophotometer, DR 2000	2776

CHEMICALS REQUIRED

Chemical	Betz Code
Chlorine Standard Solution Ampules	1219
Dechlorinating Reagent Powder Pillows	2043
DPD-2 Total Chlorine Reagent Powder Pillows	2337

INTERFERENCES

In samples containing more than 300 mg/L alkalinity or 150 mg/L acidity as $CaCO_3$, colors may not develop fully, or they may fade instantly. Neutralize these samples to a pH of 6–7 with 1 N H_2SO_4 or 1 N NaOH. Determine the amount required on a separate 25 mL sample; then add the same amount to the sample to be tested.

Bromine, iodine, ozone, and oxidized forms of manganese and chromium also react as chlorine.

PROCEDURE FOR TEST

1. Fill two clean 25 mL sample cells to the mark with 25 mL of the water sample (blank and test samples).

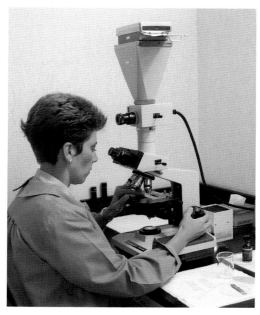

Figure 50-1. Microbiological microscopy.

2. Add the contents of one dechlorinating reagent powder pillow to one of the sample cells and swirl several times to mix. Allow to stand for 1 min. This is the blank.

3. Add the contents of one DPD-2 total chlorine reagent powder pillow to both the blank and the sample. Swirl to mix. Allow to stand for 3 min.

4. Zero the spectrophotometer at 530 nm using the blank. Immediately measure the absorbance of the sample.

CALCULATION OF RESULTS

In order to determine the concentration of total chlorine in the sample, a reading is made of the concentration of total chlorine present based on the absorbance in step 4. The reading is made from a calibration curve of absorbance versus total chlorine concentration (0.1–2.0 mg/L).

CHROMATE, HEXAVALENT

(10–100 mg/L as CrO_4)

TITRATION METHOD

Chromates were among the most widely used chemicals for the control of corrosion in industrial cooling water systems. However, only the chromate or hexavalent form of chromium salts functions as a corrosion inhibitor; trivalent chromic salts do not inhibit corrosion.

Chromates are anodic inhibitors. As such, they can intensify pitting if used in concentrations insufficient for complete prevention of corrosive attack. On the other hand, high chloride levels hinder the development of a protective film by chromate, and brines must be treated with high levels of chromate to inhibit corrosion.

The combination of phosphate and chromate, in the proper ratio and under controlled pH conditions, permits the use of relatively low chromate concentrations. Inhibition of pitting as well as general corrosion is secured at relatively low treatment concentrations.

For optimum chromate performance, it is essential that the chromate concentration be controlled within prescribed limits. The most commonly used chromate salts are the sodium and potassium salts of either chromate or dichromate. In basic solution, hexavalent chromium exists in the chromate form. In acid solution, the equilibrium shifts to the dichromate ion. Control tests are designed to measure the hexavalent chromium only and do not respond to the trivalent chromium salts.

SUMMARY OF METHOD

This method is based on the oxidizing properties of hexavalent chromium to liberate free iodine from an acidic potassium iodide solution. The liberated iodine is titrated with sodium thiosulfate solution in the presence of a starch indicator. The

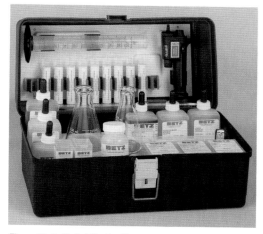

Figure 51-1. Digital titration kit.

disappearance of the blue color serves to indicate the end point.

APPARATUS REQUIRED

Apparatus	Betz Code
Buret, automatic, 25 mL	103
Cylinder, graduated, 10 mL	157
Cylinder, graduated, 100 mL	121
Flask, Erlenmeyer, 250 mL	111
Measuring Dipper, brass	113
Measuring Dipper, plastic	110

CHEMICALS REQUIRED

Chemical	Betz Code
Iodide Crystals	234
Sodium Thiosulfate, N/10	235
Starfamic Indicator	233
Sulfamic Acid	254

PROCEDURE FOR TEST

1. Transfer 50 mL of the sample to the 250 mL Erlenmeyer flask using the graduated cylinder.

2. Using the plastic dipper, add 2 scoops of sulfamic acid. Swirl the flask until all of the solid has dissolved. Allow the sample to stand for 2 min.

3. Using the brass dipper, add 1 scoop of iodide crystals. Swirl the flask until all of the solid has dissolved. Allow the sample to stand for 2 min.

4. Titrate the sample with 0.01 N sodium thiosulfate (see Note 1) until the yellow-brown iodine color almost disappears.

5. Add 1 brass dipper of starfamic indicator and swirl to dissolve.

6. Continue titrating with 0.01 N sodium thiosulfate until the blue color disappears.

CALCULATION OF RESULTS

In a 50 mL sample, the chromate concentration in milligrams per liter as CrO_4^{2-} is equal to the number of milliliters of 0.01 N sodium thiosulfate required for the titration multiplied by 7.74.

NOTES

1. The 0.01 N sodium thiosulfate used in this procedure is not stable, due to adsorption of oxygen from the air and bacterial action. Therefore, this reagent must be freshly prepared or standardized at least every 2 weeks. To 90 mL of deionized water that has been boiled and cooled to room temperature, add 10 mL of sodium thiosulfate, 0.1 N. Mix thoroughly and store in a tightly stoppered bottle. Avoid unnecessary exposure of the solution to air.

2. At times, all of the starfamic indicator will not dissolve, which may create a slight haze in the sample.

3. After the blue color disappears at the end point, disregard any reappearance of this color.

4. In some cases it may be desirable to use a sample larger than 50 mL. When this modification is required, the quantities of reagents and multiplication factor should be changed according to Table 51-1.

Table 51-1. Chromate sample size variations.

Sample Size	Sulfamic Acid	Iodide Crystals	Starfamic Indicator	Factor
100 mL	4 dippers	2 dippers	1 dipper	3.87
200 mL	8 dippers	4 dippers	1 dipper	1.94

CONDUCTANCE, SPECIFIC

The specific conductance of water is a measure of its ability to conduct an electrical current. This property is of no consequence in itself with respect to water treatment. However, from a control standpoint, the conductivity test is important as a direct measure of the total ionizable (dissolved) solids in the water. The conductivity test provides a measurement of steam purity as well as a simple control for boiler water solids. Conductivity may also be used for blowdown control in recirculating cooling water systems.

Specific conductance is inversely proportional to electrical resistance. Pure water is highly resistant to the passage of an electric current; therefore, it has a low specific conductance. If the water contains ions, it becomes a better conductor of electricity and the specific conductance is increased. Inorganic compounds, such as sodium chloride and sodium sulfate, dissociate into positive and negative ions, which conduct electricity in proportion to the number of ions present. The conductivity test, therefore, is not specific for any ion, but is a measure of the total ionic concentration.

The basic unit of electrical resistance is the ohm; because electrical conductivity is the reciprocal of resistance, the term "mho" (ohm spelled backwards) was chosen as the basic unit of conductivity. The conductivity test measures small amounts of electrical conductance, so the instrument is usually calibrated in micromhos (1 micromho = 1 millionth of a mho). Often, a conductivity meter is calibrated to read directly in milligrams per liter dissolved solids (or some specific ion or compound). This is not recommended for control testing because the conversion factor from micromhos of specific conductance to milligrams per liter varies slightly with different waters. The conductivity test provides an accurate, simple method of blowdown control. However, in alkaline samples the hydroxide ion has a disproportionately high conductance.

SUMMARY OF METHOD

Ionizable solids give water the ability to conduct electrical current. A multirange conductivity meter measures this electrical conductance. Scales are calibrated in micromhos (μmho) or the equivalent SI system microsiemens (μS). In the case of boiler water, hydroxide ions may be neutralized with a boric acid solution prior to measurement of neutralized conductivity.

APPARATUS REQUIRED

Apparatus	Betz Code
Betz/Myron L Digital Conductivity Meter, Model DC4	1592
OR	
Betz/Myron L Analog Conductivity Meter Model EP10	767
Flask, Erlenmeyer, 50 mL	389

Figure 52-1. Field conductivity meter.

CHEMICALS REQUIRED

Chemical	Betz Code
Conductivity Standard Solution, 4600 µmho	245
Conductivity Standard Solution, 575 µmho	905
Conductivity Standard Solution, 47 µmho	1594
Conductivity Standard Solution, 23 µmho	1593
Boric Acid Reagent for Neutralized Conductivity	1295

PROCEDURE FOR TEST

1. Pour approximately 45 mL of settled or filtered boiler water cooled to 50–150 °F (10–66 °C) in a rinsed 50 mL Erlenmeyer flask. Add boric acid reagent for neutralized conductivity, 1 drop at a time, until the initial pink color of the sample disappears. If no color develops after the first drop, no neutralization is necessary.

2. Rinse the permanent cell-cup that is built in-to the meter at least three times with the neutralized sample. Do not dip the meter into the water.

3. Fill the cell-cup with water to $1/4$ in. above the disc electrode. Set the range selector so that the maximum on-scale reading can be displayed.

4. Press the black button. If necessary, turn the range switch to obtain an on-scale reading.

5. Read the conductivity. In readings obtained with the EP10 conductivity meter, the neutralized conductivity of the sample equals the meter reading multiplied by the multiplier setting of the range switch. With the DC4 conductivity meter, the neutralized conductivity is equal to the digital display.

NOTES

1. In the measurement of the conductivity of neutralized boiler water, 1 µmho is approximately equal to 0.9 mg/L solids.

2. Each instrument is checked and calibrated by the manufacturer before shipment. However, a periodic check is highly recommended. Use the Betz conductivity standards that fall within the range of the samples to be measured.

3. Do not pour standard conductivity solutions back into the bottle; discard them immediately after use. If upon standardization a marked discrepancy between ranges occurs, return the instrument to the manufacturer for repair.

CONDUCTANCE OF HIGH-PURITY WATER

Samples such as makeup water, feedwater, condensed steam, and return condensate should not be neutralized prior to determining conductivity.

High-purity waters are easily contaminated and should be measured immediately. Never test a sample that has been standing. The average value for converting micromhos of conductance to dissolved solids in condensate is approximately 0.5–0.6 mg/L. In calculating the approximate solids content, gases such as ammonia and carbon dioxide must not be present because they will affect conductance.

CONDUCTANCE CORRECTIONS FOR AMMONIA AND CARBON DIOXIDE

The purpose of the conductivity determination is to obtain a measure of the solids present in the condensed steam sample as an indication of the degree of carryover of boiler water solids. To interpret the conductivity of a condensed steam or condensate sample, it is necessary to correct for the effect of any gases present. After the concentration of ammonia or carbon dioxide has been determined, the correction factor applying to that concentration is subtracted from the observed conductance.

Figure 52-2 shows the conductivity correction for ammonia in the range of 0–1.0 mg/L and carbon dioxide in the range of 0–30 mg/L. In the following example, a condensed steam sample contains 10 mg/L carbon dioxide and 0.6 mg/L ammonia as N; the conductivity correction is 6.0 µmho. This correction must be subtracted from the observed conductivity of the sample to obtain the conductivity due to the solids present.

$$\text{Specific conductance} = \frac{8.0 \ \mu\text{mho}}{\text{ammonia as N}} = 0.6 \ \text{mg/L}$$

$$\text{Free carbon dioxide as } CO_2 = 10 \ \text{mg/L}$$

$$\text{Correction due to gases} = 6.0 \ \mu\text{mho}$$

$$\text{Corrected specific conductance: } (8.0-6.0) = 2.0 \ \mu\text{mho}$$

$$\text{Solids content of sample:}$$
$$2.0 \ \mu\text{mho} \times 0.5 \ (\text{average factor}) = 1.0 \ \text{mg/L solids}$$

Figure 52-3 is useful in determining the conductivity correction where no free carbon dioxide is present. The conductivity correction for ammonia is a function of both the ammonia concentration and the pH of the sample. For example, in a sample containing 0.7 mg/L ammonia as N, with a pH of 8.0, the correction value would be 6.0 mho specific conductance.

Figure 52-4 covers a higher range of ammonia concentration. The same procedure is used to obtain the conductivity correction.

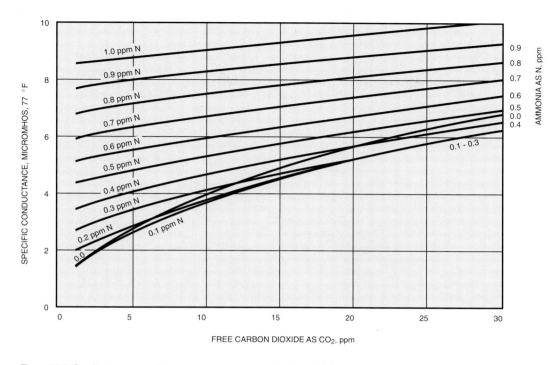

Figure 52-2. Conductance correction curves for ammonia and carbon dioxide.

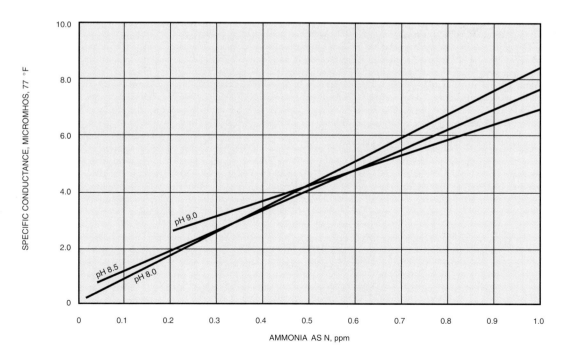

Figure 52-3. Conductance correction curves for ammonia and pH (low range).

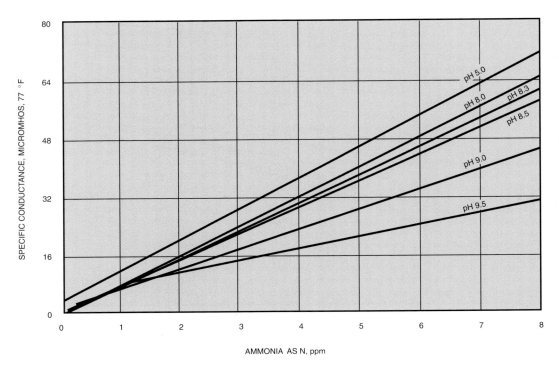

Figure 52-4. Conductance correction curves for ammonia and pH (high range).

COPPER
(0.1–5 mg/L)

BICINCHONINATE METHOD

The main reason for determining copper in industrial water is that its presence may be an indication of corrosion somewhere in the system. The chemical factors that cause corrosion of the copper alloys include low pH, ammonia, cyanide, hydrogen sulfide, and sulfur compounds. Although the corrosion rates of copper alloys in industrial waters are usually not high enough to produce significant degradation of the copper components themselves, dissolved copper can cause other operational problems. For example, dissolved copper tends to plate out on steel components in a system; this may create cell action that can cause serious pitting of the steel in both cooling and boiler systems. For these reasons, the copper test is designed to give accurate low-level measurements.

SUMMARY OF METHOD

CuVer® 2 reagent contains a bicinchoninate indicator and a reducing reagent to convert any Cu^{2+} to Cu^+. The Cu^+ then reacts with the bicinchoninate to form a purple-colored complex.

Complexed copper may then be measured with a photometer. High levels of hardness do not interfere with this determination.

APPARATUS REQUIRED

Apparatus	Betz Code
Cell, sample, 25 mL, 2.5 cm (2 required)	2601
Clippers, large	2635
Pipet, graduated, plastic, 1 mL	371
Safety Bulb, rubber	1575
Spectrophotometer, DR 2000	2776

CHEMICALS REQUIRED

Chemical	Betz Code
Copper, Standard Solution, 100 mg/L as Cu	334
CuVer® 2 Copper Reagent Powder Pillows	2346

Figure 53-1. X-ray diffraction spectrophotometer used in laboratory analysis of water-formed deposits.

INTERFERENCES

If the sample is extremely acidic (pH 2 or less), a precipitate may form. Add 8 N potassium hydroxide (Betz Code 2093) drop-by-drop with swirling to dissolve the turbidity prior to measurement. If the turbidity remains and the solution turns black, silver interference is probable. To remove silver interference, add 10 drops of saturated potassium chloride solution to 75 mL of sample. Filter through a fine or highly retentive filter and use the filtered sample in the procedure.

Cyanide interferences prevent sufficient color development but can be overcome by the addition of 0.5 mL of formaldehyde. Wait 4 min before reading.

PROCEDURE FOR TEST

1. Fill a sample cell with 25 mL of sample.

2. Add the contents of one CuVer® 2 copper reagent powder pillow to the sample cell (prepared sample). Swirl to mix.

3. Allow at least 2 min for full color development but not more than 30 min before measuring the absorbance of the sample.

4. After 2 min, fill a second sample cell with 25 mL of the original sample (blank). Place the blank into the cell holder and zero the photometer at 560 nm.

5. Remove the blank from the cell holder and place the prepared sample into the holder. Record the absorbance of the sample.

CALCULATION OF RESULTS

Determine the copper concentration of the sample using a calibration curve of absorbance versus concentration of copper standards in the range of 0.1–5 mg/L Cu.

COPPER
(5–210 µg/L)

PORPHYRIN METHOD

The Porphyrin Method is very sensitive to trace amounts of free copper. The method is free from most interferences and does not require any sample extraction or preconcentration. Interferences from other metals are eliminated by the copper masking reagent. The porphyrin indicator forms an intense, yellow-colored complex with any free copper present in the sample. This method was adapted from Ishii and Koh, *Bunseki Kagaku*, **28**, 473 (1979).

APPARATUS REQUIRED

Apparatus	Betz Code
Cell, sample, 25 mL, 2.5 cm (2 required)	2601
Clippers, large	2635
Pipet, graduated, plastic, 1 mL	371
Pipet, graduated, plastic, 10 mL	395
Safety Bulb, rubber	1575
Spectrophotometer, DR 2000	2776

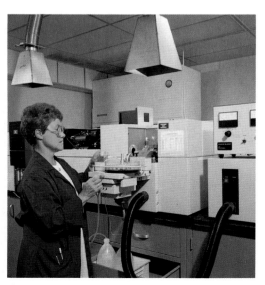

Figure 54-1. Atomic absorption spectrometer for trace metal analysis.

CHEMICALS REQUIRED

Chemical	Betz Code
Copper Masking Reagent Powder Pillows	2345
Copper Standard Solution, 100 mg/L as Cu	334
Porphyrin 1 Reagent Powder Pillows	2343
Porphyrin 2 Reagent Powder Pillows	2344

INTERFERENCES

The following may interfere when present in concentrations exceeding those listed:

Aluminum	60 mg/L
Cadmium	10 mg/L
Calcium	1,500 mg/L
Chloride	90,000 mg/L
Chromium	110 mg/L
Cobalt	100 mg/L
Fluoride	30,000 mg/L
Iron	6 mg/L
Lead	3 mg/L
Magnesium	10,000 mg/L
Manganese	140 mg/L
Mercury	3 mg/L
Molybdenum	11 mg/L
Nickel	60 mg/L
Potassium	60,000 mg/L
Sodium	90,000 mg/L
Zinc	9 mg/L

Chelating agents such as EDTA interfere at all levels unless a vigorous digestion is performed. Highly buffered samples or extreme sample pH may exceed the buffering capacity of the reagents and require sample pretreatment.

PROCEDURE FOR TEST

1. Fill two sample cells with 25 mL of sample.

2. Add the contents of one copper masking reagent powder pillow to one of the sample cells (the blank). Swirl to dissolve.

3. Add the contents of one Porphyrin 1 reagent powder pillow to each sample cell. Swirl to dissolve.

4. Add the contents of one Porphyrin 2 reagent powder pillow to each sample cell. Swirl to dissolve.

5. Allow samples to stand for 3 min.

6. Place the blank into the cell holder. Close the light shield and zero the photometer at 425 nm. Remove the blank from the cell holder.

7. Place the prepared sample into the cell holder. Close the light shield. Record the absorbance.

CALCULATION OF RESULTS

Determine the copper concentration, in micrograms per liter as Cu, using a calibration curve of the absorbance of various copper standards versus copper concentration in the range of 5–210 µg/L.

EDTA CHELANT DEMAND

TITRATION METHOD

EDTA (ethylenediaminetetraacetic acid) is a weak organic acid. When the sodium salt is applied to boiler feedwater at a relatively high pH, it is hydrolyzed to an organic anion able to form stable, soluble organometallic complexes. Metals that normally form troublesome deposits on boiler surfaces, such as calcium, magnesium, iron, and copper, remain in solution when chelated.

Although the complexes formed with EDTA are more stable than those formed with NTA (nitrilotriacetic acid), their effectiveness in a boiler treatment program is limited by the concentration of competing anions. Precipitation occurs whenever the potential of the anion to react exceeds the potential of the chelant to form complexes. The complex forming potential, also named the stability constant of the chelant, has been determined for most chelants. Except for a few special cases, EDTA is the most suitable chelating substance used for boiler water treatment to date.

Proper treatment level is necessary in attaining a successful boiler chelant program. One of the most effective ways to determine a proper EDTA feed rate is to measure the chelant demand of the feedwater.

SUMMARY OF METHOD

A standard solution of EDTA is added to a feedwater sample. Excess chelant standard is back-titrated with a standard metal solution in the presence of an indicator. This requires careful control of the pH to ensure chelation of the hardness and heavy metal cations. The amount of chelant standard needed to complex all of the cations of interest is calculated in milligrams per liter of feedwater chelant demand.

APPARATUS REQUIRED

Apparatus	Betz Code
Bottle, 2 oz., screw cap with dropper	165
Buret, micro, automatic, 2 mL	159
Buret, micro, automatic, 3 mL	176
Casserole, porcelain, 340 mL	168
Cylinder, graduated, plastic, 100 mL	490
Pipet, 1 mL (2 required)	371
Pipet, 5 mL	380
Stirring Rod, plastic	488

CHEMICALS REQUIRED

Chemical	Betz Code
Chelant Blank Reagent	369
Chelant Indicator 3	645
Chelant Liquid Buffer	646
EDTA Standard Solution	641
EDTA Titrating Solution	640
Sodium Fluoride, 3%	673
Sodium Hydroxide 1.0 N	255
Sulfuric Acid 1.0 N	283

PROCEDURE FOR TEST

Determination of Blank

1. Measure 200 mL of chelant blank reagent in a graduated cylinder and transfer to the casserole.

2. With a 1.0 mL pipet, add 1.0 mL of sodium fluoride solution and, with a clean pipet, add 1.0 mL of 1.0 N sulfuric acid. Stir.

3. Slowly add 2.0 mL of EDTA standard solution from the 2.0 mL microburet. Using the 5 mL pipet, add 1.5 mL of 1.0 N sodium hydroxide and stir.

4. With separate, clean droppers, add 1.0 mL of chelant liquid buffer and 12 drops of chelant indicator 3. Stir. A blue color appears.

5. Stirring constantly, titrate with EDTA titrating solution from the 3.0 mL microburet to the first reddish-purple end point that persists for 30 sec.

6. Record the number of milliliters of EDTA titrating solution as the feedwater blank value. Retain the titrated solution for use as a reference color when titrating the feedwater sample.

Determination of Feedwater EDTA Demand

1. Measure 100 mL of feedwater, cooled to 21–26 °C (70–80 °F), in the graduated cylinder and transfer to the second casserole.

2. Add 100 mL of chelant blank reagent and stir.

3. With the 1.0 mL pipet, add 1.0 mL of sodium fluoride solution and, with a clean pipet, add 1.0 mL of 1.0 N sulfuric acid in that order. Stir, and allow to stand for 30 sec.

4. Slowly add 2.0 mL of EDTA standard solution from the 2.0 mL microburet. Stir and allow to stand 1 min.

5. Using the 5 mL pipet, add 1.5 mL of 1.0 N sodium hydroxide and stir.

6. After 1 min, with separate, clean droppers, add 1.0 mL of chelant liquid buffer and 12 drops of chelant indicator 3. Stir. A blue color appears.

7. Stirring constantly, titrate with EDTA titrating solution from the 3.0 mL microburet to the first reddish-purple end point that persists for 30 sec. This end point should match the color obtained for the blank test.

CALCULATION OF RESULTS

Feedwater Demand

For a 100 mL feedwater sample, the EDTA feedwater demand in milligrams per liter is equal to the milliliters of titrating solution required for the blank test, minus the milliliters of titrating solution for the feedwater demand procedure, multiplied by 50.

HARDNESS

TITRATION METHOD (LIQUID BUFFER)

Hardness in water is caused by the presence of calcium and magnesium salts. Hardness of natural water varies considerably, depending on the source of the water. Some natural waters are very soft, whereas others (e.g., areas which have limestone formations) may have several hundred milligrams per liter of hardness.

In industry, high hardness is the principal source of scale formation in a variety of processes involving heat transfer, such as boiler feedwater heaters, feed lines, and economizers. Unless they are properly treated, boilers are also heavily scaled by precipitation of calcium and magnesium salts. In cooling water systems, scale develops wherever water circulates, including heat exchange equipment, engine jackets, and piping.

Hardness can be removed by means of a variety of external treatments, including lime–soda softening, ion exchange softening, hot lime/hot ion exchange, and other combinations of processes. The adverse effects of hardness are mitigated in boiler internal water conditioning by treatment with phosphate salts or chelants and polymers. Organic phosphates can be used to control hardness deposition in cooling systems. In addition, pH control can be used to retain the hardness in solution and prevent its precipitation as scale. In planning scale control programs, it is essential that the hardness levels of both untreated and treated waters be monitored.

SUMMARY OF METHOD

Titrimetric hardness is based on the determination of the total calcium and magnesium content of a sample by titration with a standard solution of chelant. Universal hardness buffer is a blend of chemicals that adjusts the sample to an alkaline pH and complexes potentially interfering ions.

Hardness indicator is red when complexed with calcium and magnesium. As the chelant titrant is added, it preferentially complexes the calcium and magnesium, the indicator regains a free form, and the sample turns blue.

APPARATUS REQUIRED

Hardness Range:
5–500 mg/L as $CaCO_3$

Apparatus	Betz Code
Buret, automatic, Teflon® stopcock	103T
Casserole, porcelain, 210 mL	156
Cylinder, graduated, 50 mL	109
Stirring Rod, plastic	488

Microhardness Range:
0.05–6 mg/L as $CaCO_3$

Apparatus	Betz Code
Buret, micro, automatic, 3 mL, Teflon© stopcock	176
Casserole, porcelain, 210 mL	156
Cylinder, graduated, 100 mL, plastic	490
Pipet, graduated, 1 mL, plastic	371
Stirring Rod, plastic	488

CHEMICALS REQUIRED

Hardness Range:
5–500 mg/L as $CaCO_3$

Chemical	Betz Code
Universal Hardness Buffer Solution	1566
Hardness Indicator	290A
Hardness Titrating Solution, 1 mL = 1 mg $CaCO_3$	292Q

Microhardness Range:
0.05–6 mg/L as $CaCO_3$

Chemical	Betz Code
Universal Hardness Buffer Solution	1566
Hardness Indicator	290A
Microhardness Titrant	834

INTERFERENCES

Hardness Range:
5–500 mg/L as CaCO₃

Large concentrations of iron, copper, or zinc may register as hardness.

Microhardness Range:
0.05–6 mg/L as CaCO₃

If polyphosphate is present in the feedwater sample, the sample must be pretreated with polyphosphate removal reagent (Betz Code 835) for accurate results.

PROCEDURE FOR TEST

Hardness Range:
5–500 mg/L as CaCO₃

1. Measure 50 mL of sample using a 50 mL graduated cylinder and transfer it to the casserole (see Note 1).

2. Add 1 mL of universal hardness buffer solution. Stir to mix.

3. Add one brass dipper of hardness indicator and stir to mix. If hardness is present, the sample will turn red.

4. While stirring, slowly add hardness titrating solution from the buret. Titrate the sample to a blue end point (see Note 2).

5. Record the total milliliters of titrant required to reach the end point.

Microhardness Range:
0.05–6 mg/L as CaCO₃

1. Transfer 100 mL of the sample to the 210 mL casserole using a 100 mL plastic graduated cylinder.

2. Transfer 1.0 mL of universal hardness buffer solution to the casserole.

3. Add 1 brass dipper of hardness indicator to the casserole and stir to mix. If hardness is present, the sample will turn pink.

4. Titrate to a blue end point using the microhardness titrant (see Note 3).

CALCULATION OF RESULTS

Hardness Range:
5–500 mg/L as CaCO₃

For the 50 mL sample, the hardness in milligrams per liter as $CaCO_3$ is equal to the milliliters of titrant employed multiplied by 20.

$$\text{Hardness (mg/L as CaCO}_3) = \text{Titrating solution (mL)} \times 20$$

Microhardness Range:
0.05–6 mg/L as CaCO₃

For a 100 mL sample, the hardness in milligrams per liter as $CaCO_3$ is equal to the milliliters of titrating solution required for the sample multiplied by 2.

$$\text{Hardness (mg/L as CaCO}_3) = \text{Microhardness titrant used (mL)} \times 2$$

NOTES

Hardness Range:
5–500 mg/L as CaCO₃

1. The pH of the water sample should be in the 7–10 pH range. If it is necessary to adjust the pH of the sample to within this range, add dilute ammonium hydroxide or dilute hydrochloric acid prior to the test.

2. When approaching the end point, the sample will begin to show some blue coloration, but a red cast can also be observed. The end point is the final discharge of the red cast. Further addition of titrant will not produce a color change.

3. In this procedure, the titrant must be added slowly because the end point is sharp and rapid. For routine hardness determination, it is recommended that 50 mL of sample be measured, but only 40–45 mL be dispensed into the casserole at the start of the test. Add the hardness buffer and indicator as described in the procedure and rapidly titrate to the end point. Then add the remainder of the sample to the casserole. The hardness present in the remainder of the sample will turn the contents of the casserole to red. Continue titrating slowly until reaching the final end point. Record the total milliliters of titrant required to reach the final end point.

Microhardness Range:
0.05–6 mg/L as CaCO₃

1. All equipment should be kept clean. Rinse repeatedly with chelant blank reagent (Betz Code 369) or an equivalent high-purity water.

2. Use a lamp during the titration to discern the subtle color changes associated with this test.

3. When approaching the end point, the sample begins to show some blue coloration, but a red cast can also be observed. The end point is the final discharge of the red cast; further addition of titrant will not produce a color change.

HYDRAZINE
(10–500 µg/L)

p-DIMETHYLAMINOBENZALDEHYDE METHOD

Hydrazine is a chemical oxygen scavenger that reacts with dissolved oxygen to produce nitrogen and water. These reaction products do not add solids to the boiler water and are not corrosive to ferrous metals. For these reasons, hydrazine was often used instead of sodium sulfite as an oxygen scavenger in high-pressure boilers.

Theoretically, 1.0 mg/L of hydrazine is required to react with 1.0 mg/L of dissolved oxygen. Unreacted hydrazine can decompose to form ammonia (which may attack copper or copper-containing alloys).

SUMMARY OF METHOD

Hydrazine in the sample reacts with the *p*-dimethylaminobenzaldehyde from the HydraVer® reagent to form a yellow color which is proportional to the hydrazine concentration. This method was adapted from the ASTM *Manual of Industrial Water*, D1385-78, 376 (1979).

APPARATUS REQUIRED

Apparatus	Betz Code
Cell, sample, 25 mL, 2.5 cm (2 required)	2601
Cylinder, graduated, plastic, 25 mL	495
Flask, Erlenmeyer, glass, 50 mL (2 required)	389
Pipet, graduated, plastic, 1 mL	371
Pipet, graduated, plastic, 10 mL	395
Safety Bulb, rubber	1575
Spectrophotometer, DR 2000	2776
Thermometer, armored, –10 to 100 °C	2636

CHEMICALS REQUIRED

Chemical	Betz Code
Deionized Water	243
HydraVer® 2 Hydrazine Reagent	2047
Hydrazine Standard Solution, 100 mg/L as N_2H_4	325

INTERFERENCES

For highly colored or turbid samples, a blank must first be prepared. Oxidize the hydrazine

Figure 57-1. Microprocessor-controlled gas chromatograph used for organic analysis.

with a 1:1 mixture of deionized water and household bleach. Add 1 drop of the mixture to 25 mL of sample in a mixing graduated cylinder and invert to mix. Use this solution in step 1, in place of deionized water, to prepare the blank. There are no other common interferences.

PROCEDURE FOR TEST

1. Pour 25 mL of deionized water into an Erlenmeyer flask.

2. Pour 25 mL of sample into a second Erlenmeyer flask.

3. Add 1 mL of HydraVer 2 hydrazine reagent to each Erlenmeyer flask. Swirl to mix. A yellow color will develop if hydrazine is present (Note: HydraVer 2 hydrazine reagent will cause a faint yellow color to appear in the blank).

4. A 12-min reaction period will begin.

5. Turn on the spectrophotometer and set the wavelength at 455 nm. Use the deionized water reagent blank to set the absorbance at zero.

6. After 12 min, transfer the reacted sample to a photometer cell and read the absorbance.

7. Convert the absorbance to concentration using a previously prepared calibration curve.

IRON, REACTIVE
(0.1–3 mg/L)

FERROVER® METHOD

Iron exists in two states in solution—the ferrous and ferric forms. The iron test described here measures the soluble and most precipitated forms of iron; both ferrous and ferric iron are detected.

Iron may be present in the water sample because it was present in the water supply, or it may be due to system corrosion. Therefore, interpretation of iron results must address the factors of both scale and corrosion. For example, iron deposits can form in a cooling system as the result of a scale problem (deposition of the natural iron in the water) or a corrosion problem (wasting away of the metal surfaces). Thus, iron determinations can reveal information on either deposition or corrosion taking place within a system.

SUMMARY OF METHOD

FerroVer® iron reagent reacts with all soluble iron and most insoluble forms of iron in the sample to produce soluble ferrous iron. This reacts with the 1,10-phenanthroline indicator in the reagent to form an orange color in proportion to the iron concentration. This method was adapted from *Standard Methods for the Examination of Water*

Figure 58-1. Inductively coupled plasma spectrometer used for multielement analysis in water.

and Wastewater, Federal Register, **45** (126), 433459 (June 27, 1980).

APPARATUS REQUIRED

Apparatus	Betz Code
Cell, sample 25 mL, 2.5 cm (2 required)	2601
Clippers, large	2635
Cylinder, graduated, plastic, 25 mL	495
Flask, Erlenmeyer, glass, 50 mL	389
Pipet, plastic, 1 mL, graduated	371
Safety Bulb, rubber	1575
Spectrophotometer, DR 2000	2776

CHEMICALS REQUIRED

Chemical	Betz Code
FerroVer® Iron Reagent Powder Pillow	2032
Iron, Standard Solution, 50 mg/L as Fe	523
RoVer® Rust Remover	2058

INTERFERENCES

The following will not interfere below the levels shown:

Chloride	185,000 mg/L
Calcium	10,000 mg/L as $CaCO_3$
Magnesium	100,000 mg/L as $CaCO_3$

A large excess of iron will inhibit color development. A diluted sample should be tested if there is any doubt about the validity of a result.

FerroVer iron reagent powder pillows contain a masking agent that eliminates potential interference from copper.

Samples containing some forms of iron oxide require digestion in an acid and peroxide solution.

Samples containing large amounts of sulfide should be treated in a fume hood or well ventilated area as follows:

1. Add 5 mL of 50% hydrochloric acid (Betz Code 247) to 100 mL of sample and boil for 20 min.

2. Adjust the pH to between 2.5 and 5.0 with 5 N sodium hydroxide (Betz Code 2003).

3. Readjust the volume to 100 mL with deionized water.

4. Analyze as described below.

   Highly buffered samples or extreme sample pH levels may exceed the buffering capacity of the reagent and require sample pretreatment.

   FerroVer iron reagent powder pillows are stable for up to 12 months, depending on storage and handling conditions. A cool, dry atmosphere is recommended for long shelf life. To check the stability, add the contents of one powder pillow to about 25 mL of water containing visual rust. If the characteristic orange color does not develop, the reagent has deteriorated beyond use and should be discarded.

PROCEDURE FOR TEST

1. Using a graduated cylinder, transfer 25 mL of sample to an Erlenmeyer flask. *Note*: If the sample is turbid, add a 0.2 g scoop of RoVer® rust remover (Betz Code 2058) to 25 mL portions of the sample and blank, and wait 5 min. Swirl to dissolve. Use the RoVer-containing blank to zero the instrument.

2. Add the contents of one FerroVer® iron reagent powder pillow and swirl to mix. An orange color will develop if iron is present. Allow at least 3 min for the color to develop fully, but do not wait more than 30 min.

3. Turn on the spectrophotometer and set the wavelength at 510 nm.

4. Transfer a portion of unreacted sample to a clean sample cell. Place the cell in the spectrophotometer and set the absorbance reading to zero.

5. Transfer the reacted sample (from step 2) to a clean sample cell, place it in the spectrophotometer, and read the absorbance.

6. Use a previously prepared calibration curve to convert the absorbance reading to a concentration value.

MOLYBDATE, MEDIUM RANGE
(0.5–20 mg/L as MoO_4^{2-})

CATECHOL METHOD

Molybdate salts are commonly used as corrosion inhibitors in cooling water systems.

SUMMARY OF METHOD

Molybdate Reagent I reacts with soluble molybdenum and forms a molybdate–catechol complex. The degree of yellow-brown color formed is proportional to the amount of molybdate in the sample. Molybdate Reagent II is added to the sample to eliminate a strong negative interference caused by iron.

APPARATUS REQUIRED

Apparatus	Betz Code
Cell, sample, 25 mL, 2.5 cm (2 required)	2601
Cylinder, graduated, 50 mL	2631
Filter Paper, prefolded, 12.5 cm*	2623
Funnel, plastic, 3 in.	366
Pipet, graduated, plastic, 10 mL (2 required)	305
Safety Bulb, rubber	1575
Spectrophotometer, DR 2000	2776
Stopper, polyethylene (2 required)	2601

*An equivalent filter paper may be substituted

CHEMICALS REQUIRED

Chemical	Betz Code
Deionized Water	243
Molybdate Reagent I	1261
Molybdate Reagent II	1262
Molybdate Standard Solution, 10 mg/L Mo (16.67 mg/L as MoO_4^{2-})	2038

INTERFERENCES

Oxidizers, such as chromate, will interfere. To eliminate the interference from the sample, add one ascorbate reagent powder pillow (Betz Code 2016).

Hydroquinone does not interfere in concentrations of up to 100 mg/L.

Chloride, phosphate, sulfate, and silica do not interfere.

PROCEDURE FOR TEST

1. Filter approximately 50 mL of sample.

2. Transfer 25 mL of filtered sample into the sample cell; this is the blank.

3. Pipet 15 mL of filtered sample to a second cell.

4. Add 1 plastic scoop of Molybdate II reagent to the 15 mL of filtered sample. This is a cationic resin intended to eliminate iron interference. Stopper the cell and agitate for 30 sec.

5. Pipet 10 mL of Molybdate Reagent I to the 15 mL of filtered sample and resin. Swirl to mix.

6. Zero the spectrophotometer at 415 nm with the filtered blank sample.

7. Place the reacted sample into the spectrophotometer and measure the absorbance.

CALCULATION OF RESULTS

From a prepared molybdate calibration curve, read the mg/L MoO_4^{2-} based on the absorbance found. Concentration versus absorbance is plotted from 0.5 to 20.0 mg/L.

NITRATE
(0.5–4.5 and 10–130 mg/L as NO$_3^-$)

CADMIUM REDUCTION METHOD

Nitrate ion is present in natural waters in relatively small quantities. Nitrogen compounds may be introduced by sewage with subsequent bacteriological oxidation to nitrate. High nitrate concentrations in drinking water appear to cause methemoglobinemia in infants (blue babies). Federal regulations have established a nitrate limit in drinking water of 10 mg/L as nitrogen. Typically, such high levels are confined to rural wells subject to surface influence.

In industrial water conditioning, the primary significance of nitrate is in relation to caustic embrittlement. Research has shown that the maintenance of certain sodium nitrate–sodium hydroxide ratios in boiler water can inhibit intercrystalline cracking (caustic metal embrittlement). Because of the widespread use of sodium nitrate to control this condition, the determination of the nitrate ion in boiler water is extremely important.

The nitrate concentration necessary to overcome embrittling tendencies in a boiler varies depending on the boiler water alkalinity and the boiler pressure. Once the proper ratio has been established, the sodium nitrate feed can be controlled on this basis. Often, the natural nitrate content of the raw water provides the necessary ratio. However, because the nitrate content of natural waters is subject to seasonal variation, it may be necessary to run nitrate control on the boiler water to ensure adequate protection.

SUMMARY OF METHOD

Cadmium metal reduces nitrates present in the sample to nitrite. The nitrite ion reacts in an acidic medium with sulfanilic acid to form an intermediate diazonium salt. This salt couples to gentisic acid to form an amber-colored product. This chemistry can be used to accommodate a high and low range. Adjusting the wavelength of the spectrophotometer and zeroing with a reagent

blank establishes the lower detection range. This method was adapted from the APHA-AWWA *Standard Methods for the Examination of Water and Wastewater.*

APPARATUS REQUIRED

Apparatus	Betz Code
Cell, sample, 25 mL, 2.5 cm (2 required)	2601
Clippers, large	2635
Spectrophotometer, DR 2000	2776
Stopper, polyethylene	2601A

CHEMICALS REQUIRED

Chemical	Betz Code
Deionized Water	243
Nitrate Standard, 100 mg/L as NO$_3^-$	279
NitraVer® 5 Nitrate Reagent Powder Pillows	2336

INTERFERENCES

Strong oxidizing and reducing substances adversely affect the cadmium reducing agent and also the nitrate–nitrite balance. High concentrations of nitrite may reduce the accuracy of nitrate determinations. Ferric iron causes high results and must be absent. Chloride concentrations above 100 mg/L cause low results. The test may be used at high chloride levels (e.g., seawater), but a calibration must be performed using stan-

Figure 60-1. Ion chromatograph used for analysis of anions and cations in industrial water systems.

dards prepared with the same chloride concentration. Highly buffered samples or extreme sample pH ranges may exceed the buffering capacity of the reagents and require sample pretreatment.

PROCEDURE FOR TEST
Nitrate Range: 0.5–4.5 mg/L as NO_3^-

1. Fill a sample cell with 25 mL of sample (the prepared sample).

2. Fill a second sample cell with 25 mL of deionized water (the blank).

3. Add the contents of one NitraVer® 5 nitrate reagent powder pillow to each cell. Stopper the cells.

4. Shake both cells for 1 min (see Note 1).

5. Allow the samples to develop for 5 min (see Note 2).

6. Place the blank into the spectrophotometer and zero the meter at 400 nm.

7. Place the prepared sample into the photometer and read the absorbance.

Nitrate Range: 10–130 mg/L as NO_3^-

1. Fill a sample cell with 25 mL of sample.

2. Add the contents of one NitraVer 5 nitrate reagent powder pillow to the cell (the prepared sample). Stopper and shake vigorously for 1 min (see Note 1).

3. Allow the sample to develop for 5 min (see Note 2).

4. Fill a second sample cell with 25 mL of sample (the blank).

5. Place the blank into the photometer and zero the meter at 500 nm.

6. Place the prepared sample into the photometer and read the absorbance.

CALCULATION OF RESULTS
Nitrate Range: 0.5–4.5 mg/L as NO_3^-

The nitrate in milligrams per liter is read off a calibration curve of absorbance versus concentration from 0.5 to 4.5 mg/L of nitrate.

Nitrate Range: 10–130 mg/L as NO_3^-

The nitrate in milligrams per liter is read off a calibration curve of absorbance versus concentration from 10 to 130 mg/L of nitrate.

NOTES

1. Shaking time and technique influence color development. When preparing the calibration curve, test each standard in triplicate.

2. A deposit of unoxidized metal will remain after the NitraVer 5 nitrate reagent powder dissolves and will have no effect on test results.

3. If the sample has a nitrate concentration in between the ranges of these tests, dilute the sample with an equal volume of deionized water and repeat the 0.5–4.5 mg/L test. Calculate the nitrate concentration by multiplying the test result by 2.

NITRITE
(2–500 mg/L as NO$_2^-$)

DIRECT TITRATION METHOD

Sodium nitrite is a good steel corrosion inhibitor which can be used in open recirculating cooling systems. However, due to problems with nitrifying bacteria that oxidize nitrite to nitrate, it is most often applied in closed systems. Although nitrites are often regarded as reducing agents, their oxidizing properties actually inhibit corrosion, by promoting the formation of a passive film on the steel surface.

To establish a protective iron oxide film, it is sometimes necessary to feed nitrite initially at a greater concentration than normally required. The feed rate can be reduced to normal once the protective film has been established. Oxidizing and reducing agents destroy nitrite. These decomposition problems and the problem of bacteria-induced oxidation make it difficult to maintain effective nitrite concentrations in open circulating systems without using excessive feed rates. However, where nitrite concentrations can be continuously maintained at an effective level, especially in closed systems, satisfactory protection of ferrous metals is obtained.

SUMMARY OF METHOD

This test is based on the determination of the nitrite content of a sample by titration in an acid medium with a standard oxidizing agent, potassium permanganate. Nitrite is quantitatively oxidized to nitrate. The permanganate titrant is purple, but forms a colorless product in the reaction with nitrite. Thus, the persistence of a definite pink color for 1 min, which indicates excess permanganate, is taken as the end point.

APPARATUS REQUIRED

Apparatus	Betz Code
Cylinder, graduated, 10 mL	858
Buret, automatic, 25 mL	103T
Casserole, porcelain, 210 mL	156
Flask, volumetric, glass, 500 mL	934
Pipet, volumetric, glass, 50 mL	118
Safety Bulb, rubber	1575
Stirring Rod, plastic	488

CHEMICALS REQUIRED

Chemical	Betz Code
Potassium Permanganate, 0.1 N	263
Sulfuric Acid, 5%	256

INTERFERENCES

This method is affected by any oxidizable substances in the sample such as organic matter, sulfides, hydrogen sulfide, and mercaptans. If present, these substances will interfere by reacting with the titrant, potassium permanganate, yielding an erroneously high nitrite concentration.

PROCEDURE FOR TEST

1. Measure a 10 mL sample into the graduated cylinder and cool to room temperature. Transfer to the casserole.

2. Measure 3 mL of the 5% sulfuric acid into the 10 mL graduate and add to the sample casserole.

3. Add the standard 0.01 N potassium permanganate (see Note) from the buret, drop-by-drop, to the sample in the casserole, stirring constantly, until a pink color starts to persist.

4. At this point, add the permanganate *1 mL* at a time with constant stirring.

5. Continue to add 1 mL increments of the permanganate until a definite pink color persists for 1 min.

CALCULATION OF RESULTS

Formula:

$$\text{mg/L as NO}_2^- = \text{mL of 0.01 N KMnO}_4 \times 23$$

NOTE

The titrating solution, 0.01 N potassium permanganate, has limited stability and must be freshly prepared each month. To prepare, pipet 50 mL of 0.1 N potassium permanganate into a 500 mL volumetric flask and dilute to 500 mL with deionized water. Protect the titrating solution from light by storing it in amber-colored bottles.

OXYGEN, DISSOLVED

When water has been in contact with a gas, a small amount of the gas can be found in the water. For example, the atmosphere consists mainly of nitrogen and oxygen, and both are somewhat soluble in water. Nitrogen is an inert gas and of minor importance in water treatment, but the control of dissolved oxygen in industrial waters is essential for efficient operation.

The term "dissolved oxygen" refers only to the oxygen gas which is dissolved in the water and in no way refers to combined oxygen present in the water molecule.

In a given water sample, the solubility of the gas is determined not only by the pressure of the gas and temperature of the water but also by the composition of the solution. For example, mineral constituents in the water affect the solubility, and distilled water absorbs more oxygen than waters containing dissolved solids.

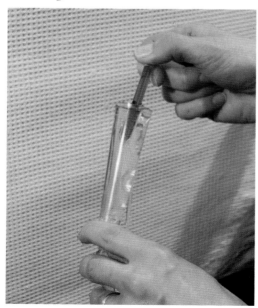

Figure 62-1. Chemet used to measure dissolved oxygen in a flowing water sample.

In addition to the natural presence of oxygen in water supplies, aeration is frequently used to remove other gases, such as carbon dioxide and hydrogen sulfide. Efficient aeration results in saturation of the water with dissolved oxygen.

Dissolved oxygen is undesirable in industrial water because of its corrosive effect on iron and copper alloys. Increased temperatures and low pH values intensify corrosion due to oxygen.

Removal of oxygen from boiler feedwater is especially important because of the accelerated effect oxygen has at the elevated temperatures of boiler systems. In boiler systems, corrosion may occur in feed lines, heaters, economizers, boilers, and condensate systems. The tests described below focus on measurement or detection of the very low oxygen levels typically associated with boiler operation.

SUMMARY OF METHOD

Several methods can be used to measure dissolved oxygen. These methods differ in their sensitivity to various levels of dissolved oxygen, their reliability, and their degree of difficulty.

Classically, milligram per liter levels of dissolved oxygen have been measured by the Winkler iodometric method or by membrane-detecting electrometric instruments.

The need to measure to low microgram per liter levels of dissolved oxygen accurately has necessitated special sampling techniques, colorimetric methods, and newly designed membrane-detecting on-line instruments.

In many industrial plants, the measurement of microgram per liter levels of dissolved oxygen is accomplished with on-line analyzers. In support of or in place of these on-line analyzers, the use of reagent containing evacuated ampules, such as the Chemetrics Chemet® and Vacu-vial®, provides a convenient and reasonably accurate test.

Dissolved oxygen Chemets® contain a measured quantity of Rhodazine D dye sufficient to react with a level of oxygen slightly higher than the range of the test. In its reduced form the dye is a pale yellow, but contact with oxygen changes the color to a deep rose. In the test, the tip of the Chemet is snapped in a sample stream flowing at 500–1000 mL/min in a conical-shaped funnel. The reagent mixes with the sample to form the deep rose color.

APPARATUS REQUIRED

Apparatus	Betz Code
Comparator, dissolved oxygen, 0–40 µg/L	1244
Comparator, dissolved oxygen, 0–100 µg/L	1275

CHEMICALS REQUIRED

Chemical	Betz Code
Dissolved Oxygen Chemets®, 0–40 mg/L and 0–100 µg/L as O_2	1276

INTERFERENCES

Sulfite and hydrazine are known not to interfere in the test. However, organic oxygen scavengers may give positive interferences. If these are used, it is necessary to employ an on-line membrane analyzer to verify the ampule values.

PROCEDURE FOR TEST

1. Purge the sampling tube and funnel free of air bubbles with a flowing water sample.

2. Insert a Chemet into the funnel and press to snap the tip.

3. Remove the Chemet, cover the tip with a finger cot, and mix the contents by rotating the vial.

4. Compare to either the 0–40 mg/L or 0–100 µg/L comparators to obtain the level of dissolved oxygen.

NOTES

Chemet and Vacu-vial are registered trademarks of Chemetrics, Inc.

pH MEASUREMENT

The pH of most natural waters is in the range of 6.0–8.0. However, more acidic conditions (and therefore a lower pH) may be present where the water contains high concentrations of free carbon dioxide or oxides of nitrogen or sulfur, or where the water originates from acid mine drainage. A pH above 8.0 is seldom encountered except when the water has been polluted by alkaline wastes or treated by chemical processes.

Because pH is a measure of relative acidity or alkalinity of water, it is a most important factor influencing scale-forming and corrosion tendencies. Low-pH water increases the potential for corrosion of equipment. High-pH water can cause precipitation of calcium carbonate on the surfaces of pipelines, heat exchangers, condensers, and other components. The Langelier Index is a measure of the calcium carbonate scaling tendencies of water. In the calculation of this index, determination of pH is required. In clarification processes, pH is an important determination. The control of coagulation is optimized through pH control because every coagulant possesses an optimum pH range in which it provides the most efficient performance.

Control of pH is an important factor in such processes as iron removal, recarbonation, and acid treatment. The pH value of a boiler water is usually adjusted to control both deposition and corrosion.

In summary, pH is a very important factor in industrial water treatment, and any analysis of a water for scale-forming or corrosive tendencies is incomplete without the determination of pH.

SUMMARY OF METHOD

Solution pH is a measure of the active acid or base in a solution. pH is defined as the negative logarithm of the hydrogen ion concentration. Neutral water has a pH of 7. Values below 7 approaching 0 are increasingly acid while values above 7 are increasingly basic.

Figure 63-1. pH meter and combination electrode.

A pH measurement system has four parts: a pH sensing electrode, electronic circuits that translate the signal into a form the user can read, a reference electrode, and the sample being measured. pH is conveniently measured with a pH meter (electronics and readout device) and an electrode which combines the reference and sensing portions in one housing.

APPARATUS REQUIRED

Apparatus	Betz Code
pH Electrode, Combination Calomel, BNC	397A
pH Meter, pH/mV/temp, digital	1285
Adapter, AC	1291

CHEMICALS REQUIRED

Chemical	Betz Code
Buffer, pH 4.0	623
Buffer, pH 7.0	624
Buffer, pH 10.0	625

INTERFERENCES

To compensate for temperature, calibrate the meter at the given temperature of the sample or

...cause pH actually changes with tem-
buffe in many instances it is important to
per the pH at the desired temperature.
...are the least likely part of the pH measuring
...n to cause a problem. In a majority of cases,
...lectrode system is the cause of incorrect or
...gish response.

...ROCEDURE FOR TEST

1. Fully extend the hinged cover, open the
battery compartment, and connect the 9 V
battery. Close the compartment and adjust the
cover to the desired viewing angle.

2. Press the "on" switch down and set the "mode"
switch to the "pH" position. This will activate
the liquid crystal display (LCD).

3. Remove the wetting cap from the electrode tip
and connect the cable to the BNC input.
Condition the new electrode in buffer 7.0 until
the LCD is stable (30 sec). Give a vigorous
stirring action to the electrode each time it is
placed in a new solution.

4. Let the buffers and samples come to ambient
temperature. Adjust the temperature control
to read this ambient value. This electrode
operates in the range of 5–60 °C (41–140 °F).

5. Adjust the calibration control so that the LCD
shows the value of buffer 7.0 in Table 63-1.

6. Rinse the electrode with buffer 4.0, blot dry,
and immerse in buffer 4.0. When the LCD is

Table 63-1. Effects of temperature on buffer solutions.

Temperature (°C)	Buffer 7.00	Buffer 4.00
5	7.08	3.99
10	7.06	3.99
20	7.01	3.99
25	7.00	4.00
30	6.99	4.01
40	6.98	4.03
50	6.97	4.05
60	6.96	4.08

stable (30 sec), adjust the slope control to
make the LCD show the value of buffer 4.0 in
Table 63-1. The system is now calibrated to
read samples with pH 0–14.

7. Rinse the electrode with a small portion of
sample, blot dry, and immerse in the sample.
When the LCD is stable (30 sec), it will read
the pH of the sample. Repeat this step for
continued samples. Periodically check calibration, steps 5 and 6.

8. When all samples are completed, unplug the
electrode and rinse with buffer 4.0. Put a small
portion (1 mL) in the wetting cap and place
over the electrode tip until the next pH measurement. Closing the hinged cover shuts off
the system.

PHOSPHATE, ORTHO
(1–20 mg/L as PO_4^{3-})

AMINO ACID METHOD

Various forms of phosphate are used in the treatment of boiler and cooling systems. Orthophosphate ion, the simplest form, has the formula PO_4^{3-}. It also exists in water as $H_2PO_4^-$ or HPO_4^{2-}. Orthophosphate is used to control calcium carbonate deposition in boiler systems. It is also used to control corrosion in boiler and cooling systems.

SUMMARY OF METHOD

In a highly acidic solution, ammonium molybdate reacts with orthophosphate to form molybdophosphoric acid. This complex is then reduced by the amino acid reagent to yield an intensely colored molybdenum blue compound. This method was adapted from the APHA-AWWA *Standard Methods for the Examination of Water and Wastewater*.

APPARATUS REQUIRED

Apparatus	Betz Code
Cell, sample 25 mL, 2.5 cm (2 required)	2601
Clippers, large	2635
Cylinder, graduated, plastic, 50 mL	486
Filter Paper	2623
Flask, Erlenmeyer, glass, 50 (2 required)	389
Funnel, plastic, 3 in.	366
Pipet, graduated, plastic, 10 mL	395
Safety Bulb, rubber	1575
Spectrophotometer, DR 2000	2776

CHEMICALS REQUIRED

Chemical	Betz Code
Amino Acid Reagent Powder Pillows	2012
Molybdate Reagent for Phosphate	2044
Phosphate Standard, 50 mg/L as PO_4^{3-}	2061
Sulfuric Acid, 10 N	2033

INTERFERENCES

For best results, the sample should be from 21 to 26 °C (70 to 80 °F). Sulfides interfere by forming a blue color directly with the molybdate reagent. Nitrites bleach the blue color.

The following may interfere when present in concentrations exceeding those listed:

Chloride	150,000 mg/L as Cl^-
Calcium	10,000 mg/L as $CaCO_3$
Magnesium	40,000 mg/L as $CaCO_3$

In waters containing high salt levels, low results may be obtained. To eliminate this interference, dilute the sample until two successive dilutions yield approximately the same results.

As the concentration of phosphate increases, the color changes from blue to green, then to yellow, and finally to brown. The brown color may suggest a concentration as high as 100,000 mg/L PO_4^{3-}. If a color other than blue is formed, dilute the sample and retest.

Figure 64-1. Automated photometric analyzer.

PROCEDURE FOR TEST

1. Filter the sample.

2. Transfer two 25 mL portions of filtered sample into separate Erlenmeyer flasks.

3. Add 1 mL of 10 N sulfuric acid to each flask. Swirl to mix.

4. Add 1 mL of molybdate reagent to one of the flasks. This is the prepared sample. Swirl to mix.

5. Add the contents of one amino acid powder pillow to the flask containing the prepared sample. Swirl to mix.

6. A 10-min reaction period will begin.

7. Turn on the spectrophotometer and set the wavelength at 530 nm. Transfer the sample blank containing only the sulfuric acid to a clean sample cell.

8. Set the spectrophotometer at zero absorbance with the sample blank.

9. After the 10-min reaction period, transfer the prepared sample to a clean sample cell.

10. Read the absorbance of the prepared sample.

11. Convert the absorbance reading to phosphate concentration using a previously prepared calibration curve.

PHOSPHATE, TOTAL INORGANIC
(1–20 mg/L as PO$_4^{3-}$)

HYDROLYSIS TO ORTHOPHOSPHATE

Complex forms of inorganic phosphate, or poly-phosphates, are used for both scale and corrosion control in water distribution and cooling water systems. Examples of complex inorganic phosphates are pyrophosphate and hexametaphosphate.

SUMMARY OF METHOD

This procedure lists the steps necessary to convert condensed forms of phosphate (meta-, pyro-, or other polyphosphate) to orthophosphate before analysis, using acid and heat to hydrolyze the sample. Organic phosphates are not readily converted to orthophosphate by this process, and a very small fraction of them may be unavoidably included in the result. Thus, the "acid hydrolyzable" phosphate results are primarily a measure of inorganic phosphorus. The amino acid method is used to determine the amount of orthophosphate in the sample. This method was adapted from the APHA-AWWA *Standard Methods for the Examination of Water and Wastewater.*

APPARATUS REQUIRED

Apparatus	Betz Code
Cell, sample 25 mL, 2.5 cm (2 required)	2601
Clippers, large	2635
Cylinder, graduated, plastic, 50 mL	486
Filter Paper	2623
Flask, Erlenmeyer, 125 mL (2 required)	126
Funnel, plastic, 3 in.	366
Hot Plate	2646
Spectrophotometer, DR 2000	2776

CHEMICALS REQUIRED

Chemical	Betz Code
Amino Acid Reagent Powder Pillows	2012
Deionized Water	243
Molybdate Reagent for Phosphate	2044
Sulfuric Acid, 10 N	2033

INTERFERENCES

This method measures orthophosphate along with phosphates hydrolyzable under the conditions of the test. A high concentration of chromate (over 150 mg/L) may interfere. High concentrations of organic materials also may interfere.

PROCEDURE FOR TEST

1. Filter at least 50 mL of sample.

2. Transfer 25 mL of filtered sample into each Erlenmeyer flask.

3. Add 1 mL of 10 N sulfuric acid solution to each flask. Swirl to mix.

4. Place one flask (the prepared sample) onto a hot plate. Boil gently for 30 min.

5. Cool the prepared sample to room temperature (21–26 °C, 70–80 °F).

6. Pour the prepared sample into a graduated cylinder. Add deionized water to return the volume to 25 mL.

7. Turn on the spectrophotometer and set the wavelength at 530 nm. Transfer the unboiled sample blank to a clean cell and set the spectrophotometer at zero absorbance using the sample blank.

8. Transfer the prepared sample back into the Erlenmeyer flask. Add 1 mL of molybdate reagent. Swirl to mix.

9. Add the contents of one amino acid powder pillow to the prepared sample. Swirl to mix.

10. After 10 min, transfer the prepared sample to a clean cell.

11. Read the absorbance of the prepared sample.

12. Convert the absorbance reading to phosphate concentration using a previously prepared calibration curve.

PHOSPHATE, TOTAL ORGANIC AND INORGANIC
(1–20 mg/L as PO_4^{3-})

ACID PERSULFATE DIGESTION METHOD (EPA APPROVED)

Modern water treatment practices make use of a variety of phosphate ions with different organic functional groups attached. A broad class of these compounds, known as phosphonates, is useful for scale control. Phosphonates function by retarding or preventing crystal growth and by dispersing scale-forming particles in solution. Examples of typical phosphonates include the di-, tri-, and tetraphosphonic acids. Because of the large number of titratable protons for each of these acids, the ionic forms present in aqueous media strongly depend on the system pH.

SUMMARY OF METHOD

Phosphates present in organic and condensed inorganic forms (meta-, pyro-, or other polyphosphates) must be converted to reactive orthophosphate before analysis. Pretreatment of the sample with acid and heat provides the conditions for hydrolysis of the condensed inorganic forms.

Figure 66-1. Superconducting NMR spectrometer.

Organic phosphates are heated with persulfate for conversion to orthophosphate. For indirect determination of organically bound phosphates, the result of an acid hydrolyzable phosphate test is subtracted from the total phosphate result.

The amino acid method is used to determine the content of orthophosphate in the sample. This method was adapted from the APHA-AWWA *Standard Methods for the Examination of Water and Wastewater.*

APPARATUS REQUIRED

Apparatus	Betz Code
Cell, sample, 25 mL, 2.5 cm (2 required)	2601
Clippers, large	2635
Cylinder, graduated, plastic, 50 mL	486
Filter Paper	2623
Flask, Erlenmeyer, 125 mL (2 required)	126
Funnel, plastic	366
Hot Plate	2646
Spectrophotometer, DR 2000	2776

CHEMICALS REQUIRED

Chemical	Betz Code
Amino Acid Reagent Powder Pillows	2012
Deionized Water	243
Molybdate Reagent for Phosphate	2044
Potassium Persulfate Reagent Powder Pillows	2004
Sulfuric Acid, 10 N	2033

INTERFERENCES

The method measures orthophosphate, hydrolyzable phosphate, and all organic forms which can be reverted under the conditions of the test. High concentrations of chromate may interfere.

PROCEDURE FOR TEST

1. Filter at least 50 mL of sample.

2. Transfer 25 mL of filtered sample into each Erlenmeyer flask.

3. Add 1 mL of 10 N sulfuric acid solution to each flask. Swirl to mix.

4. Add the contents of one potassium persulfate powder pillow to one flask (the prepared sample). Swirl to mix.

5. Place the prepared sample flask onto the hot plate and boil gently for 30 min.

6. Cool the prepared sample to room temperature (21–26 °C, 70–80 °F).

7. Pour the prepared sample into a graduated cylinder. Add deionized water to return the volume to 25 mL.

8. Turn on the spectrophotometer and set the wavelength at 530 nm. Transfer the unboiled sample blank to a clean cell and set the spectrophotometer at zero absorbance using the unboiled sample blank.

9. Transfer the prepared sample back into the Erlenmeyer flask. Add 1 mL of molybdate reagent. Swirl to mix.

10. Add the contents of one amino acid powder pillow to the prepared sample. Swirl to mix.

11. After 10 min, transfer the prepared sample to a clean cell.

12. Read the absorbance of the prepared sample.

13. Convert the absorbance readings to phosphate concentration using a previously prepared calibration curve.

SILICA
(0.005–1.6 and 2–100 mg/L as SiO₂)

HETEROPOLY BLUE/ SILICOMOLYBDATE METHOD

The element silicon exists in nature mainly as silicates of various metals or as silica (SiO_2) in quartz and sand. Silica and silicates make up 50–75% of most common rocks throughout the world.

Natural weathering processes bring surface waters into contact with these silicate rocks and, invariably, a small portion is dissolved. The silica content of natural waters depends to a considerable extent on locality. Well waters from certain aquifers are also rich in silica. It is not uncommon in some regions of the United States to find water supplies that contain silica of 50 mg/L or more.

In water, silica takes many forms, the most common of which is the simple silicate ion, SiO_3^{2-}. A variety of complex silicate ions also occurs in natural waters. Structurally, these are somewhat analogous to the different complex phosphates. Additionally, agglomerated silicates can reach the size, in solution, where they are no longer ions but are actually colloidal particles that can remain in suspension indefinitely. This poses a difficult problem in operation and analysis. The particles impart a milky appearance to water and are very hard to remove by filtration.

Reverse osmosis systems are used for this purpose. Coagulation rarely helps. Larger silicate particles become visible to the naked eye, and these can be filtered, coagulated, or allowed to settle by themselves in time.

Silica testing takes advantage of the formation of a silicomolybdate complex ion which can be reduced to the molybdenum blue complex. As in the case of phosphate testing, only the simplest ion, SiO_3^{2-}, reacts with molybdic acid. Fortunately, in most instances 90–100% of the silica in solution

Figure 67-1. X-ray flourescence spectrometer used for analysis of metals in deposits.

is active to the molybdate. No simple procedure exists for converting complex silicate ions to the simple form. Field control methods are limited to determining soluble silicate, although there are modern laboratory techniques for determining total silica. These techniques are also highly responsive to colloidal silicates.

In operating systems, silica can form calcium, magnesium, iron, and aluminum silicates. The presence of silica in boiler feedwaters is particularly objectionable because the dense scales formed have very poor heat transfer properties. In high-pressure boilers, the possibility exists for silica in the boiler water to vaporize and be carried into the turbine, where it deposits on the blades. A serious loss in turbine efficiency can result.

SUMMARY OF METHOD

Silica and phosphate in the sample react with molybdate ion under acid conditions to form a yellow color due to the silicomolybdic and phosphomolybdic acid complexes. Citric acid is added, which preferentially destroys the phospho-molybdic acid complex. For large amounts of silica, the remaining yellow color is intense enough to be measured directly. For low concentrations, an amino acid reducing agent is used to convert the faint yellow color to a dark heteropoly blue species. The color formed in either case is directly proportional to the amount of silica present in the sample. These methods were adapted from APHA-AWWA *Standard Methods for the Examination of Water and Wastewater.*

APPARATUS REQUIRED

Apparatus	Betz Code
Beakers, plastic, 50 mL	2627
Cell, sample, 25 mL, 2.5 cm (2 required)	2601
Clippers, large	2635
Cylinder, plastic, 25 mL	495
Spectrophotometer, DR 2000	2776
Stirring Rod, plastic	488

CHEMICALS REQUIRED

Heteropoly Blue Method (0.005–1.6 mg/L as SiO_2)

Chemical	Betz Code
Amino Acid F Reagent Powder Pillows	2351
OR	
Amino Acid F Reagent Solution	2380
Citric Acid F Reagent	2350
Molybdate 3 Reagent	2322A
Silica Standard Solution, 1 mg/L	2065

Silicomolybdate Method (2–100 mg/L as SiO_2)

Acid Reagent Powder Pillows	2011
Citric Acid Powder Pillows	2013
Molybdate Reagent Powder Pillows	2010
Silica Standard Solution, 50 mg/L	253

INTERFERENCES

Color and turbidity interferences are eliminated when the instrument is zeroed with the blank. Sulfides and large amounts of iron interfere. Phosphate does not interfere in quantities below 50 mg/L phosphate. At 60 mg/L phosphate, an interference of −2% is observed. At 75 mg/L, the interference is −11%.

Occasionally, a sample contains forms of silica that are very slow to react with molybdate. The exact nature of these forms is not known. A pretreatment with sodium bicarbonate, and with sulfuric acid, renders these forms reactive to molybdate. This pretreatment is given in the APHA-AWWA *Standard Methods for the Examination of Water and Wastewater,* "Silica— Digestion with Sodium Bicarbonate." A longer reaction time of the sample with the molybdate reagent before addition of the citric acid is often helpful in lieu of the bicarbonate pretreatment.

PROCEDURE FOR TEST

Heteropoly Blue Method (0.005–1.6 mg/L as SiO_2)

1. Transfer 25 mL of sample to each of two plastic beakers.

2. Add 0.5 mL of molybdate 3 reagent to each beaker. Stir to mix. Wait 4 min.

3. Add 0.5 mL of citric acid F reagent to each beaker. Stir to mix. Wait 1 min for the destruction of possible phosphate interference.

4. Add the contents of one amino acid F powder pillow or 0.5 mL of amino acid F reagent solution to one beaker. Stir to mix. Wait 1 min. This is the prepared sample.

5. Transfer the solution from the beaker without the added amino acid to a sample cell. Place the cell into the photometer and zero the meter at 815 nm.

6. Transfer the prepared sample to a second sample cell. Place the cell into the photometer and read the absorbance.

Silicomolybdate Method (2–100 mg/L as SiO_2)

1. Transfer 25 mL of sample to a beaker (the prepared sample).

2. Add the contents of one molybdate reagent powder pillow for high-range silica. Stir to dissolve.

3. Add the contents of one acid reagent powder pillow for high-range silica. Stir to dissolve. Wait 10 min.

4. Add the contents of one citric acid powder pillow to the beaker. Stir to dissolve. Wait 2 min for the destruction of possible phosphate interference.

5. Transfer a portion of the original sample to a sample cell (the blank). Place the blank into the photometer and zero the meter at 450 nm.

6. Transfer the prepared sample to a second sample cell. Place the cell into the photometer and read the absorbance.

CALCULATION OF RESULTS
Heteropoly Blue Method
(0.005–1.6 mg/L as SiO$_2$)

The silica in milligrams per liter is read off a calibration curve of absorbance versus concentration from 0.005 to 1.600 mg/L silica.

Silicomolybdate Method
(2–100 mg/L as SiO$_2$)

The silica in milligrams per liter is read off a calibration curve of absorbance versus concentration from 2 to 100 mg/L silica.

SULFATE
(2–65 mg/L as SO_4^{2-})

SULFAVER® 4 METHOD

The principal problem posed by sulfate ions in water is the possibility of calcium sulfate scale formation. Precipitation of calcium sulfate can occur when high concentrations of calcium and sulfate exist simultaneously. There are three primary areas of water treatment in which calcium sulfate precipitation may occur: boilers, cooling systems, and ion exchangers regenerated with sulfuric acid.

In boilers, calcium sulfate scale formation has become rare due to modern treatment practices. The low feedwater hardness levels and normal boiler treatments largely preclude calcium sulfate boiler scale.

By contrast, cooling systems are subject to calcium sulfate scale deposition, because calcium is high in the makeup water and the recirculating water is treated with sulfuric acid for pH control. Calcium sulfate scale deposition can result unless the cycles of concentration are properly controlled and the proper scale control chemicals are used.

Cation ion exchangers regenerated with sulfuric acid can also have major calcium sulfate scaling problems. Calcium sulfate precipitation can be prevented through proper control of the acid concentration and flow rate through the exchanger.

SUMMARY OF METHOD

Sulfate ions in the sample react with barium chloride and form insoluble barium sulfate turbidity. The amount of turbidity formed is proportional to the sulfate concentration. This method was adapted from the APHA-AWWA *Standard Methods for the Examination of Water and Wastewater.*

APPARATUS REQUIRED

Apparatus	Betz Code
Cell, sample, 25 mL, 2.5 cm (2 required)	2601
Clippers, large	2635
Cylinder, graduated, mixing, 50 mL	2631
Spectrophotometer, DR 2000	2776

CHEMICALS REQUIRED

Chemical	Betz Code
Sulfate Standard Solution, 100 mg/L	333
SulfaVer® 4 Sulfate Reagent Powder Pillows	2019

INTERFERENCES

Silica and calcium may interfere at levels above 500 mg/L and 20,000 mg/L, respectively. Chloride and magnesium do not interfere at levels up to at least 40,000 and 10,000 mg/L, respectively.

PROCEDURE FOR TEST

1. Fill a sample cell with 25 mL of sample. Filter highly colored or turbid samples. Use the filtered sample here and in step 3.

2. Add the contents of one SulfaVer 4 sulfate reagent powder pillow to the sample cell (the prepared sample). Swirl to dissolve. Wait 5 min. The sample should not be disturbed during this development period.

3. Fill a second sample cell with 25 mL of sample (the blank). Place the blank into the photometer and zero the meter at 450 nm.

4. Place the prepared sample into the photometer and read the absorbance.

CALCULATION OF RESULTS

The sulfate, in milligrams per liter, is read off a calibration curve of absorbance versus concentration in the range of 2–65 mg/L sulfate.

SULFITE
$(1–100 \ \mathrm{mg/L} \ \mathrm{as} \ SO_3^{2-})$

TITRATION METHOD

The determination of sulfite usually is made only on boiler waters or on waters that have been treated with catalyzed sodium sulfite for corrosion prevention. Generally, sulfite is not present in natural waters.

In boiler feedwater conditioning, sodium sulfite is fed to remove dissolved oxygen and thereby prevent pitting. For the reaction between sulfite and oxygen to proceed rapidly and completely, it is necessary to maintain an excess sulfite concentration at an elevated temperature.

Theoretically, 7.88 lb of chemically pure sodium sulfite is required to remove 1 lb of oxygen. The efficiency of the oxygen removal is somewhat lower because of feed solution oxidation from contact with air, blowdown losses, and other factors. Therefore, approximately 10 lb of commercial sodium sulfite is required for each pound of oxygen removed (or 10 mg/L sulfite per 1 mg/L dissolved oxygen).

When water contains large amounts of dissolved oxygen, it is usually passed through a deaerator and later treated with sulfite. This combination of mechanical and chemical processes is more economical than chemical removal of all of the oxygen in the water.

To prevent corrosion and pitting in feed lines, closed heaters, and economizers, it is necessary to feed the sulfite continuously to the deaerator storage compartment. Reaction between sulfite and oxygen is not instantaneous; as much contact time as possible should be provided to complete the reaction.

Catalyzed sodium sulfite reacts with dissolved oxygen even at cold water temperatures. For this reason, the use of catalyzed sulfite has increased in the treatment of closed cooling water, process water, and distribution systems for prevention of oxygen corrosion.

SUMMARY OF METHOD

The titration of sulfite is based on the reaction of sulfite with iodine in acidic solution. The titrant used is a standard solution of iodide–iodate, which generates iodine in an acid solution. At the end point, excess iodine combines with the indicator to form a blue color.

APPARATUS REQUIRED

Apparatus	Betz Code
Buret, automatic, 25 mL	103T
Casserole, porcelain, 210 mL	156
Cylinder, graduated, 50 mL	486
Stirring Rod, plastic	488
Measuring Dipper, plastic	110

CHEMICALS REQUIRED

Chemical	Betz Code
Phenolphthalein Indicator	212
Potassium Iodide–Iodate, 1 mL = 0.5 mg SO_3^{2-}	237
Sulfite Indicator	219

Figure 69-1. Automatic titrator used for multiple sample analysis.

INTERFERENCES

This method is affected by any oxidizable substances in the water sample such as organic matter, sulfides, and nitrites. The presence of these interfering substances causes erroneously high results.

PROCEDURE FOR TEST

1. Measure 50 mL of sample with the graduated cylinder and transfer to the casserole (see Note 1).

2. Add 3–4 drops of phenolphthalein indicator to the sample.

3. Use the plastic dipper to add sulfite indicator, one measure at a time, to the sample, stirring thoroughly between each measure, until the red color disappears. Add one more measure of sulfite indicator and stir (see Note 2).

4. Titrate with potassium iodide–iodate solution until a faint, permanent bluish-purple color develops in the sample.

CALCULATION OF RESULTS

For a 50 mL sample, sulfite in milligrams per liter as SO_3^{2-} is equal to the milliliters of potassium iodide–iodate required multiplied by 10.

NOTES

1. The water sample should be freshly obtained with as little contact with air as possible. Do not filter. Cool to room temperature (21–26 °C, 70–80 °F).

2. The additional dipper of sulfite indicator is essential in achieving the proper pH drop. At the wrong pH, the end point achieved is a faint, permanent brownish color. At the right pH, each drop of titrant looks brownish when hitting the sample surface, but the end point is a bluish purple. If left to stand, this end point color begins to fade toward the brownish color achieved at the wrong pH.

TURBIDITY
(5–450 FTU)

ABSORPTOMETRIC METHOD

Turbidity, caused by the presence of suspended matter in a finely divided state, can be interpreted as a lack of cleanness or brilliance in a water. Turbidity should not be confused with color, because a water may have a dark color and yet be clear and not turbid. Clay, silt, finely divided organic and inorganic matter, and microscopic organisms can all cause turbidity.

Turbidity is a measure of relative light transmission through a sample. Light not transmitted may be either absorbed or reflected by the particles in the sample. Although closely allied with suspended matter, which is the absolute quantity of matter in a water sample that can be removed by filtration, turbidity is not synonymous with suspended matter.

The turbidity of industrial water, especially boiler and cooling system feedwaters, should be as low as possible. The small particles in suspension concentrate in the boiler water and may settle out in the form of a heavy sludge, or "mud." Turbidity in boiler feedwater also limits the cycles of concentration that can be carried in the boiler. Finely divided particles can be responsible for a foaming and priming condition in the boiler.

Turbid makeup water to cooling water systems may cause plugging and overheating where solids settle out on heat exchange surfaces. Corrosive action may also be increased under such deposits, which hinder penetration of corrosion inhibitors.

Although turbidity may be partially removed by settling, it is usually necessary to use coagulation and filtration to eliminate it from industrial water.

SUMMARY OF METHOD

The turbidity test measures an optical property of the water sample that results from the scattering and absorbing of light by the particulate matter present. The amount of turbidity registered is dependent on such variables as the size, shape, and refractive properties of the particles. No direct relationship exists between the turbidity of a water sample and the weight concentration of the matter present.

Two common units in use for turbidity measurement are FTU (Formazin Turbidity Unit) and NTU (Nephelometer Turbidity Unit). An FTU is equivalent to an NTU when readings are made on a nephelometer. A nephelometer measures the intensity of light scattered at an angle of 90° to the incident light. A photometer measures the reduction in light intensity caused by scattering and absorption as light passes through the sample. This photometric test procedure is designed using a formazin turbidity standard calibration curve and the readings are in terms of FTU. This method was adapted from FWPCA *Methods for Chemical Analysis of Water and Wastes*, 275 (1969).

APPARATUS REQUIRED

Apparatus	Betz Code
Cell, sample, 25 mL, 2.5 cm (2 required)	2601
Spectrophotometer, DR 2000	2776
Volumetric Flask, 100 mL	2680

CHEMICALS REQUIRED

Chemical	Betz Code
Formazin Turbidity Standard (See calibration curve preparation)	–
Turbidity Blank Solution	2637

PROCEDURE FOR TEST

1. Fill one sample cell with turbidity blank solution (the blank).

2. Place the blank into the photometer and zero the meter at 450 nm.

3. Fill a second sample cell with 25 mL of sample. Place the cell into the photometer and read the absorbance.

CALIBRATION CURVE PREPARATION

A calibration curve should be prepared for this test. Formazin turbidity standards can be purchased from a local supplier or prepared by the following procedure:

1. Dissolve 1.000 g of hydrazine sulfate (98%, ACS grade) in deionized water and dilute to the 100 mL mark in a volumetric flask.

2. Dissolve 10.00 g of hexamethylene tetramine (95%, analytical grade) in deionized water and dilute to the 100 mL mark in a second volumetric flask.

3. Mix 5.0 mL of each solution in a 100 mL volumetric flask and allow to stand undisturbed for 24 hr at 77 °F (25 °C).

4. Dilute the mixture to the 100 mL mark with deionized water and mix.

The turbidity of this standard is 400 FTU. It should be prepared monthly. Appropriate dilutions of the standard should be made to prepare the calibration curve.

CALCULATION OF RESULTS

The turbidity in Formazin Turbidity Units (FTU) is read off a calibration curve of absorbance versus Formazin Turbidity Units of standards from 5 to 450 FTU.

ZINC
(0.02–2.0 mg/L as Zn)

ZINCON METHOD

The principal purpose of zinc analysis is to maintain proper treatment levels of zinc-containing corrosion inhibitors. Zinc corrosion inhibitors are used in water distribution and cooling water systems.

Zinc functions as a cathodic inhibitor by forming a film of zinc hydroxide or zinc phosphate at the cathodic site and causing a localized deposition of these compounds that stifles the cathodic reaction. Proper levels of zinc are essential for obtaining optimum corrosion protection.

SUMMARY OF METHOD

Zinc and other metals in the sample are complexed with cyanide. The addition of cyclohexanone causes a selective release of zinc. The zinc then reacts with 2-carboxy-2-hydroxy-5-sulfoformazylbenzene (zincon) indicator. The resulting blue color is proportional to the zinc concentration. This method was adapted from *Standard Methods for the Examination of Water and Wastewater,* Federal Register, **45** (105), 36166 (May 29, 1980).

APPARATUS REQUIRED

Apparatus	Betz Code
Cell, sample, 25 mL, 2.5 cm (2 required)	2601
Clippers, large	2635
Cylinder, graduated, mixing, 50 mL	2631
Spectrophotometer, DR 2000	2776
Pipet, graduated, plastic, 1 mL	371
Safety Bulb, rubber	1575

CHEMICALS REQUIRED

Chemical	Betz Code
Cyclohexanone	2018
Zinc Standard P1 Solution, 100 mg/L as Zn	530
ZincoVer® 5 Zinc Reagent Powder Pillow	2017

INTERFERENCES

The following may interfere when present in concentrations exceeding those listed:

Aluminum	6 mg/L
Cadmium	0.5 mg/L
Copper	5 mg/L
Iron (ferric)	7 mg/L
Manganese	5 mg/L
Nickel	5 mg/L

Large amounts of organic material may interfere. Perform a digestion to eliminate this interference.

Highly buffered samples or extreme sample pH may exceed the buffering capacity of the reagents and require sample pretreatment.

PROCEDURE FOR TEST

1. Fill a 50 mL mixing graduated cylinder to the 50 mL mark with sample.
2. Add the contents of one ZincoVer® 5 reagent powder pillow. Stopper and invert several times to dissolve the powder completely.
3. Measure 25 mL of the solution into a 2.5 cm sample cell (the blank).
4. Add 1.0 mL of cyclohexanone to the remaining solution in the cylinder.
5. Stopper the cylinder (the prepared sample). Shake for 30 sec and transfer solution to a second sample cell.
6. Wait 3 min.
7. Place blank sample into the photometer and zero meter at 620 nm.
8. Place the prepared sample cell into the photometer and read the absorbance.

CALCULATION OF RESULTS

The zinc in milligrams per liter is read off a calibration curve of absorbance versus concentration from 0.02 to 2.0 mg/L zinc.

COVERSION FACTORS/
TABLES OF MEASUREMENT

WATER ANALYSIS

		ppm	mg/L	parts/ 100,000	grains/ U.S. gal	grains/ Imperial gal
1 ppm	=	1.0	1.0	0.10	0.058	0.07
1 part/100,000	=	10.0	10.0	1.00	0.585	0.70
1 grain/U.S. gal	=	17.1	17.1	1.71	1.000	1.20
1 grain/Imperial gal	=	14.3	14.3	1.43	0.833	1.00

HARDNESS AS CaCO$_3$

		ppm	mg/L	grains/ U.S. gal	Clark degrees	French degrees	German degrees
1 ppm	=	1.0	1.0	0.058	0.07	0.10	0.056
1 grain/U.S. gal	=	17.1	17.1	1.000	1.20	1.71	0.958
1 Clark degree	=	14.3	14.3	0.833	1.00	1.43	0.800
1 French degree	=	10.0	10.0	0.583	0.70	1.00	0.560
1 German degree	=	17.9	17.9	1.044	1.24	1.78	1.000

SI PREFIXES

Prefix	Symbol	Multiplication Factor
exa	E	10^{18}
peta	P	10^{15}
tera	T	10^{12}
giga	G	10^9
mega	M	10^6
kilo	k	10^3
hecto	h	10^2
deka	da	10
deci	d	10^{-1}
centi	c	10^{-2}
milli	m	10^{-3}
micro	μ	10^{-6}
nano	n	10^{-9}
pico	p	10^{-12}
femto	f	10^{-15}
atto	a	10^{-16}

CONVERSION FACTORS[a]

AREA

1 ft^2	= 144 in.^{2}
	= 0.0929 m^2
1 in.^{2}	= 6.45 cm^2

LENGTH

1 statue mile	= 5280 ft
	= 1.609 km
1 ft	= 12 in.
	= 30.48 cm
1 in.	= 25.40 mm
	= 1000 mils
1 km	= 1000 m
	= 0.621 statue mile
1 m	= 100 cm
	= 1000 mm
	= 1.094 yd
	= 3.281 ft
	= 39.37 in.
1 cm	= 1 x 10^8 Å
1 µm	= 0.001 mm
	= 0.000039 in.

VELOCITY

100 ft/min	= 0.508 m/sec
1 m/sec	= 196.9 ft/min
1 mil/yr	= 0.0254 mm/yr

WEIGHT

1 U.S. long ton	= 2240 lb
	= 1016 kg
1 U.S. short ton	= 2000 lb
	= 907 kg
1 lb	= 16 oz
	= 7000 grains
	= 0.454 kg
1 oz	= 0.0625 lb
	= 28.35 g
1 grain	= 64.8 mg
	= 0.0023 oz
1 metric ton (tonne)	= 1000 kg
	= 0.984 U.S. long ton
	= 1.102 U.S. short ton
	= 2205 lb

1 kg	= 1000 g
	= 2.205 lb
1 g	= 1000 mg
	= 0.03527 oz
	= 15.43 grains

FLOW

1 ft^3/min	= 1.699 m^3/hr
	= 7.48 gpm
1 m^3/hr	= 0.589 ft^3/min
	= 4.4 gpm
1 ft^3/sec	= 646,316 gpd
	= 448.83 gpm
1 gpm	= 0.00144 million gpd
	= 0.227 m^3/hr

VOLUME

1 yd^3	= 27 ft^3
	= 0.765 m^3
1 ft^3	= 1728 in.^{3}
	= 28.32 L
	= 7.48 U.S. gal
1 in.^{3}	= 16.39 cm^3
1 Imperial gal	= 277.4 in.^{3}
	= 4.55 L
1 U.S. gal	= 0.833 Imperial gal
	= 3.785 L
	= 231 in.^{3}
	= 0.1337 ft^3
1 U.S. barrel (petroleum)	= 42 U.S. gal
	= 35 Imperial gal
1 m^3	= 1000 L
	= 35.31 ft^3
1 L	= 1000 mL
	= 0.2200 Imperial gal
	= 0.2642 U.S. gal
	= 61.0 in.^{3}

1 board ft	= 12 in. x 12 in. x 1 in. thick
	= 144 in.^{3}

DENSITY

Weight/Volume

1 ft^3/lb	= 0.0624 m^3/kg
1 lb/ft^3	= 16.02 kg/m^3
	= 0.016 g/cm^3
1 grain/ft^3	= 2.288 g/m^3
1 grain/U.S. gal	= 17.11 g/m^3
	= 17.11 mg/L
1 m^3/kg	= 16.02 ft^3/lb
1 kg/m^3	= 0.0624 lb/ft^3
1 g/m^3	= 0.437 grain/ft^3
	= 0.0584 grain/U.S. gal
1 g/mL	= 62.4 lb/ft^3
1 g/L	= 58.4 grains/U.S. gal
1 kg/m^3	= 1 g/L
	= 1 ppth
1 g/m^3	= 1 mg/L
	= 1 ppm
1 ppm	= 8.33 lb/million gal
1 grain/gal	= 143 lb/million gal
1 lb/million gal	= 0.12 ppm
1 lb/million gal	= 0.007 grain/gal
1 lb/thousand gal	= 120 ppm

Water at 62 °F (16.7 °C)

1 ft^3	= 62.3 lb
1 lb	= 0.01604 ft^3
1 U.S. gal	= 8.34 lb

Water at 39.2 °F (4 °C): Maximum Density

1 ft^3	= 62.4 lb
1 lb	= 0.01602 ft^3
1 m^3	= 1000 kg
1 L	= 1.0 kg

CONVERSION FACTORS[a] (CONTINUED)

PRESSURE

1 atm	= 760 mm (29.92 in.) mercury with density 13.595 g/mL = 14.696 lb/in.² = 1.033 kg/cm² = 103.3 kPa = 1.03 bars		
1 atm (metric)	= 1 kg/cm² = 10,000 kg/m² = 10 m head of water = 14.22 lb/in.² = 1 bar = 100 kPa		

1 lb/ft²	= 0.1924 in. of water = 4.88 kg/m²
1 lb/in.²	= 2.036 in. head of mercury = 2.309 ft head of water = 0.0703 kg/cm² = 0.0690 bar = 7.03 kPa
1 ton/in.²	= 1.406 kg/mm²
1 in. head of water	= 5.20 lb/ft² = 0.25 Pa
1 ft head of water	= 0.433 lb/in.²

1 m head of water	= 0.1 kg/cm² = 10 kPa = 0.1 bar
1 in. head of mercury	= 0.491 lb/in.² = 3.387 kPa
1 m head of mercury	= 1.360 kg/cm²
1 kg/m²	= 1 mm head of water = 0.2048 lb/ft²
1 kg/cm²	= 735.5 mm of mercury = 14.22 lb/in.² = 1 bar = 100 kPa
1 kg/mm²	= 0.711 ton/in.²

[a] In these conversions, inches and feet of water are measured at 62 °F (16.7 °C), millimeters and meters of water at 39.2 °F (4 °C), and inches, millimeters, and meters of mercury at 32 °F (0 °C).

TEMPERATURE[a]

	Fahrenheit	Celcius	Absolute (Kelvin)	Absolute (Rankine)
Boiling point	212 °F	100 °C	373 °K	672 °R
Freezing point	32 °F	0 °C	273 °K	492 °R

$$°F = \frac{9}{5} °C + 32 \qquad °R = °F + 460$$

$$°C = \frac{5}{9} (°F - 32) \qquad A(°K) = °C + 273$$

[a] The standards for these scales are the freezing and boiling points of water at 1 atm pressure.

CYLINDRICAL TANK CAPACITIES

Diameter (ft)	ft³/ft of depth	U.S. gal/ft of depth
1.0	0.785	5.87
1.5	1.767	13.22
2.0	3.142	23.50
2.5	4.909	36.72
3.0	7.069	52.88
3.5	9.621	71.97
4.0	12.57	94.00
4.5	15.9	119.0
5.0	19.63	146.9
5.6	23.76	177.7
6.0	27.28	211.5
6.6	33.18	248.2
7.0	38.48	287.9
7.5	44.18	330.5
8.0	50.27	376.0
8.5	56.75	424.5
9.0	63.62	475.9
9.5	70.88	530.2
10.0	78.54	587.5
10.5	86.59	647.7
11.0	95.03	710.9
11.6	103.9	777.0
12.0	113.1	846.0

ENERGY AND HEAT

1 J	= 0.2390 cal
1 Btu	= 1.055 kJ
	= 0.252 kcal
1 Btu/ft²	= 11.360 kJ/m²
	= 2.71 kcal/m²
1 Btu/ft²/hr	= 3.152 W/m²
1 Btu/ft²/hr/°F	= 5.674 W/m² · °K
1 Btu/ft²/hr/°F/in.	= 0.1441 W/m · °K

POWER

1 Btu/hr	= 0.2931 W
1 kW	= 1000 J/sec
	= 0.2390 cal/sec
	= 0.9478 Btu/sec

CONDUCTIVITY

1 µmho	= 1 µS

CALCIUM CARBONATE EQUIVALENTS

		Conversion Factor	
Ion	Equivalent Weight	Substance to $CaCO_3$ Equivalent	$CaCO_3$ Equivalent to Substance
Cations			
Al^{3+}	9.0	5.56	0.18
NH_4^+	18.0	2.78	0.36
Ca^{2+}	20.0	2.50	0.4
Fe^{2+}	27.9	1.79	0.56
Fe^{3+}	18.6	2.69	0.37
H^+	1.0	50.0	0.02
Mg^{2+}	12.2	4.1	0.24
Mn^{2+}	27.5	1.82	0.55
K^+	39.1	1.28	0.78
Na^+	23.0	2.18	0.46
Zn^{2+}	32.7	1.53	0.65
Anions			
HCO_3^-	61.0	0.82	1.22
CO_3^{2-}	30.0	1.66[a]	0.60
Cl^-	35.5	1.41	0.71
NO_3^-	62.0	0.81	1.24
OH^-	17.0	2.94	0.34
PO_4^{2-}	31.7	1.58	0.63
SiO_2	30.0	0.83	0.60
SO_4^{2-}	48.0	1.04	0.96

[a] For ion exchange calculations, assume that carbonate exchanges as the monovalent ion and the conversion factor is 0.83.

PERIODIC TABLE OF THE ELEMENTS

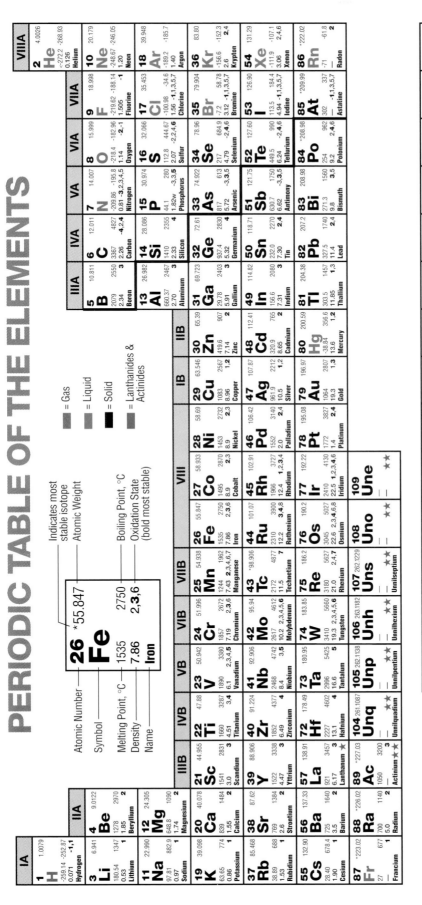

Legend (key example):

- Atomic Number — 26
- Symbol — Fe
- Melting Point, °C — 1535
- Density — 7.86
- Name — Iron
- Indicates most stable isotope — *
- Atomic Weight — *55.847
- Boiling Point, °C — 2750
- Oxidation State (bold most stable) — 2,3,6

= Gas
= Liquid
= Solid
= Lanthanides & Actinides

*Asterisked numbers are mass numbers of most stable isotope of that element.

INDEX

Downtime, boiler systems, 92–95
Drift, 222, 223
Drift eliminators, 222, 241
Drift loss, 223
Dry polymers, 261–262
  in volumetric feeders, 253
Dry storage of boilers, 93–94

E

Economizers, corrosion/erosion in, 80
Economizer tubes, 113–114
EDTA chelant demand, water analysis of, 335–336
Effective particle size, of filter media, 31
Electrode-type conductance probes, 266
Electrodialysis and Electrodialysis reversal, 68–70
Electrostatic precipitators, 9
  in handling gaseous fuel, 152
Elemental mapping, 123
Embrittlement, 115, 119–120
Embrittlement detector, 85, 115
Emulsion polymers, 261, 262–263
Energy conservation/loss, in boiler blowdown, 107–109
Entrainment, boiler water, 140
Environmental considerations, 8–12
  history/scope of, 8
Environmental impact, of wastewater chemicals, 228
Environmental regulation, 8–12
  for wastewater, 275, 290
Equivalents per million (EPM), 302–303
Erosion,
  in boiler system components, 79–80
  by steam solids, 130
  of steam turbine blades, 143
  in high solids systems, 295
Erosion corrosion, 143, 172, 175
Ethylene glycol, antifreeze, 245
Ethylenediaminetetraacetic acid (EDTA), 100–101
  as cleaner, 127
  inhibited, 129
  in scale control, 182
Evaporation, 219, 222
  heat rejection by, 219
Evaporation factor, 222
Evaporative condenser, 219
Evaporator, component in refrigeration system, 239
Extracellular polymeric substance (EPS), slime component, 186
Extractives, wood components, 210

F

Facultative lagoons, 280
Fast rinse, in softener regeneration cycle, 53

Fatigue, 115, 116
Fatigue cracking, 86, 113, 114
Feed methods, 259
Feed points, 258, 296
Feed verification, 257, 261
Feedback control, 255, 256
Feedforward control, 254
Feedwater lines, erosion of, 113
Ferrous compounds, as inorganic coagulants, 24
Ferrous metal piping, corrosion of, 215
Fiber optics, inspection by, 273
Field cleaning, 125
Fill-and-soak cleaning, *see* Acid cleaning
Film, protective, 86, 87, 89
  formation in cooling systems, 235–236
  (*see also* Iron oxide, Magnetite)
Film fill, cooling tower, 221
Filming amines, 150–151
  feeding of, 260
Film-type deaerators, 77
Filter beds/filter media,
  diatomaceous earth, 37
  mixed, 32
  types of, 31, 32
Filters, 31–37
  gravity, 32, 34
  in air washers, 243
  pressure, 32
  typical construction, 31
  upflow, 34
  washing/backwashing of, 33, 34, 35–36
Filtration,
  activated carbon, *see* Activated carbon filtration
  downflow, 32
  mechanisms in, 31
  media in, 31–33
  precoat, 36, 37, 62
  wastewater, 31, 278
Filtration, occurences of,
  in air cooling systems, 238
  in in-line clarification, 28–29
  natural, in groundwater, 4
  in softening, 38–41
Fireside,
  additives, in volumetric feeders, 253
  corrosion/corrosion control, 159–160
  deposition/deposit control, 157–159
  fouling, treatment for, 158–159
  preboiler systems, 152–156
  slagging, 158
  storage of boilers, 95
Fixed media oxidation, 284
Flame emission spectroscopy/Flame spectrophotometer testing, 138
Flash steam, from blowdown flash tanks, 109, 110
Flocculants, 24, 26, 29
  as coagulant aids, 26
  polyelectrolytes as, 24, 26